AF902912

PRODUCTS AND APPLICATIONS OF BIOPOLYMERS

Edited by
Casparus Johannes Reinhard Verbeek

Products and Applications of Biopolymers
Edited by Casparus Johannes Reinhard Verbeek

Published by InTech
Janeza Trdine 9, 51000 Rijeka, Croatia

Copyright © 2012 InTech
All chapters are distributed under the Creative Commons Attribution 3.0 license, which allows users to download, copy and build upon published articles even for commercial purposes, as long as the author and publisher are properly credited, which ensures maximum dissemination and a wider impact of our publications. After this work has been published by InTech, authors have the right to republish it, in whole or part, in any publication of which they are the author, and to make other personal use of the work. Any republication, referencing or personal use of the work must explicitly identify the original source.

As for readers, this license allows users to download, copy and build upon published chapters even for commercial purposes, as long as the author and publisher are properly credited, which ensures maximum dissemination and a wider impact of our publications.

Notice
Statements and opinions expressed in the chapters are these of the individual contributors and not necessarily those of the editors or publisher. No responsibility is accepted for the accuracy of information contained in the published chapters. The publisher assumes no responsibility for any damage or injury to persons or property arising out of the use of any materials, instructions, methods or ideas contained in the book.

Publishing Process Manager Romana Vukelic

Technical Editor Teodora Smiljanic

Cover Designer InTech Design Team

Additional hard copies can be obtained from orders@intechopen.com

Products and Applications of Biopolymers, Edited by Casparus Johannes Reinhard Verbeek
p. cm.
ISBN 978-953-51-0226-7

We are IntechOpen, the world's leading publisher of Open Access books

Built by scientists, for scientists

4,100+
Open access books available

116,000+
International authors and editors

120M+
Downloads

151
Countries delivered to

Our authors are among the
Top 1%
most cited scientists

12.2%
Contributors from top 500 universities

CLARIVATE ANALYTICS
BOOK CITATION INDEX
INDEXED

WEB OF SCIENCE™

Selection of our books indexed in the Book Citation Index
in Web of Science™ Core Collection (BKCI)

Interested in publishing with us?
Contact book.department@intechopen.com

Numbers displayed above are based on latest data collected.
For more information visit www.intechopen.com

For EU product safety concerns:
IN TECH d.o.o., Prolaz Marije Krucifikse Kozulić 3, 51000 Rijeka, Croatia,
info@intechopen.com or visit our website at intechopen.com.

Meet the editor

Dr Verbeek obtained his PhD in Engineering and is currently a Senior Lecturer and the convener of the Chemical & Biological Engineering program at the University of Waikato. He has also won and been nominated for various awards, such as the Bayer Innovators Award in the Agriculture & Environment category. His research into the use of bloodmeal to make bioplastics has been picked up as a commercial business opportunity material can be used in agricultural-type applications such as seedling trays, biodegradable plant pots, pegs and vine clips. He has published several papers on protein plastics and polymer composites and his recent paper on protein extrusion was one of the most downloaded papers in the journal Macromolecular Materials and Engineering.

Contents

Preface

It is interesting to consider that biopolymers are by no means new to this world. It is only because of our fascination with petrochemical products that these wonderful materials have been neglected for so long. In fact, natural or biopolymers have been considered in the 1940s and Henry Ford has used these biopolymers in the construction of a car. However, with the discovery of petrochemical polymers, the low cost of these quickly over shadowed natural materials.

Today we face a different challenge. Environmental pressure is pushing away from synthetic or petro-chemically derived products, while economic factors are pulling back from often more expensive "green" options. This paradox is what fuels research into materials and products whose properties fulfill environmental and economic requirements.

This book presents two aspects of biopolymers; products and applications. In the first section, *Products of Biopolymers*, a collection of chapters is presented covering topics such as antioxidant properties of one of the oldest known natural polymers, gum arabic, the use of poly(lactic acid) nano-composites in packaging, utilizing sweet sorghum as a novel biomass source to produce polyhydroxyalkanoates, extracting xylan from corn cobs, hydrodynamic properties of gelatin which can be used to make films and the role of biopolymers in "green" nano-technology. These chapters have been written by prominent researchers, who are experts in their fields. Section 1 aim at highlighting the diverse range of products and research in this field, which represents many industrial applications from pharmaceutical to possibly automotive and construction.

In section 2, *Applications of Biopolymers*, the use and significance of biopolymers in biotechnology are described in a collection of chapters covering topics on stable isotope analysis in bone collagen which can be used to gather information on historical living conditions, biopolymer-capped nano-particles in the emic-conductor industry, protection of DNA against UV damage and the role of biopolymers in plant defense systems. It shows that biopolymer research is not limited to products made of biopolymers, but that these polymers could play a vital role in other areas of research too.

The diverse field of biopolymers outlined in this book does underline its importance today. We can expect to see biopolymers in packaging, medicine, construction, in fact in almost every part of life – just as synthetic plastics are ubiquitous with everyday life at the moment.

Dr. Casparus Johannes Reinhard Verbeek
School of Engineering
Faculty of Science and Engineering
University of Waikato
Hamilton
New Zealand

Part 1

Biopolymer Products

1

Gum Arabic: More Than an Edible Emulsifier

Mariana A. Montenegro[1], María L. Boiero[1], Lorena Valle[2] and Claudio D. Borsarelli[2]
[1]Departamento de Química, Universidad Tecnológica Nacional- Facultad Regional Villa María, Córdoba
[2]Laboratorio de Cinética y Fotoquímica, Instituto de Química del Noroeste Argentino (INQUINOA-CONICET) Universidad Nacional de Santiago del Estero, Santiago del Estero Argentina

1. Introduction

Gum Arabic (GA) or *Acacia* gum is an edible biopolymer obtained as exudates of mature trees of *Acacia senegal* and *Acacia seyal* which grow principally in the African region of Sahe in Sudan. The exudate is a non-viscous liquid, rich in soluble fibers, and its emanation from the stems and branches usually occurs under stress conditions such as drought, poor soil fertility, and injury (Williams & Phillips, 2000).

The use of GA dates back to the second millennium BC when the Egyptians used it as an adhesive and ink. Throughout the time, GA found its way to Europe and it started to be called "gum arabic" because was exported from Arabian ports.

Chemically, GA is a complex mixture of macromolecules of different size and composition (mainly carbohydrates and proteins). Today, the properties and features of GA have been widely explored and developed and it is being used in a wide range of industrial sectors such as textiles, ceramics, lithography, cosmetics and pharmaceuticals, encapsulation, food, etc. Regarding food industry, it is used as a stabilizer, a thickener and/or an emulsifier agent (e.g., soft drink syrup, gummy candies and creams) (Verbeken et al., 2003).

In the pharmaceutical industry, GA is used in pharmaceutical preparations and as a carrier of drugs since it is considered a physiologically harmless substance. Additionally, recent studies have highlighted GA antioxidant properties (Trommer & Neubert, 2005; Ali & Al Moundhri, 2006; Hinson et al., 2004), its role in the metabolism of lipids (Tiss et al., 2001, Evans et al., 1992), its positive results when being used in treatments for several degenerative diseases such as kidney failure (Matsumoto et al., 2006; Bliss et al., 1996; Ali et al., 2008), cardiovascular (Glover et al., 2009) and gastrointestinal (Wapnir et al., 2008; Rehman et al., 2003).

Therefore, there is substantial evidence that GA can play a positive health-related role in addition to its well-known properties as an emulsifier. Therefore, the aim of this chapter is to describe general aspects of the source, composition, and already known uses of GA as well as some new aspects of its antioxidant capacity against some reactive oxygen substances (ROS), and its antimicrobial activity (AMA).

2. GA sources and processing

Acacia senegal and *Acacia seyal* trees are the main sources of GA. These species grow naturally in the semi-arid sub-Saharan regions of Africa. There are over 1000 species of acacia and a summary of their botanical classification was reported by Phillips and Williams, 1993. Sudan has traditionally been the main GA producer and its derivatives until the early 60s with a production of 60 kTn/year. Nevertheless, such a production decreased from 60 kTn/year to 20 kTn/year in the '70s and '80s due to extensive drought and unstable governments. These facts prompted new GA-producing countries such as Chad and Nigeria which produce mainly *Acacia seyal* (Abdel Nour, 1999). Europe and U.S. are the most important GA markets importing 40 kTn/year, on average, while Japan, the largest Asian consumer, imports about 2 kTn/year.

The crude exudate of GA is processed differently according to the quality finally required for it to be marketed. Air drying is the easiest method to be applied which, together with mechanical milling (kibbling), are used in order to produce a granular material that is much more soluble than the raw product. Other processing methods are spray drying and roller drying. These methods involve dissolving exudate in water under controlled heating conditions and constant stirring. Heating must be mild to avoid distortion of the gum which could have a detrimental effect on its functional properties. After removing the insoluble material by decantation or filtration, the solution is pasteurized and subjected to spray or roller drying. Spray drying involves spraying the solution into a stream of hot air. The water completely evaporates and the dry powder is separated from air by a cyclone, resulting in 50 to 100 mμ particles. During the roller-drying, the solution is passed to the hot rollers and the water is evaporated by the air flow. The thickness of the resulting GA film is controlled by adjusting the distance between the rollers. The film is separated from the roll by scraping blades giving way to particle scales of several hundred μm in size. GA samples produced by spray drying and drying rollers have an advantage over raw gum as they are virtually free of microbial contamination and dissolve much faster.

3. Chemical composition and structure

In recent years, several investigations have been conducted in order to reveal the molecular structure of GA and relate it to its exceptional emulsifying and rheological properties. The chemical composition of GA is complex and consists of a group of macromolecules characterized by a high proportion of carbohydrates (~97%), which are predominantly composed of D-galactose and L-arabinose units and a low proportion of proteins (<3%) (Islam et al., 1997). The chemical composition of GA may vary slightly depending on its origin, climate, harvest season, tree age and processing conditions, such as spray dying (Al-Assaf, et al., 2005 (a,b); Flindt et al., 2005; Hassan et al., 2005; Siddig et al., 2005). Therefore, there are some differences between the chemical composition of the GA taken from *Acacia senegal* and *Acacia seyal*. In fact, both gums have the same sugar residues but *Acacia seyal* gum has a lower content of rhamnose and glucuronic acid and a higher content of arabinose and 4-O-methyl glucuronic acid than *Acacia senegal* gum. Instead, *Acacia seyal* gum contains a lower proportion of nitrogen, and specific rotations are also completely different. The determination of the latter parameters may clearly spot the difference between the two species (Osman et al., 1993).

Table 1 presents the chemical composition and some properties of both gums reported by Osman et al., 1993 and Williams & Phillips, et al., 2000. Despite having different protein content, amino acid composition is similar in both gums. Recently, Mahendran et al., 2008, reported the GA amino acid composition in *Acacia Senegal*, being rich in hydroxyproline, serine, threonine, leucine, glycine, histidine, **Table 2.**

Parameter	*Acacia senegal*	*Acacia seyal*
% Rhamnose	14	3
% Arabinose	29	41
% Galactose	36	32
% Glucuronic acid	14.5	6.5
% Nitrogen	0.365	0.147
% Protein	2.41	0.97
Specific rotation (degrees)	-30	+ 51
Average molecular mass (kDa)	380	850

Table 1. Comparative chemical composition and some properties of Gum arabic taken from *Acacia senegal* and *Acacia seyal* trees (Osman et al., 1993 and Williams& Phillips, et al., 2000).

Aminoacid	(nmol/ mg) GA	% Aminoacid
Hydroxyproline	54.200	0.711
Serine	28.700	0.302
Threonine	15.900	0.208
Proline	15.600	0.180
Leucine	15.100	0.198
Histidine	10.700	0.166
Aspartic acid	10.600	0.141
Glutamic acid	8.290	0.122
Valine	7.290	0.085
Phenylalanine	6.330	0.105
Lysine	5.130	0.075
Alanine	5.070	0.045
Isoleucine	2.380	0.031
Tyrosine	2.300	0.042
Arginine	2.120	0.037
Methionine	0.110	0.002
Cysteine	0.000	0.000
Tryptophan	0.000	0.000

Table 2. Aminoacid content in Gum Arabic taken from *Acacia senegal* (Mahendran et al., 2008).

Table 3 shows some physicochemical properties used as international GA quality parameters, for example: moisture, total ash content, volatile matter and internal energy, with reference to gums taken from *Acacia senegal* species in Sudan (FAO, 1990, Larson & Bromley, 1991). The physicochemical properties of GA may vary depending on the origin and age of trees, the exudation time, the storage type, and climate. The moisture content facilitates the solubility of GA carbohydrate hydrophilic and hydrophobic proteins. The total ash content is used to determine the critical levels of foreign matter, insoluble matter in acid, calcium, potassium and magnesium (Mocak et al., 1998). The compositions of cations in the ash residue are used to determine the specific levels of heavy metals in the gum arabic quality (FAO, 1990, 1996). The volatile matter determines the nature and degree of polymerization of the compositions contained in sugar (arabinose, galactose and rhamnose) which exhibits strong binding properties to act as emulsifiers and stabilizers in the manufacture of cough syrups in the pharmaceutical industry (Phillips & Williams, 2001). The GA internal energy is the required energy to produce an amount of carbon by heating at 500 °C to release carbon dioxide. Optical rotation is used to determine the nature of GA sugars as well as to identify the source of production.

Property	Value
Moisture (%)	13 - 15
Ash content (%)	2 - 4
Internal energy (%)	30 - 39
Volatile Matter (%)	51 - 65
Optical rotation (degrees)	(-26) - (-34)
Nitrogen content (%)	0.26 - 0.39
Cationic composition of total ash at 550 °C	
Copper (ppm)	52 - 66
Iron (ppm)	730 - 2490
Manganese (ppm)	69 - 117
Zinc (ppm)	45 - 111

Table 3. International specifications of Gum Arabic quality (FAO, 1990).

Gel permeation chromatography studies using both refractive index and UV(260 nm) absorption detections have confirmed that both *Acacia senegal* and *Acacia seyal* gums consist of three main components (Islam et al., 1997, Idris et al., 1998, Williams & Phillips, 2000, Al Assaf 2006):

i. A main fraction (88-90%) of a polysaccharide of β-(1→3) galactose, highly branched with units of rhamnose, arabinose and glucuronic acid (which is found in nature like salts of magnesium, potassium and calcium). This fraction is called Arabinogalactan (AG) and contains a low protein content (~0.35%) and MW≈ 300 kDa (Renard et al., 2006, Sanchez et al., 2008);
ii. A secondary fraction constituting ~10% of the total, with a protein content of 11% and a molecular weight of 1400 kDa, corresponding to a complex Arabinogalactan-Protein (AGP) (Goodrum et al., 2000), and finally,
iii. A smaller fraction (1% of total) composed by a glycoprotein (GP) consisting of the highest protein content (50 wt%) with an amino acid composition different from the complex AGP (Williams et al., 1990).

Although the total content of carbohydrate fractions of the three components is similar, as reported by Williams et al., 1990, it was found that protein-rich fractions have a significantly lower glucuronic acid content. Circular dichroism studies conducted on different GA fractions showed that only the AGP and GP components have a secondary structure (Renard et al., 2006). The AGP fraction was isolated by gel filtration chromatography and subjected to deglycosylation with hydrofluoric acid (HF) to separate the protein (Qi et al., 1991). About 400 amino acids were contained by the AGP protein fraction (~33% are hydroxyproline residues). In addition, it was shown that the AGP fraction is composed of blocks of carbohydrates attached to the polypeptide chain by covalent bonds through serine and hydroproline residues (Mahendran et al., 2008). Further SDS-PAGE studies conducted after deglycosylation with HF indicated the presence of two proteins moieties, one with a mass of about 30 kDa corresponding to a polypeptide chain of approximately 250 amino acids, and the second one with about 45 amino acids (~5 kDa). This minor protein fraction is thought to be associated with the main AG fraction. It was proposed that in the structure of AGP, the polypeptide chain of 400 amino acids acts as "cable connector" of the blocks of carbohydrates (≤ 40 kDa) which are covalently linked to the protein ("wattle blossom" model) (Fincher et al., 1983; Mahendran et al., 2008).

4. Physicochemical properties

The GA is a heterogeneous material having both hydrophilic and hydrophobic affinities. GA physicochemical responses can be handled depending on the balance of hydrophilic and hydrophobic interactions. GA functional properties are closely related to its structure, which determines, for example, solubility, viscosity, degree of interaction with water and oil in an emulsion, microencapsulation ability, among others.

4.1 Solubility and viscosity

GA has high water solubility and a relatively low viscosity compared with other gums. Most gums cannot dissolve in water in concentrations above 5% due to their high viscosity. Instead, GA can get dissolved in water in a concentration of 50% w/v, forming a fluid solution with acidic properties (pH ~4.5). The highly branched structure of the GA molecules leads to compact relatively small hydrodynamic volume and, consequently GA will only become a viscous solution at high concentrations. Solutions containing less than 10% of GA have a low viscosity and respond to Newtonian behavior (Williams et al., 1990). However, steric interactions of the hydrated molecules increase viscosity in those solutions containing more than 30% of GA resulting in an increasingly pseudoplastic behavior. Its high stability in acidic solutions is exploited to emulsify citrus oils. The viscosity of GA solutions can be modified by the addition of acids or bases as these ones change the electrostatic charge on the macromolecule. In very acidic solutions, acid groups neutralize so inducing a more compact conformation of the polymer which leads to a decreased viscosity; while a higher pH (less compact molecule) results in maximum viscosity around pH 5.0-5.5. In very basic solutions, the ionic strength increment reduces the electrostatic repulsion between GA molecules producing a more compact conformation of the biopolymer and thus reducing the viscosity of the solution (Anderson et al., 1990; Williams et al., 1990).

4.2 Emulsifying properties

GA is well recognized as emulsifier used in essential oil and flavor industries. Randall et al., 1998, reported that the AGP complex is the main component responsible for GA ability to stabilize emulsions, by the association of the AGP amphiphilic protein component with the surface of oil droplets, while the hydrophilic carbohydrate fraction is oriented toward the aqueous phase, preventing aggregation of the droplets by electrostatic repulsion. However, only 1-2% of the gum is absorbed into the oil-water interface and participates in the emulsification; thus, over 12% of GA content is required to stabilize emulsions with 20%

orange oil (Williams et al., 1990). If there is not enough GA amount to cover all the gum drops, unstable emulsion is formed and flocculation and coalescence occurs.

4.3 Molecular association

It is well known the tendency of polysaccharides to associate in aqueous solution. These molecular associations can deeply affect their function in a particular application due to their influence on molecular weight, shape and size, which determines how molecules interact with other molecules and water. There are several factors such as hydrogen bonding, hydrophobic association, an association mediated by ions, electrostatic interactions, which depend on the concentration and the presence of protein components that affect the ability to form supramolecular complexes.

Al-Assaf et al., 2007, showed that molecular associations in GA can lead to an increase in molecular weight in the solid state by maturation under controlled heat and humidity. The process does not involves change in the basic structural components and, while the maturation takes place, the level of association increases giving way to AGP with higher molecular weight and protein content. This process mimics the biological process which produces more AGP throughout the tree growth, and gets maturation to continue during the storage of GA after harvest. Subsequently, Al-Assaf et al., 2009, analyzed the role of protein components in GA to promote molecular association when the gum is subjected to different processing treatments such as maturation, spray drying and irradiation. Results demonstrate the ability of protein components to promote hydrophobic associations that influence the size and proportion of the high molecular weight component AGP. When GA undergoes maturation (solid state heat treatment) there is an increase in the hydrophobic nature of the gum and therefore an increase of its emulsifying properties. Spray drying involves not only the aggregation through hydrophobic associations but also changes in the surface properties of peptide residues increasing GA hydrophilicity compared with the association promoted by the treatment of maturity in the solid state. Ionizing radiation in both aqueous solutions and solid state induces cross-linking between polysaccharide blocks by the formation of -C-C- bonds.

It was also reported that, by using mild UV-radiation, it is possible to induce GA cross-linking (Kuan et al., 2009). The process reduced the solution viscosity and improved emulsification properties. This GA modification can be used in food products requiring better reduced viscosity emulsifying properties such as dressings, spreads, and beverages, as well as in other nonfood products such as lithographic formulations, textiles, and paper manufacturing.

5. Pharmacological action

Although GA is being widely used as an experimental vehicle for drugs in physiological and pharmacological experiments, and it is supposed to be an inert substance, recent reports have confirmed that GA has some biological properties as an antioxidant (Trommer & Neubert, 2005; Ali & Al Moundhri, 2006, Hinson et al., 2004) on the metabolism of lipids (Tiss et al., 2001, Evans et al., 1992), positive contribution in treating kidney, (Matsumoto et al., 2006; Bliss et al., 1996, Ali et al., 2008), cardiovascular (Glover et al., 2009) and gastrointestinal diseases (Wapnir et al., 2008, Rehman et al., 2003).

GA has been extensively tested for its properties as non-digestible polysaccharide which can reach the large intestine without digestion; in the small intestine, it can be classified as dietary fiber. Due to its physical properties, it reduces glucose absorption, increases fecal mass, bile acids and has the potential to beneficially modify the physiological state of humans (Adiotomre et al., 1990). GA is slowly fermented by the bacterial flora of the large intestine producing short chain fatty acids (Annison et al., 1995). Therefore, its tolerance is excellent and can be consumed in high daily doses without intestinal complications. In addition, GA is able to selectively increase the proportion of lactic acid bacteria and biphidus bacteria in healthy subjects.

A daily intake of 25 and 30 g of GA for 21 to 30 days reduced total cholesterol by 6 and 10.4%, respectively (Ross et al., 1983, Sharma 1985). The decrease was limited only to LDL and VLDL, with no effect on HDL and triglycerides. However, Topping et al. (1985) reported that plasma cholesterol concentrations were not affected by the supply of GA, but triglyceride concentration in plasma was significantly lower than in controls.

Various mechanisms have been proposed to explain the hypocholesterolemic effect of GA (Annison et al., 1995; Tiss et al., 2001). Some studies have suggested that the viscosity of fermentable dietary fiber contributes substantially to the reduction of lipids in animals and humans (Gallaher et al., 1993; Moundras et al., 1994). However, other studies suggested that this property is not related to plasma lipids (Evans et al., 1992). The mechanism involved is clearly linked to increased bile acid excretion and fecal neutral sterol or a modification of digestion and absorption of lipids (Moundras et al., 1994).

6. Applications

GA is being widely used for industrial purposes such as a stabilizer, a thickener, an emulsifier and an encapsulating in the food industry, and to a lesser extent in textiles, ceramics, lithography, cosmetic, and pharmaceutical industry (Verbeken et al., 2003). In the food industry, GA is primarily used in confectionery, bakery, dairy, beverage, and as a microencapsulating agent.

6.1 Confectionery and baking

GA is employed in a variety of products including gum, lozenges, chocolates, and sweets. In these products, GA performs two important functions: to delay or to prevent sugar crystallization, and to emulsify fat to keep it evenly distributed throughout the product. In baking, GA is extensively used for its low moisture absorption properties. GA solubility in cold water allows greater formation of clear solutions than in sugar solutions. It has also favorable adhesive properties to be used in glace and meringues, and it provides softness when used as an emulsion stabilizer. Baking properties of wheat and rye flours can be improved by adding a small amount of GA since its capacity for retaining moisture reduces the hardening of bread.

6.2 Dairy products

GA is used as a stabilizer in frozen products like ice-cream due to its water absorption properties. The role of GA in these products is to cause a fine texture and growth by inhibiting the formation of ice crystals which is achieved by combining a large amount of water and holding it as water of hydration, being its higher melting point the main attraction of ice-cream.

6.3 Beverages

GA is used as an emulsifier in beverages such as citrus juices, beer, and cola drinks. GA ability to stabilize foams is used in the manufacture of beer and soft drinks. Besides, it can be used for clarifying wines.

6.4 Microencapsulation

In the food industry, microencapsulation is an important process to improve the chemical stability of sensitive compounds, to provide the controlled release of microencapsulated compounds and to give a free flowing powder with improved handling properties (Anandaraman and Reineccius, 1986; Sheu and Rosenberg, 1993). The encapsulating material must preserve and protect the encapsulated compounds during manufacture, storage, and handling to release them into the final product during manufacture or consumption.

Solubility and low viscosity emulsion properties have facilitated the use of GA as an encapsulating agent for retention and protection of chemically reactive and volatile commercial food flavoring. Reineccius (1988) has reported on the encapsulation of orange oil using GA as wall material. Its main drawback is its cost for the oversubscription. However, due to its efficacy with regard to other wall materials such as maltodextrin (Krishnan et al., 2005) and modified starch, reported by various studies (Reineccius, 1989), the cost may not be relevant as long as extra protection or stability are achieved for microencapsulated high-value products, and in food or pharmaceutical fields.

GA is mainly used for fat microencapsulation because it produces stable emulsions in the case of most oils in a wide pH range, and it has the ability to form films (Kenyon, 1995). Barbosa et al., 2005 studied the photostability of the microencapsulated carotenoid bixin in different edible polysaccharide. They found out that microencapsulated bixin in GA was three to four times more stable than the one microencapsulated with maltodextrin, and about ten-fold than in homogeneous solvents.

7. Antioxidant action

Several reports suggest that GA has antioxidant capacity. However, there are controversial results of it, mainly *in vivo* studies. For example, GA has been reported to exert a protective effect against gentamicin and cisplatin nephrotoxicity (Al-Majed et al., 2002, 2003), and doxorubicin cardiotoxicity (Abd-Allah et al., 2002) used as biological models in rats. However, Ali et al., (2003) reported that treatment of rats with GA causes only a slight palliative effect of gentamicin nephrotoxicity. Later, Trommer & Neubert (2005) studied lipid peroxidation antioxidant and reducing effects in vitro of various polysaccharides (including GA). They found that GA reduces lipid peroxidation of skin in a dose-dependent. In contrast, Ali (2004) reported that administration of GA at concentrations of 2.5%, 5.0% and 10.0% in drinking water for eight consecutive days to rats did not significantly alter the concentrations of free radical scavenger's glutathione (GSH) and acid ascorbic acid (AA), and superoxide dismutase (SOD), or lipid peroxidation.

Consequently, the antioxidant activity of GA in biological systems is still an unresolved issue, and therefore it requires a more direct knowledge of the antioxidant capacity of GA that can be obtained by *in vitro* experiments against different types of oxidant species. The total antioxidant activity of a compound or substance is associated with several processes that include the scavenging of free radical species (eg. HO•, ROO•), ability to quench reactive excited states (triplet excited states and/or oxygen singlet molecular $^1O_2$), and/or sequester of metal ions (Fe^{2+}, Cu^{2+}) to avoid the formation of HO• by Fenton type reactions. In the following sections, we will discuss the *in vitro* antioxidant capacity of GA for some of these processes.

7.1 Quenching of electronically excited states

From the viewpoint of food and biological systems, Vitamin B_2 (riboflavin, RF) is a widely distributed molecule. **Table 4** shows the most relevant spectroscopic and photophysical properties in aqueous solution (Valle et al., 2011). As many isoalloxasine derivatives, RF is a well known photosensitizer by absorbing near UV radiation and visible light (blue-edge), generating both reactive excited states and reactive oxygen species (ROS) in the presence of ground state molecular oxygen ($^3O_2$), **Scheme 1**.

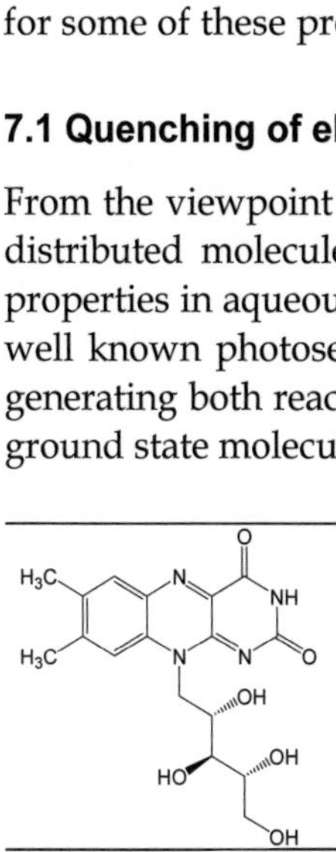	**λ_{ab}^{max} (nm) [$\varepsilon_\lambda^{max}$×$10^{-3}$ ($M^{-1}cm^{-1}$)]:**	267 [2.94]; 375 [0.96]; 445 [1.12]
	λ_{fl}^{max} (nm); τ_{fl} (ns); Φ_{fl}; E_S (kJ/mol):	520; 4.7; 0.2; 239
	λ_T^{max} (nm); τ_T (μs); Φ_T; E_T (kJ/mol):	720; 25; 0.58; 209

Table 4. Chemical structure and spectroscopical and photophysical properties of riboflavin (RF) in aqueous phosphate buffer pH 7.4 (Valle et al, 2011).
λ_{ab}^{max}, λ_{fl}^{max}, λ_T^{max} are the spectral maximum of UV-Vis absorption, emission, and triplet state absorption, respectively. ε_λ is the molar extinction coefficient. τ_{fl} and τ_T, Φ_{fl} and Φ_T, E_S and E_T; are the lifetime, quantum yield, and energy content of the singlet and triplet excited states of RF, respectively.

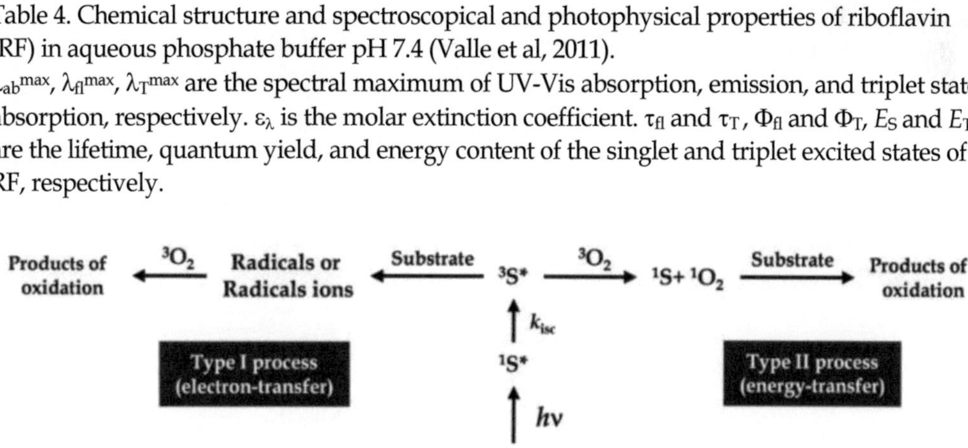

Scheme 1. Photosensitized Type I and Type II oxidative processes involving a sensitizer molecule S.

Depending on the relative concentration of reactive substrate and dissolved molecular oxygen ($^3O_2$), RF is able to induce photosensitized oxidation of molecular targets by either Type I (electron-transfer) or Type II (energy-transfer) processes (Foote, 1991). In Type I process, the excited triplet state of RF, i.e. $^3RF^*$, can react with aminoacids residues (A) of peptides and proteins to produce radicals ions pair $RF^{\bullet-}$ and $A^{\bullet+}$ (eqns 1 and 2). The semi reduced radical anion $RF^{\bullet-}$ undergoes secondary reactions that can subsequently generate ROS, such as superoxide anion $O_2^{\bullet-}$, hydrogen peroxide H_2O_2, eqns 3-6. In the presence of heavy metal cations, e.g. Fe^{2+}, H_2O_2 can produce the very reactive hydroxyl radical $HO^{\bullet}$ by Fenton reaction, eqn. 7.

$$RF \xrightarrow{h\nu} {}^1RF^* \xrightarrow{k_{isc}} {}^3RF^* \tag{1}$$

$$^3RF^* + A \xrightarrow{k2} A^{\bullet+} + RF^{\bullet-} \tag{2}$$

$$RF^{\bullet-} + H^+ \xrightarrow{k3} RFH^{\bullet} \tag{3}$$

$$2RFH^{\bullet} \xrightarrow{k4} RFH_2 + RF \tag{4}$$

$$RF^{\bullet-} + {}^3O_2 \xrightarrow{k5} RF + O_2^{\bullet-} \tag{5}$$

$$RFH_2 + {}^3O_2 \xrightarrow{k6} RF + H_2O_2 \tag{6}$$

$$H_2O_2 \xrightarrow{M^{n+}/\text{Fenton}} 2HO^{\bullet} \tag{7}$$

On the other hand, Type II process competes efficiently with the electron-transfer pathway in aerobic environments where the concentration of ground triplet state molecular oxygen is relatively high (~0.27 mM), and singlet molecular oxygen ($^1O_2$) is the most abundant ROS generated under these conditions, with a quantum yield $\Phi_\Delta \approx 0.48$ (Valle et al., 2011), eqn. 8. It is also possible an electron-transfer reaction from $^3RF^*$ to $^3O_2$ to form anion superoxide, but this reaction occurs with very low efficiency <0.1% (Lu et al., 2000).

$$^3RF^* + {}^3O_2 \xrightarrow{k8} RF + {}^1O_2 \tag{8}$$

In turn, $^1O_2$ is a very electrophilic excited state species of molecular oxygen that interacts efficiently with electron-rich molecules, such as aminoacid residues of proteins like histidine, metionine, tryptophan, tyrosine, etc., by both physical and chemical quenching processes, eqns. 9 and 10 (Davies, 2003; Bisby et al., 1999).

$$^1O_2 + A \xrightarrow{kq} {}^3O_2 + A \tag{9}$$

$$^1O_2 + A \xrightarrow{kr} AO_2 \text{ (oxidation product)} \tag{10}$$

A larger k_q/k_r ratio for the molecule A, better is its ability as catalytic quencher, since physical quenching of $^1O_2$ (eqn. 9) does not consume the antioxidant molecule (Montenegro et al, 2002; Morán Vieyra et al., 2009).

Therefore, it is a very relevant issue the evaluation of molecules and macromolecules that can efficiently act as quenchers of electronically excited states, such as $^3RF^*$ and $^1O_2$ as examples, to avoid the formation of ROS and/or eliminate them (Wondrak et al., 2006).

Figure 1a shows that the addition of GA (Powdered food grade MW = 3.5 x 10^5 g/mol of Colloids Naturels Brazil, Saõ Paulo, Brazil) in RF aqueous solutions produces additive absorption changes, since GA showed a slight scattering effect (dashed line spectrum) that distorted the absorption spectrum of RF at the UV edge. This effect did not modify the fluorescence emission spectra of RF obtained by excitation at 445 nm. These results indicate that no strong interactions are occurring with GA either in the ground or singlet excited state of RF.

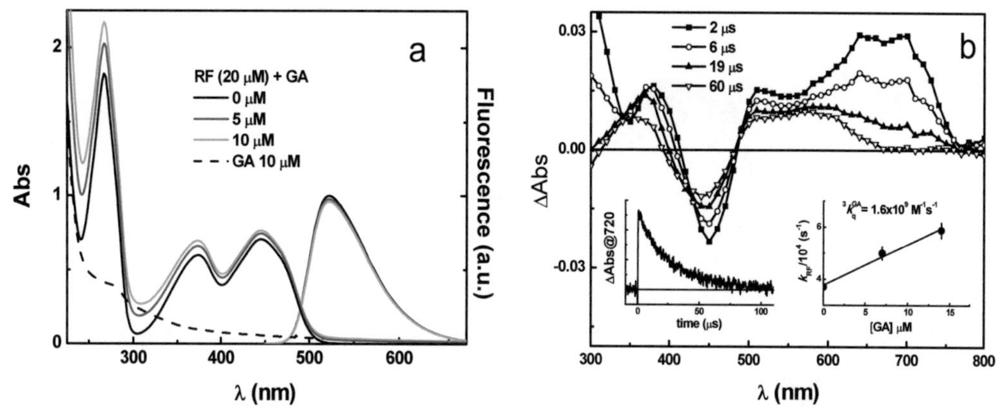

Fig. 1. **a)** UV-Vis absorption and fluorescence emission spectra of riboflavin (RF, 20 μM) and Gum Arabic aqueous solutions at pH 7 (phosphate buffer 100 mM). **b)** Transient absorption spectra of RF (35 μM) in N_2-saturated MeOH-Water (1:1) solution. The insets show the transient decay at 720 nm for the $^3RF^*$ species and the Stern-Volmer plot for the quenching of $^3RF^*$ by GA, eqn 11.

Figure 1b shows the transient absorption spectra of RF (i.e. the difference between the ground singlet and excited triplet states) obtained by laser-flash photolysis using a Nd:Yag pulsed laser operating at 355 nm (10 ns pulse width) as excitation source. At short times after the laser pulse, the transient spectrum shows the characteristic absorption of the lowest vibrational triplet state transitions (0 ← 0) and (1 ← 0) at approximately 715 and 660 nm, respectively. In the absence of GA, the initial triplet state decays with a lifetime around 27 μs in deoxygenated solutions by dismutation reaction to form semi oxidized and semi reduced forms with characteristic absorption bands at 360 nm and 500-600 nm and (Melø et al., 1999). However, in the presence of GA, the $^3RF^*$ is efficiently quenched by the gum with a bimolecular rate constant $^3k_q^{GA} = 1.6\times10^9$ $M^{-1}s^{-1}$ calculated according to eqn 11, where k_{RF} and $k_{RF}{}^0$ represent the observed rate constant for the triplet decay in presence and absence of GA, respectively.

$$k_{RF} = k_{RF}{}^{0} + {}^{3}k_q^{GA}[GA] \qquad (11)$$

The value of $^3k_q^{GA}$ was similar to those obtained for the quenching of $^3RF^*$ by aminoacids such as histidine (His) and tyrosine (Tyr) with $^3k_q^{His} = 3.8\times10^8$ $M^{-1}s^{-1}$ y $^3k_q^{Tyr} = 1.8\times10^9$ $M^{-1}s^{-1}$, respectively. The values of $^3k_q^{His}$ and $^3k_q^{Tyr}$ obtained were in very good agreement with 2.0×10^8 $M^{-1}s^{-1}$ and 1.4×10^9 $M^{-1}s^{-1}$ reported by Huvaere and Skibsted (2009) and Cardoso et al.,

(2004), respectively. Thus, the quenching of $^3RF^*$ can be associated with the protein moiety of GA, which is rich in His and Tyr, see **Table 2**. Further evidence of the reaction of $^3RF^*$ with aminoacid residues of GA, is the blue-shift of about 20 nm observed for the long-lived transient band at the UVA (~350 nm) with the increment of the concentration of GA, **Figure 2a**. The same behavior was previously reported by Lu & Liu (2002) for the reaction of the $^3RF^*$ with Tyr or Trp due to the formation of the neutral radicals $RFH^•$ and $Tyr^•/Trp^•$ by one-electron transfer coupled with proton-transfer.

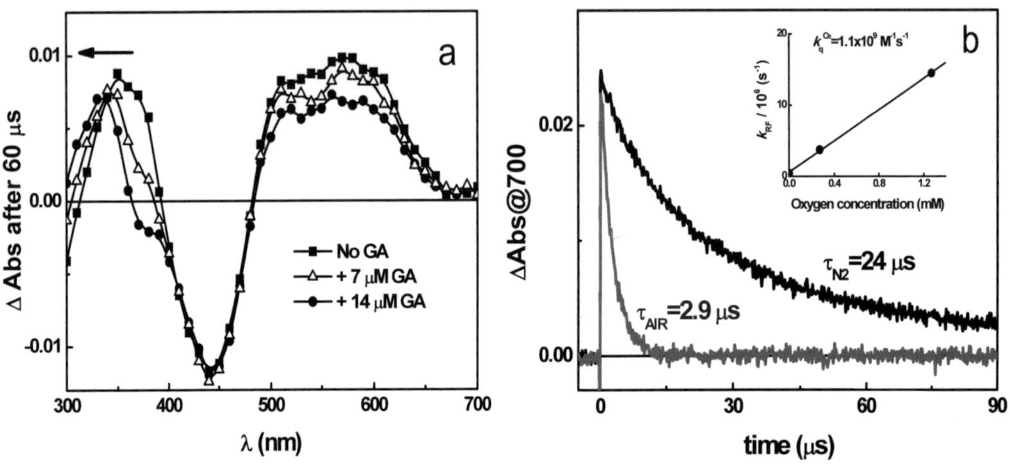

Fig. 2. **a)** Transient absorption spectra of RF (35 μM) in N_2-saturated MeOH-Water (1:1) solution observed after 60 μs of the laser pulse as a function of the concentration of GA. **b)** Effect of dissolved molecular oxygen ($^3O_2$) on the decay of the $^3RF^*$ at 700 nm. Inset: Stern-Volmer plot for the quenching of $^3RF^*$ by $^3O_2$.

Nevertheless, in aerobic media, molecular oxygen $^3O_2$ competes efficiently with GA for the interaction with $^3RF^*$, as showed by the large decreases of the decay time of $^3RF^*$ in air-saturated aqueous solutions, **Figure 2b.** The bimolecular quenching rate constant of $^3RF^*$ by $^3O_2$, i.e. $k_q^{O_2}$ = 1.1×10^9 $M^{-1}s^{-1}$ (inset of **Figure 2b**). This value is almost the same that the obtained by the quenching of $^3RF^*$ by GA, see above. Thus, the predominant quenching process will modulated by the relative molar concentration between GA and $^3O_2$. Normally, the concentration of dissolved oxygen in air-saturated water at standard conditions is about 270 μM (Murov et al., 1997), and the concentration of GA it will depend on the type of food and pharmacological preparations. In any case, the combinatory effect will quench efficiently the harmful riboflavin excited triplet state. However, according to eqn 8, th quenching if $^3RF^*$ by $^3O_2$ produces mainly singlet molecular oxygen, $^1O_2$, and ground state RF. In a previous work, we reported that the total (physical + chemical) quenching rate constant of $^1O_2$ by GA, i.e. k_t^{GA} = 2.7×10^7 $M^{-1}s^{-1}$, as obtained by using time-resolved near-infrared phosphorescence detection of $^1O_2$ (Faria et al., 2010). In order to separate the contribution of physical and chemical quenching, we monitored the consumption of dissolved $^3O_2$ by GA using a FOXY-R oxygen-sensitive luminescent sensor coupled with to CCD detector from OceanOptics, **Figure 3a**. The dye methylene blue was used as sensitizer

($\Phi_\Delta \approx 0.52$) and histidine as actinometer, since this aminoacid reacts completely with $^1O_2$ with $k_r^{His} = 4 \times 10^7$ $M^{-1}s^{-1}$, (Bisby et al., 1999).

Under photostationary conditions, the slopes of the linear plots of the consumption of dissolved oxygen are the observed pseudo-first order rate constant of the chemical quenchers, k_{obs} (Criado et al., 2008), and the rate constant for the reactive quenching of $^1O_2$ by GA is calculated with eqn. 12.

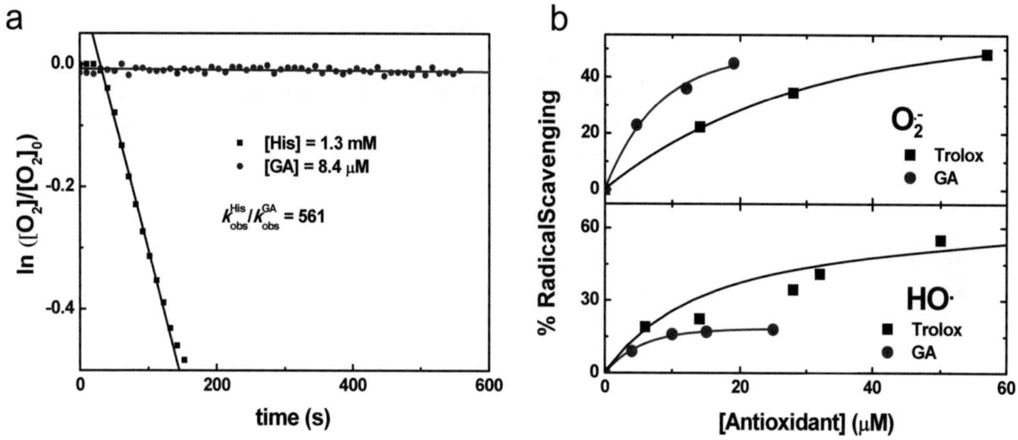

Fig. 3. a) First order plot of oxygen uptake in the Methylene-blue (MB)-sensitized photooxidation of GA 8.4 μM and 1.3 mM histidine (control) in phosphate buffer pH 7. b) Percentage radical scavenging activity for the control molecule Trolox and GA at pH 7.4 in phosphate buffer 10 mM (hydroxyl radical) and pH 10 in sodium carbonate buffer 50 mM (anion superoxide radical).

$$k_r^{GA} = k_r^{His} \times \frac{k_{obs}^{GA}[His]}{k_{obs}^{His}[GA]} \quad (12)$$

In the present case, $k_r^{GA} = 1.1 \times 10^7$ $M^{-1}s^{-1}$, a typical value for the reaction of aminoacid moieties with $^1O_2$ (Michaeli & Feitelson, 1994; Bisby et al., 1999). By comparison with the total quenching rate constant, $k_t^{GA} = 2.7 \times 10^7$ $M^{-1}s^{-1}$, it can be concluded that almost 60% of the interaction with $^1O_2$ is through physical quenching and about 40% of the reactive moieties of GA are oxidized by $^1O_2$.

In addition, Montenegro et al., (2007) determined that the photosensitized RF-mediated degradation of vitamins A, D_3, and RF itself in skimmed milk was strongly reduced by the addition of small amounts of lycopene-gum arabic-sucrose microcapsules, prepared by spray-drying. Under these conditions, the bulk properties of the skimmed milk were unmodified. The main photoprotection mechanism of the milk vitamins was the efficient quenching of the $^3Rf^*$ by the protein moiety of GA. Small contributions (<5%) to the total photoprotection percentage was due to both inner filter effect and $^1O_2$ quenching by the microencapsulated lycopene.

These results show the functional ability of GA to act as quencher of electronically excited states in food systems either as non-processed gum or spray-drying microencapsulated preparations.

7.2 Scavenging of free radical species

The capacity of GA to scavenge *in vitro* the chemically generated free radicals hydroxyl ($HO^{\bullet}$) and superoxide anion ($O_2^{\bullet -}$) was determined by the Trolox Equivalent Antioxidant Capacity (TEAC) assay (Huang, 2005; Gliszczyńska-Świgł, 2006). The $HO^{\bullet}$ was generated by Fenton reaction at pH 7.4 (Aruoma, 1994), and the reaction was monitored spectrophotometrically at 532 nm, color created by the adduct formed between thiobarbituric acid with malonaldehyde, the product of oxidation of deoxyribose by $HO^{\bullet}$ (Gutteridge & Halliwell, 1988). The superoxide anion was detected by using the Nitro blue tetrazolium (NBT) method, as described by Sabu & Ramadasan, 2002. In this method the generation of $O_2^{\bullet -}$ is performed by auto-oxidation of hydroxylamine hydrochloride in presence of NBT, which gets reduced to nitrite, which in presence of EDTA gives a color measured at 560 nm. The radical scavenging (RS) activity of GA was reported as the percentage of inhibition of color formation, and calculated according to eqn. 13:

$$\%RS = \frac{A_0 - A_t}{A_0} \times 100 \qquad (13)$$

where A_0 in the absorbance value obtained for the control solution without GA or Trolox (TX), A_t is the absorbance value in the presence of these molecules. All the tests were performed by triplicate and, **Figure 3b** shows the increment of radical scavenging capacity with the TX or GA concentration. By comparing the initial linear slopes between the GA and TX curves (Re et al., 1999), obtained by fitting of experimental data with a second-order polynomial function, the TEAC value of 1.82 and 0.71 was calculated for the scavenging of $O_2^{\bullet -}$ and $HO^{\bullet}$, respectively. These results are consistent with the results reported by Liu et al., 1997, for the scavenging of $O_2^{\bullet -}$ and $HO^{\bullet}$ by polysaccharides extracts of mushrooms. These authors found that the radical scavenging efficiency by the polysaccharides was higher for $O_2^{\bullet -}$, and also increased by the protein content of the polysaccharide-protein complex.

8. Antimicrobial action

As for the antimicrobial activity of GA, few studies have been performed, mainly reporting growth inhibitory activity of certain periodontal pathogenic species (causal agent of tooth decay or agent involved in the plaque), such as *Prophyromonas gingivalis* and *Prevotella intermedia* (Clark et al., 1993). These results suggested that GA could inhibit the formation of plaque and improve dental remineralization, acting as a potential preventive agent in the formation of caries (Onishi et al., 2008). Such effects are attributed to the high salt content of Ca^{2+}, Mg^{2+} and K^{+} of polysaccharides in GA, and the effect of the gum in the metabolism of Ca and possibly phosphate. It is also known that cyanogenic glucosides and GA contains many types of enzymes such as oxidases, peroxidases, and pectinases, some of which have antimicrobial properties (Tyler et al., 1997; Kirtikar & Basu, 1984).

Saini et al., 2008 studied the antimicrobial effect of different acacia species, including *A. senegal*, using different plant parts (pods, bark, etc.), against three strains of Gram positive (*Bacillus cereus, Escherichia coli, Salmonella typhi*), two gram negative (*Pseudomonas aeruginosa, Staphylococcus aureus*) and three fungal strains (*Candida albicans, Aspergillus niger* and *Microsporum canis*). The study revealed that methanol extracts of the species *A. catechu* and *A. nilotica* showed the highest antimicrobial activity. This is due to the presence of hydrophilic components such as polyphenols, polysaccharides and tannins present in one or more parts of the plant. The hexane extracts of these species also showed significant activity. As for *A. senegal*, using the bark, determined that the hexane extract showed antimicrobial activity (AMA) against *S. aureus* and the fungus *C. albicans*, while the methanol extract showed AMA against *E. coli, B. cereus*, and fungi *C. albicans* and *A. niger*.

However, to date, no AMA studies have been conducted against spore-forming microorganisms. It is very important to search for a compound having action on the development of such organisms as *Bacillus subtilis* and *B. cereus*, since these microorganisms are able to withstand pasteurization conditions and contain hydrolytic enzymes, which generate off-flavor in the food.

For this reason, we evaluated the AMA of GA obtained from *Acacia* trees against "off-flavor" microbial producers in food such as *Pseudomonas aeruginosa, Bacillus subtilis* and *Micrococcus luteus*. **Figure 4a** shows the growth inhibition diameters of the three organisms analyzed by solutions with increasing concentration from 5 to 50 μM, as determined by the agar diffusion method (Ferreira et al., 2004). These results indicate that GA exerts only a moderated AMA against *P. aeruginosa*, while *Micrococcus luteus* and *Bacillus sutilis* were unaffected by the presence of GA. However, the in the presence of GA the inhibition zones measured were translucent, indicating that a bacteriostatic effect occurs, meaning that growth slows, but does not destroy the bacteria.

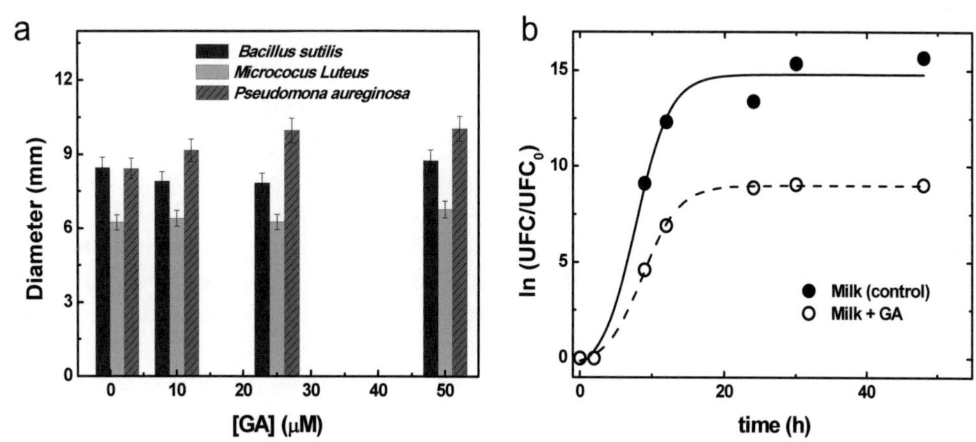

Fig. 4. a) Inhibition of microbial growth by GA, from *Bacillus subtilis, Micrococcus luteus* and *Pseudomonas aeruginosa*. b) Growth curve of *Bacillus subtilis* at 32 ° C (●) Milk, (O) milk with added GA 47 μM.

Additionally, the AMA of GA against *Bacillus subtilis* was evaluated in milk. The growth rates (h^{-1}) of the microbial culture in milk with and without added GA stored at 32 °C were determined from the slopes of the linear portion of the growth curves of the organism, which were built by standard plate counts determined according to the IDF standard No 100 B (IDF, 1991), **Figure 4b**. The growth rate determined for *Bacillus subtilis* in milk was 0.92 h^{-1}, whereas with the addition of GA in 47 μM the growth rate decreased almost 40% (0.56 h^{-1}), confirming a bacteriostatic effect on *Bacillus subtilis* by GA.

9. Conclusions

GA is a natural biopolymer with wide industrial use as a stabilizer, a thickener, an emulsifier and in additive encapsulation not only in food industry but also in textiles, ceramics, lithography, cosmetic and pharmaceutical industry (Verbeken et al., 2003).

Besides all the sensory and texturizing properties, GA has interesting antioxidant properties such as an efficient capacity for deactivation of excited electronic states and moderated radical scavenging capacity. There is increasing experimental evidence that associate the antioxidant function with its protein fraction, mainly by amino acid residues such as histidine, tyrosine and lysine, which are generally considered as antioxidants molecules (Marcuse, 1960, 1962; Park et al., 2005).

In summary, besides its use as texturing additive, the combined functionality of GA is an advantage over other edible biopolymers that do not show antioxidant activities

10. Acknowledgments

The authors thank the Argentinian Funding Agencies CICyT-UNSE, CONICET, and ANPCyT. CDB thanks also Alexander von Humboldt Foundation for a Georg Foster Fellowship.

11. References

Abd-Allah, A.R.; Al-Majed, A.A.; Mostafa, A.M.; Al-Shabanah, O.A.; Din, A.G. & Nagi, M.N. (2002). Protective effect of arabic gum against cardiotoxicity induced by doxorubicin in mice: a possible mechanism of protection. *Journal of Biochemical and Molecular Toxicology*, Vol.16, No.5, (August 2002), pp. 254–259, 1095-6670.

Abdel Nour, H. O. Gum arabic in Sudan: production and socio-economic aspects. In: Medicinal, Culinary and Aromatic Plants in the Near East. Food and Agriculture Organization of The United Nations, Cairo, 1999. Internet reference fao.org/docrep/x5402e/x5402e12.htm.

Adiotomre, J.; Eastwood, M. A.; Edwards, C. A. & Brydon, W. G. (1990). Dietary fiber: in vitro methods that anticipate nutrition and metabolic activity in humans. *American Journal of Clinical Nutrition*, Vol.52, No.1, (January 1990), pp. 128–134, ISSN 0002-9165.

Al-Assaf, S.; Phillips, G.O. &. Williams, P.A. (2005 a). Studies on acacia exudate gums. Part I: the molecular weight of Acacia senegal gum exudate. Food Hydrocolloids, Vol.19, No.4, (July 2005), pp. 647-660, ISSN: 0268-005X.

Al-Assaf, S.; Phillips, G.O. &. Williams, P.A. (2005 b). Studies on Acacia exudate gums: part II. Molecular weight comparison of the Vulgares and Gummiferae series of Acacia gums. Food Hydrocolloids, Vol.19, No.4, (July 2005), pp. 661-667, ISSN: 0268-005X.

Al-Assaf, S.; Phillips, G. O. & Williams, P. A. (2006). Controlling the molecular structure of food hydrocolloids. Food Hydrocolloids, Vol.20, No.2-3, (March-May 2006), pp. 369–377, ISSN: 0268-005X.

Al-Assaf, S.; Phillips, G.O.; Aoki, H. & Sasaki, Y. (2007). Characterization and properties of Acacia senegal (L.) Willd. var. senegal with enhanced properties (Acacia (sen) SUPER GUM™): Part 1-Controlled maturation of Acacia senegal var. senegal to increase viscoelasticity, produce a hydrogel form and convert a poor into a good emulsifier *Food Hydrocolloids*, Vol.21, No.3, (May 2007), pp. 319-328, ISSN: 0268-005X.

Al-Assaf, S.; Sakata, M.; McKenna, C.; Aoki, H. &. Phillips, G.O. (2009). Molecular associations in acacia gums. *Structural Chemistry*, Vol.20, No.2 (April 2009), pp. 325–336, ISSN: 1040-0400.

Al-Majed, A.A.; Abd-Allah, A.R.; Al-Rikabi, A.C.; Al-Shabanah, O.A. & Mostafa, A.M. (2003). Effect of oral administration of Arabic gum on cisplatin-induced nephrotoxicity in rats. *Journal of Biochemical and Molecular Toxicology*, Vol.17, No.3, (January 2003), pp. 146–153, ISSN: 1099-0461.

Al-Majed, A.A.; Mostafa, A.M.; Al-Rikabi, A.C. & Al-Shabanah, O.A. (2002). Protective effects of oral arabic gum administration on gentamicin-induced nephrotoxicity in rats. *Pharmacological Research*, Vol.46, No.5, (November 2002), pp. 445–451, ISSN: 1043-6618.

Ali, B.H.; Al-Qarawi, A.A.; Haroun, E.M. & Mousa, H.M. (2003). The effect of treatment with G.A. on gentamicin nephrotoxicity in rats: a preliminary study. Renal Failure, Vol.25, No.1, (January 2003), pp.15–20, ISSN: 0886-022X.

Ali, B.H. (2004). Does GA have an antioxidant action in rat kidney? Renal Failure, Vol.26, No.1, (January 2004), pp. 1-3, ISSN: 0886-022X.

Ali, B.H. & Al Moundhri, M.S. (2006). Agents ameliorating or augmenting the nephrotoxicity of cisplatin and other platinum compounds: a review of some recent research. Food and Chemical Toxicology, Vol. 44, No.8, (August 2006), pp. 1173–1183, ISSN: 0278-6915.

Ali, A.A.; Ali, K.E.; Fadlalla, A. & Khalid, K.E. (2008). The effects of GA oral treatment on the metabolic profile of chronic renal failure patients under regular haemodialysis in Central Sudan. Natural Product Research, Vol.22, No.1, (January 2008), pp.12–21, ISSN: 1478-6419.

Anandaraman, S. & Reineccius, G. A. (1986). Stability of encapsulated orange peel oil. *Food Technology*, Vol.40, No.11, (nd), pp. 88-93, ISSN: 0015-6639.

Anderson, D. M. W.; Brown Douglas, D. M.; Morrison, N. A. & Weiping, W. (1990). Specifications for gum arabic (Acacia Senegal); analytical data for samples collected between 1904 and 1989. *Food Additives and Contaminants*. Vol. 7, No.3, (June 1990), pp. 303-321, ISSN: ISSN 0265-203X.

Annison, G.; Trimble, R. P. & Topping, D. L. (1995). Feeding Australian acacia gums and gum arabic leads to non-Starch polysaccharide accumulation in the cecum of rats. *Journal of Nutrition*, Vol. 125, No.2, (February 1995), pp. 283–292, ISSN 0022-3166.

Aruoma, O. I. (1994). Deoxyribose assay for detecting hydroxyl radicals. Methods in Enzymology, Vol. 233, pp. 57-66, ISBN: 978-0-12-182148-7.

Barbosa, M.I.; Borsarelli, C.D. & Mercadante, A. Z. (2005). Light Stability of Spray-Dried Bixin Encapsulated with Different Edible Polysaccharide Preparations. Food Research International,Vol. 38, No. 8-9, (October-November 2005), pp 989–994, ISSN: 0963-9969.

Bisby, R. H.; Morgan, C. G.; Hamblet, I, & Gorman, A. A. (1999). Quenching of Singlet Oxygen by Trolox C, Ascorbate, and Amino Acids: Effects of pH and Temperature. *Journal of Physical Chemistry A.*, Vol.103, No. 37, (August 1999), pp. 7454–7459, ISSN 1089-5639.

Bliss, D.Z.; Stein, T.P.; Schleifer, C.R. & Settle, R.G. (1996). Supplementation with G A fiber increases fecal nitrogen excretion and lowers serum urea nitrogen concentration in chronic renal failure patients consuming a low-protein diet. *The American Journal of Clinical Nutrition,* Vol. 63, No.3, (March 1996), pp. 392–398, ISSN 0002-9165.

Cardoso, D. R.; Franco, D. W.; Olsen K.; Andersen, M. L. & Skibsted, L. H. (2004). Reactivity of Bovine Whey Proteins, Peptides, and Amino Acids toward Triplet Riboflavin as Studied by Laser Flash Photolysis. *Journal of Agricultural and Food Chemistry*, Vol. 52, No. 21, (October 2004), pp. 6602-6606, ISSN: 0021-8561.

Clark, D.T.; Gazi, M.I.; Cox, S.W.; Eley, B.M. & Tinsley, GF. (1993). The effects of Acacia arabica gum on the in vitro growth and protease activities of periodontopathic bacteria. *Journal of Clinical Periodontology*, Vol.20, No.4, (April 1993), pp. 238–43, ISSN: 0303-6979.

Criado, S.; Escalada, J. P.; Pajares, A. & García, N. A. (2008). Singlet molecular oxygen [O2(1Δg)]-mediated photodegradation of tyrosine derivatives in the presence of cationic and neutral micellar systems. Amino Acids, Vol. 35, No. 1, (June 2008), 201-208, ISSN: 0939-4451

Davies, M. J. (2003). Singlet oxygen-mediated damage to proteins and its consequences. *Biochemical and Biophysical Research Communication*. Vol. 305, No.3, (June 2003), pp. 761-770, ISSN: 0006-291X.

Evans, A.J.; Hood, R.L.; Oakenfull, D.G. & Sidhu, G.S. (1992). Relationship between structure and function of dietary fibre: a comparative study of the effects of three galactomannans on cholesterol metabolism in the rat. *British Journal of Nutrition*, Vol.68, No.1, (July 1992), pp. 217-229, ISSN: 0007-1145.

FAO (1996). A Review of Production, Markets and Quality Control of Gum Arabic in Africa. FAO, ROME. Forestry Dept., 191 p.

FAO. (1990). Specifications for identity and purity of certain food additives; *Food and Nutrition Paper*, 49; FAO: Rome.

Faria, A.F; Mignone, R.A.; Montenegro, M.A.; Mercadante, A.Z. & Borsarelli, C.D. (2010). Characterization and singlet oxygen quenching ability of spray-dried microcapsules of edible biopolymers containing antioxidant molecules. *Journal of Agricultural and Food Chemistry*, Vol. 58, No.13, (June 2010), pp. 8004-8011, ISSN: 0021-8561.

Ferreira, I.C.F.R.; Calhelha, R.C.; Estevinho, L.M. &. Queiroz, M.J.R.P. (2004). Screening of antimicrobial activity of diarylamines in the 2,3,5-trimethylbenzo[b]thiophene series: a structure–activity evaluation study . Bioorganic & Medicinal Chemistry Letters, Vol.14, No.23, (December 2004), pp. 5831-5833, ISSN: 0960-894X

Fincher, G. B.; Stone, B. A., & Clarke, A. E. (1983). Arabinogalactan–protein: Structure, synthesis and function. *Annual Review of Plant Physiology*, Vol.34, (June 1983), pp. 47-70, ISSN 0066-4294.

Flindt, C.; Al-Assaf, S.; Phillips, G.O. & Williams, P.A. (2005). Studies on acacia exudate gums. Part V. Structural features of Acacia seyal. *Food Hydrocolloids*, Vol.19, No.4, (July 2005), pp. 687-701, ISSN: 0268-005X.

Foote, C. S. (1991). Definition of Type I and Type II Photosensitized Oxidation. Photochemistry and Photobiology, Vol. 54, No.5, (November 1991), pp. 659, ISSN: 0031-8655.

Gallaher, D.D.; Hassel, C.A.& Lee, K.J. (1993). Relationships between viscosity of hydroxypropyl methylcellulose and plasma cholesterol in hamsters. *Journal of Nutrition*, Vol. 123, No.10, (October 1993), pp. 1732–1738, ISSN 0022-3166.

Gliszczyńska-Świgł, A. (2006). Antioxidant activity of water soluble vitamins in the TEAC (trolox equivalent antioxidant capacity) and the FRAP (ferric reducing antioxidant power) assays. Food Chemistry, Vol.96, No.1, (May 2006), pp. 131-136, ISSN 0308-8146.

Glover, D.A.; Ushida, K.; Phillips, A. O. & Riley. S. G. (2009) Acacia(sen) SUPERGUM™ (Gum arabic): An evaluation of potential health benefits in human subjects. Food Hydrocolloids, Vol. 23, No. 8, (December 2009), pp. 2410–2415, ISSN: 0268-005X.

Goodrum, L. J.; Patel, A.; Leykam, J. F. & Kieliszewski, M. J. (2000). Gum arabic glycoprotein contains glycomodules of both extension and arabinogalactan-glycoproteins. *Phytochemistry*, Vol.54, No. 1, (May 2000), pp. 99-106, ISSN: 0031-9422.

Gutteridge, J.M. & Halliwell, B. (1988). The deoxyribose assay: an assay both for 'free' hydroxyl radical and for site-specific hydroxyl radical production. *Biochemical Journal*, Vol. 253, (April 1988), pp. 932-933, ISSN 0264-6021.

Hassan, E. A.; Al Assaf, S.; Phillips, G.O. &. Williams, P.A. (2005). Studies on Acacia gums: Part III molecular weight characteristics of Acacia seyal var. seyal and Acacia seyal var fistula. *Food Hydrocolloids*, Vol. 19, No. 4, (July 2005), pp. 669-677, ISSN: 0268-005X.

Hinson, J.A.; Reid, A.B.; McCullough, S.S. & James, L.P. (2004). Acetaminophen-induced hepatotoxicity: role of metabolic activation, reactive oxygen/nitrogen species, and mitochondrial permeability transition. *Drug Metabolism Reviews*, Vol.36, No. 3-4, (January 2004), pp. 805–822, ISSN: 0360-2532.

Huang, D.; Ou, B. & Prior, R. L. (2005). The Chemistry behind Antioxidant Capacity Assays. *Journal of Agricultural and Food Chemistry*, Vol. 53, No. 6, (February 2006), pp. 1841-1856, ISSN 0021-8561.

Huvaere, K., and Skibsted, L. H. (2009) Light-Induced Oxidation of Tryptophan and Histidine. Reactivity of Aromatic N-Heterocycles toward Triplet-Excited Flavins, Journal of American and Chemical Society, Vol. 131, (May 2009) pp. 8049-8060, ISSN: 0002-7863.

IDF (1991). Enumeration of microorganisms; colony count technique at 30 °C, Standard 100B.

Idris, O.H.M.; Williams, P.A. & Phillips, G.O. (1998). Characterization of gum from Acacia senegal trees of different age and location using multidetection gel permeation chromatography. *Food Hydrocolloids*, Vol. 12, No. 4, (October 1998), pp. 379–388, ISSN: 0268-005X.

Islam, A. M.; Phillips, G. O.; Sljivo, A.; Snowden, M. J. & Williams, P. A. (1997). A review of recent developments on the regulatory, structural and functional aspects of gum arabic. *Food Hydrocolloids*, Vol. 11, No. 4, (October 1997), pp. 493-505, ISSN: 0268-005X.

Kenyon, M.M. (1995). Modified starch, maltodextrin, and corn syrup solids as wall materials for food encapsulation, American Chemical Society symposium series, Vol. 590, pp 42–50, ISBN: 0841231648.

Kirtikar, K.R. & Basu, B.D. (1984). Indian medicinal plants, 2nd ed., vol. II. Allahabad: Lalit Mohan Basu; pp. 919–35.

Kuan, Y. H.; Bhat, R.; Senan, C.; Williams, P. A. & Karim, A. A. (2009). Effects of Ultraviolet Irradiation on the Physicochemical and Functional Properties of Gum Arabic. *Journal of Agricultural and Food Chemistry*, Vol. 57, No.19, (September 2009), pp. 9154-9159, ISSN: 0021-8561.

Larson, B.A., & Bromley, D.W. (1991). Natural resource prices, export policies and deforestation: the case of Sudan. *World Development*, Vol.19, No.10, (October 1991), pp.1289-1297, ISSN: 0305-750X.

Liu, F., Ooi, V.E. & Chang, S.T. (1997). Free radical scavenging activities of mushroom polysaccharide extracts. *Life Science*, Vol.60, No.10, (nd), pp. 763-771, ISSN 0024-3205.

Lu, C.; Lin, W.; Wang, W.; Han, Z.; Yao, S. & Lin, N. (2000). Riboflavin-(VB2) photosensitized oxidation of 2′-deoxyguanosine-5′-mono-phosphate (dGMP) in aqueous solution: A transient intermediates study. *Physical Chemistry Chemical Physics*, Vol.2, (January 2000), pp.329-334, ISSN: 1463-9076.

Lu, C. Y. & Lui, Y.Y. (2002). Electron transfer oxidation of tryptophan and tyrosine by triplet states and oxidized radicals of flavin sensitizers: a laser flash photolysis study. *Biochimica et Biophysica Acta (BBA) - General Subjects*, Vol. 1571, No.1, (May 2002), pp. 71-76, ISSN: 0304-4165

Mahendran, T.; Williams, P. A.;. Phillips, G. O.; Al-Assaf, S. & Baldwin, T. C. (2008). New Insights into the Structural Characteristics of the Arabinogalactan-Protein (AGP) Fraction of Gum Arabic. *Journal of Agricultural and Food Chemistry*, Vol. 56, No.19, (September 2008), pp. 9269–9276, ISSN: 0021-8561.

Marcuse, R. (1960). Antioxidative effect of amino-acids. Nature, Vol.186, No.4728, (June 1960), pp. 886-887, ISSN: 0028-0836.

Marcuse, R. (1962). The effect of some amino acids on the oxidation of linoleic acid and its methyl ester. *Journal of the American Oil Chemists' Society*, Vol. 39, No.2 (February 1962) pp. 97-103, ISSN: 0003-021X.

Matsumoto, N.; Riley, S.; Fraser, D.; Al-Assaf, S.; Ishimura, E; Wolever, T.; Phillips, G.O. & Phillips, A.O. (2006). Butyrate modulates TGF-beta1 generation and function: potential renal benefit for Acacia (sen) SUPERGUM (G.A.)? *Kidney International*, Vol.69, No.2 (January 2006), pp. 257–265, ISSN: 0085-2538.

Melø, T. B.; Ionescu, M. A.; Haggquist, G. W. & Naqvi, K. R. (1999). Hydrogen abstraction by triplet flavins. I: time-resolved multi-channel absorption spectra of flash-irradiated riboflavin solutions in water Spectrochimica Acta, Part A: Molecular and Biomolecular Spectroscopy, Vol.55, No.11, (September 1999), pp. 2299-2307, ISSN 1386-1425.

Michaeli, A. & Feitelson, J. (1994). Reactivity of singlet oxygen toward amino acids and peptides. *Photochemistry and Photobiology*, Vol.59, No.3, (March 1994), pp. 284-289, ISSN 0031-8655.

Mocak, J.; Jurasek, P.; Phillips, G.O, Vargas, S.; Casadei, E. & Chikamai, B.N. (1998). The classification of natural gums. X. Chemometric characterization of exudate gums that conform to the revised specification of the gum arabic for food use, and the identification of adulterants. Food Hydrocolloids, Vol. 12, No. 2, (April 1998), pp 141-150, ISSN: 0268-005X.

Montenegro, M.A.; Nazareno, M.A; Durantini, E.N. & Borsarelli, C.D. (2002). Singlet oxygen quenching ability of carotenoids in a reverse micelle membrane mimetic system. *Photochemistry and Photobiology*, Vol. 75, No. 4, (April 2002), pp.353-361, ISSN: 0031-8655

Montenegro, M. A.; Nunes, I.; Mercadante, A. Z. & Borsarelli, C. D. (2007). Photoprotection of Vitamins in Skim Milk by Aqueous Soluble Lycopene - Gum Arabic Microencapsulated. *Journal of Agricultural and Food Chemistry*, Vol. 55, No. 2, (January 2007), pp. 323-329, ISSN 0021-8561.

Morán Vieyra, F.E.; Zampini, I.C.;. Ordoñez, R.M.; Isla, M.I.; Boggetti, H.J.; De Rosso, V.; Mercadante, A.Z.; Alvarez, R.M.S. & Borsarelli, C.D. (2009). Singlet oxygen quenching and radical scavenging capacities of structurally related flavonoids present in *Zuccagnia punctata* Cav. *Free Radical Research*, Vol.43, No.6, (January 2009), pp. 553-564, ISSN 1071-5762

Moundras, C.; Behr, S.R.; Demigné, C.; Mazur, A. & Rémésy, C. (1994). Fermentable polysaccharides that enhance fecal bile acid excretion lower plasma cholesterol and apolipoprotein E-rich HDL in rats. *Journal of Nutrition*, Vol. 124, No.11, (November 1994), pp. 2179–2188, ISSN 0022-3166.

Murov, S. L.; Carmichael, I.; Hug, G. L. *Handbook of Photochemistry*, 2nd ed.; Marcel Dekker:New York (NY), 1993, 420 p, Chapter 13.

Onishi, T.; Umemura, S.; Yanagawa, M.; Matsumura, M.; Sasaki, Y.; Ogasawara, T. & Ooshima, T. (2008). Remineralization effects of gum arabic on caries-like enamel lesions. Archives of Oral Biology, Vol.53, No.3, (March 2008), pp. 257-260, ISSN: 0003-9969.

Osman, M.E.; Williams, P. A.; Menzies, A. R. & Phillips, G. O. (1993) Characterization of Commercial Samples of Gum Arabic. *Journal of Agricultural and Food Chemistry*, Vol. 41, No. 1, (January 1993), pp. 71-77, ISSN: 0021-8561.

Park, E. Y.; Murakami, H. & Matsumura, Y. (2005). Effects of the addition of amino acids and peptides on lipid oxidation in a powdery model system. *Journal of Agricultural and Food Chemistry*, Vol. 53, No. 21, (September 2005), pp. 8334-8341 7, ISSN: 0021-8561.

Phillips, G.O. & Williams, P.A. (1993) In Nishinari, K. and Doi, E. (eds), Food Hydrocolloids; Structures, Properties and Functions. Plenum Press, New York, pp. 45-61.

Phillips, G.O. & Williams, P.A. (2001). Tree exudate gums: natural and versatile food additives and ingredients. *Food Ingredients and Analysis International*, Vol.23, (nd),pp. 26- 28, ISSN: 0968-574X.

Qi, W.; Fong, C. & Lamport, D. T. A. (1991). *Plant Physiology*, Vol.96, No. 3, (July 1991), pp. 848–855, ISSN 0032-0889.

Randall, R. C.; Phillips, G. O. & Williams, P. A. (1988).The role of the proteinaceous component on the emulsifying properties of gum arabic. *Food Hydrocolloids*, Vol. 2, No.2, (June 1988), pp. 131–140, ISSN: 0268-005X.

Re, R.; Pellegrini, N.; Proteggente, A.; Pannala, A.S.; Yang, M. & Rice-Evans, C.A. (1999). Antioxidant activity applying an improved ABTS radical cation decolorization assay. *Free Radical Biology and Medicine*, Vol.26, No.9-10, (May 1999), pp.1231-1237, ISSN: 0891-5849.

Rehman, K.U.; Wingertzahn, M.A.; Teichberg, S.; Harper, R.G. & Wapnir, R.A. (2003). Gum arabic. (GA) modifies paracellular water and electrolyte transport in the small intestine. *Digestive Diseases and Sciences*, Vol. 48, No.4, (April 2003), pp. 755–760, ISSN: 0163-2116.

Reineccius GA. (1988). Spray drying of food flavors.In: Risch S.J.& Reineccius G.A., editors. Flavor encapsulation.Washington DC: American Chemistry Society, Symposium Series, Vol. 370, pp 55-66, ISBN: 0-8412-1482-4.

Reineccius, G.A. (1989). Flavor Encapsulation. Food Reviews International, Vol.5, No. 2 (March 1989), pp 147-176, ISSN: 8755-9129.

Renard, D.; Lavenant-Gourgeon, L.; Ralet, M. & Sanchez, C. (2006). Acacia senegal Gum: Continuum of Molecular Species Differing by Their Protein to Sugar Ratio, Molecular Weight, and Charges. *Biomacromolecules*, Vol.7, No.9, (Augudt 2006), pp. 2637–2649, ISSN: 1525-7797.

Ross, A.H.; Eastwood, M.A.; Brydon, W.G.; Anderson, J.R. & Anderson, D.M. (1983). A study of the effects of dietary gum arabic in humans. *American Journal of Clinical Nutrition*, Vol. 37, No.3, (March 1983), pp. 368–375, ISSN: 0002-9165.

Sabu, M.C. & Ramadasan, K. (2002). Anti-diabetic activity of medicinal plants and its relationship with their antioxidant property. *Journal of Ethnopharmacology*, Vol. 81, No.2, (July 2002), pp. 155-160, ISSN: 0378-8741.

Saini, M.; Saini, R.; Roy, S. & Kumar, A. (2008). Comparative pharmacognostical and antimicrobial studies of acacia species (Mimosaceae). *Journal of Medicinal Plants Research*, Vol. 2, No.12, (December 2008), pp. 378-386, ISSN 1996-0875.

Sanchez, C.; Schmitt, C.; Kolodziejczyk, E'.; Lapp, A.; Gaillard, C. & Renard, D. (2008). The Acacia Gum Arabinogalactan Fraction Is a Thin Oblate Ellipsoid: A New Model Based on Small-Angle Neutron Scattering and Ab Initio Calculation. *Biophysical Journal*, Vol. 94, No. 2, (January 2008), pp.629–639, ISSN: 0006-3495.

Sharma, R.D. (1985). Hypocholesterolaemic effect of gum acacia in men. *Nutrition Research*, Vol. 5, No.12, (December 1985), pp. 1321–1326, ISSN: 0271-5317.

Sheu, T. Y. & Rosenberg, M. (1993). Microencapsulating Properties of Whey Proteins. 2. Combination of Whey Proteins with Carbohydrates. *Journal of Dairy Science*, Vol.76, No. 1, (October 1993), pp.2878-2885, ISSN: 0022-0302.

Siddig, N.E.; Osman, M.E.; Al-Assaf, S.; Phillips, G.O. & Williams, P.A. (2005). Studies on acacia exudate gums, part IV. Distribution of molecular components in Acacia seyal in relation to Acacia senegal. *Food Hydrocolloids*, Vol. 19, No.4, (July 2005), pp. 679-686, ISSN: 0268-005X.

Tiss, A.; Carrière, F. & Verger, R. (2001). Effects of gum arabic on lipase interfacial binding and activity. *Analytical Biochemistry*, Vol. 294, No. 1, (July 2001) pp. 36–43, ISSN: 0003-2697.

Topping, D.; Illman, R.J. & Trimble, R.P. (1985). Volatile fatty acid concentrations in rats fed diets containing gum Arabic and cellulose separately and a mixture. *Nutrition Reports International*, Vol. 32, (nd) pp. 809–814, ISSN: 0029-6635.

Trommer, H. & Neubert, R.H. (2005). The examination of polysaccharides as potential antioxidative compounds for topical administration using a lipid model system. *International Journal of Pharmaceutics*, Vol. 298, No. 1, (July 2005), pp.153–163, ISSN: 0378-5173.

Tyler, V.; Brady, L. & Robbers, J. Pharmacognosy. 7th ed. Philadelphia: Lea & Febiger; 1977. p. 66–8.

Valle, L.; Morán Vieyra, F.E. & Borsarelli, C.D. (2011). Microenvironmental modulation of photophysical properties of flavins. Submitted to Photophysical and Photochemical Sciences.

Verbeken, D.; Dierckx, S. & Dewettinck. (2003). Exudate gums: Occurrence, production, and applications. Applied Microbiology and Biotechnology, Vol.63, No. 1, (November 2003), pp. 10–21, ISSN: 0175-7598.

Wapnir, R.A.; Sherry, B.; Codipilly, C.N.; Goodwin, L.O. & Vancurova, I. (2008). Modulation of rat intestinal nuclear factor NF-kappaB by gum arabic. Rat small intestine by gum arabic. Digestive Diseases and Sciences, Vol. 53, No. 1, (May 2007), pp 80–87, ISSN: 0163-2116.

Williams, P. A. Phillips, G. O. In Handbook of Hydrocolloids; Williams, P. A., Phillips, G. O., Eds.; CRC Press: Cambridge, 2000; p 155-168.

Williams, P. A.; Phillips, G. O. & Stephen, A. M. (1990). Spectroscopic and molecular comparisons of three fractions from Acacia senegal gum. *Food Hydrocolloids*, Vol.4, No.4, (December 1990), pp. 305-311, ISSN: 0268-005X.

Wondrak, G. T.; Jacobson, M. K. & Jacobson, E. L. (2006). Endogenous UVA photosensitizers: mediators of skin photodamage and novel targets for skin photoprotection. *Photochemical & Photobiological Sciences*, Vol.5, (nd), (August 2005), pp. 215 - 237, ISSN 1474-905X.

2

Poly(Lactic Acid) as a Biopolymer-Based Nano-Composite

Emad A. Jaffar Al-Mulla[1,2] and Nor Azowa Bt Ibrahim[1]
[1]*Department of Chemistry, Faculty of Science,*
Universiti Putra Malaysia, Serdang, Selangor
[2]*Department of Chemistry, College of Science,*
University of Kufa, An-Najaf,
[1]*Malaysia*
[2]*Iraq*

1. Introduction

Petrochemical-based polymer technology has created a lot of benefits to soceity. One of these benefits is the use of plastics in packaging. The most important factors determining rapid growth in the use of plastics in packaging industries are convenience, safety, low price and good aesthetic qualities. However, petrochemical-based polymers are produced from fossil fuel, consumed and discarded into the environment, ending up as undegradable waste. Increasing undegradable wastes are significantly disturbing and damaging the environment. Environmental specialists do not have a clear answer about dealing with these undegradable wastes yet. Incineration of these wastes produces large amounts of carbon dioxide that will contribute to global warming.

These environmental issues, created a dire need for the development of green polymeric materials, which would not involve the use of toxic and noxious component in their manufacture and could be degradable in nature. For these reasons, through the world today, the development of biodegradable materials with controlled properties has been a subject of great research challenge for the community of material scientists and engineers.

The importance of renewable products for industrial applications has become extremely clear in recent years with increasing emphasis on environmental issues such as waste disposal and depleting non-renewable resources. Renewable resources can create a platform to substitute petroleum-based polymers through innovative bio-based polymers which can compete with or even surpass existing petroleum-based materials on a cost performance basis with the added advantage of eco-friendliness. There is a growing urgency to develop and commercialize new bio based products and other innovative technologies that can reduce widespread dependence on fossil fuel and the same time would enhance national security, the environment and the economy (Miyagawa *et al.*, 2005).

Biodegradable polymers are polymers that undergo microbially induced chain scission leading to mineralization. Biodegradable polymers may not been produced from bio-source only, but it can be derived from the petroleum source (Ray and Bousmina, 2005). Efforts have been made to improve mechanical properties by the addition of reinforcement particles or fibres to polymer matrices optimising them for engineering applications (Kulinski and Piorkowska, 2005). Flax has been considered as cost-effective alternative to glass in composites, since new technology and separation techniques have lower the cost to produce fibres that are more uniform in colour, strength, length and fineness and thus better suited to composites (Foulk *et al.*, 2004). Clay is an abundant natural mineral so it has been used as filler for rubber and plastics for many years, but its reinforcing ability is poor so it can only be used for conventional micro-composites. A new way has been found to improve the reinforcing ability of clay; clay can be chemically modified to produce clay complexes with organic monomers and polymers (Usuki *et al.*, 1993).

2. Poly(lactic acid) (PLA)

PLA is a thermoplastic aliphatic polyester, biodegradable and derived from renewable resources by means of a fermentation process using sugar from corn, followed by either ring-opening polymerization or by condensation polymerization of lactic acid (Scheme 1). It is one of the most important biocompatible and biodegradable polymers in a group of degradable plastics. In addition to its application in textile industries, automotive and clinical uses, PLA represents a good candidate to produce disposable packaging due to its good mechanical properties and processability (Murariu *et al.*, 2008).

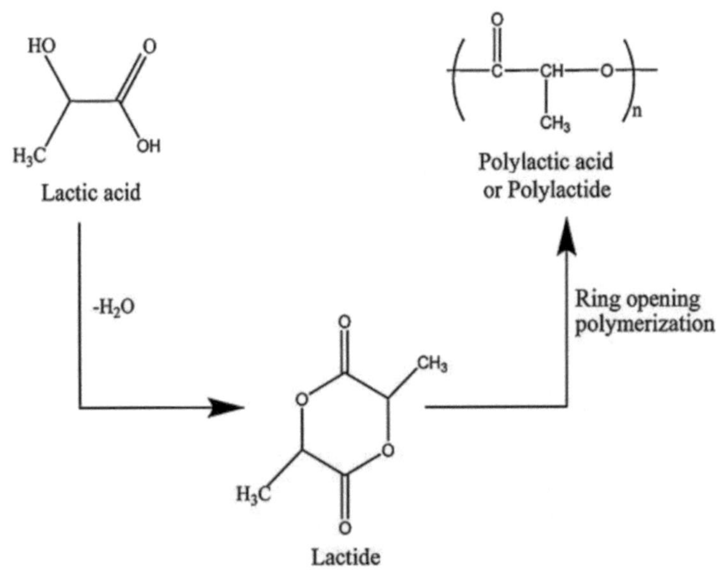

Scheme 1. Reaction scheme to produce PLA (Linnemann *et al.*, 2003)

3. Composites

Composites are combinations of two or more materials with the properties shown by individual components. They are made to perform as a single material. Nature made the first composite in living things. Wood is a composite of cellulose fibers held together with a matrix of lignin. Most sedimentary rocks are composites of particles bonded together by nature cement and many metallic alloys are composites of several quite different constituents. Steel reinforced concrete and medical pills are composite materials that are homogenous on a macro scale. The term composite was used in the reinforced plastic industry during the 1940s (Donald and Dominick, 1994).

4. Addition of fillers

Some polymers ignite quite easily when exposed to a flame. In particular acrylates, which are commonly used for extruded sheets or molded, paints and shellacs, are highly combustible and very difficult to render flame retardant, even with the addition of large amounts of conventional flame retardant (FR) agents, such as halogenated compounds, phosphorus and inorganic materials. Recently several groups have reported that addition of a small amount of organo-clay can significantly decrease the heat release and mass loss rate, as measured by cone calorimetry (Zanetti *et al.*, 2001; Alexandre and Dubois, 2000; Gilman *et al.*, 2000).

Clays have long been used as fillers in polymer systems because of low cost and the improved mechanical properties of the resulting polymer composites. If all other parameters are equal, the efficiency of a filler to improve the physical and mechanical properties of a polymer system is sensitive to its degree of dispersion in the polymer matrix (Krishnamoorti *et al.*, 1996). In the early 1990s, Toyota researchers (Okada *et al.*, 1990) discovered that treatment of montmorillonite (MMT) with amino acids allowed dispersion of the individual 1 nm thick silicate layers of the clay scale in polyamide on a molecular. Their hybrid material showed major improvements in physical and mechanical properties even at very low clay content (1.6 vol %). Since then, many researchers have performed investigations in the new field of polymer nano-composites. This has lead to further developments in the range of materials and synthesizing methods available.

5. Structure of clay

Natural clays such as MMT belong to the 2:1 layered silicate group (Figure 1). Their crystal lattice consists of two tetrahedral silica sheets sandwiching an edge shared octahedral sheet of either aluminium or magnesium hydroxyl. A regular Van der Waal's space between the layers, called an interlayer or gallery arises due to stacking of the layers (Alexandre and Dubois, 2000). The gallery is normally occupied by cations such as Na^{+}, Ca^{2+} and Mg^{2+}. The starting clay materials are easily available and cheap. MMT is chosen as the filler due to its suitable layer charge density, high cation exchange capacity (70-150 meq/100g) and its ability to show extensive interlayer expansion.

Due to the relatively weak forces between the layers of MMT, water and other polar molecules can enter between the unit layers, causing the lattice to expand in the thickness direction. The charge deficiency on the sheet surface is typically balanced by exchangeable cations adsorbed between the unit layers and around their edges because of the substitution of ions of different valence.

X-ray was discovered in 1895 by W.K. Rontgen. After its discovery, studies of this radiation were expanded when in 1912 Laue and Friedrich found that the atoms in crystals diffracted

X-rays. This was followed by the mathematical solution of crystal structure from X-ray diffraction data in 1913 by Bragg. Since that, many applications of X-ray were found including structure determination of fine-grained materials, like soils and clays, which had been previously thought to be amorphous. Since then, crystals structures of the clay minerals were well studied (Ray and Okamoto, 2003).

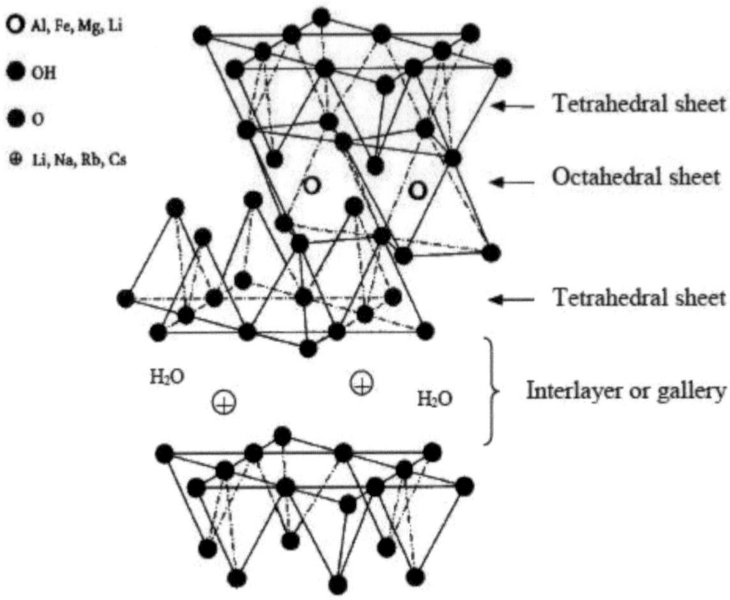

Fig. 1. Structure of 2:1 phyllosilicates (Giannelis *et al.*, 1999).

MMT, saponite and hectorite are the most commonly used layered silicates (LSs). LSs have two types of structures; tetrahedral and octahedral substituted. In the case of tetrahedrally substituted LSs, the negative charge is located on the surface of silicate layers, and hence, the polymer matrices can interact more readily with these than with octahedrally substituted material. Two particular characteristics of LSs are generally considered. The first is the silicate particles ability to disperse into individual layers. The second characteristic is modification ability for their surface chemistry through ion exchange reactions with organic and inorganic cations. These two characteristics of LSs are interrelated since the degree of dispersion of a layered silicate in a particular polymer matrix depends on the interlayer cation (Ray and Okamoto, 2003).

6. Modification of clay

MMT, and other layered silicate clays, are naturally hydrophilic. This property makes them poorly suited for mixing and interacting with most polymer matrices. Moreover, electrostatic forces hold tightly the stacks of clay platelets together. The counter ions can be shared by two platelets, resulting in stacks of platelets that are tightly held together (Giannelis, 1996).

Therefore, the clay should be modified before it can be used to make a nano-composite. After all, these stacks of clay platelets are larger than one nanometer in every dimension.

Making a composite using unmodified clay would not be a very effective use of material, because most of the clay is unable to interact with the matrix and would be stuck inside. A popular and relatively easy method of clay modification, making it compatible with an organic matrix, is ion exchange. The cations are not strongly restricted to the clay surface, so small organic cations can replace the cations present on clay (Giannelis, 1996).

If the cations were quaternary alkylammonium ions with long chains, clay would be much more compatible with an organic matrix. By exchanging sodium ions with various organic cations, MMT can be compatibilized with several different matrix polymers. They can be more easily intercalated and exfoliated because this process helps to separate the clay platelets. Nano-composites can then be formed by incorporating the intercalated or exfoliated clay in a matrix. The first commercial clay nano-composite was prepared via an ion exchanging process (Ray and Okamoto, 2003; Giannelis, 1996). Recently, many studies were reported for MMT modification using alkylammonium ions produced from vegetable oils (Al-Mulla *et al.*, 2009; Al-Mulla *et al.*, 2010a; Hoidy *et al.*, 2010a; Hoidy *et al.*, 2010b).

7. Nano-composites

Performance of polymers during use is a key feature of any composite material, which decides the real fate of products in outdoor applications. Whatever the application, there is concern regarding the durability of polymers, partly because of their useful lifetime, maintenance and replacement. The deterioration of these materials depends on the duration and extent of interaction with the environment (Homminga *et al.*, 2005).

Nano-composites (NCs) are materials that comprise a dispersion of particles of at least one of their dimentions is 100 nm or less in a matrix. The matrix may be single or multicomponent. It may include additional materials that add other functionalities to the system such as reinforcement, conductivity and toughness (Alexandre and Dubois, 2000). Depending on the matrix, NCs may be metallic (MNC), ceramic (CNC) or polymeric (PNC) materials. Since many important chemical and physical interactions are governed by surface properties, a nanostructured material could have substantially different properties from large dimensional material of the same composition (Hussain *et al.*, 2007).

8. Polymer layered-silicate nano-composites

Polymer layered-silicate clay nano-composites (PLCN) attracted lately major interests into the industry and academic fields, since they usually show improved properties with comparison by virgin polymers or their conventional micro and macro-composites. Improvements included increase in strength, heat resistance (Giannelis, 1998), flammability (Gilman, 2000) and a decrease in gas permeability (Xu *et al.*, 2001) as well as an increase in biodegradability (Sinha *et al.*, 2002).

However, the field of polymer clay silicate has only started to speed up recently, mixing the appropriate modified layered silicate with synthetic layered silicates has long been known (Theng, 1979). The interest in these materials came from two important findings, first has been reported by Toyota research group of a Nylon-6 (N6)/Na-MMT nano-composites (Okada *et al.*, 1990) where very small amounts of layered silicate loadings resulted in the improvements of thermal and mechanical properties and second the findings of Vaia *et al.* (1993) about the possibility to melt- mix polymers with layered silicates without using organic solvents. Worldwide effort in applying this technology, as evident from using several polymer matrices such as PLA (Ljungberg *et al.*, 2005), ethylene vinyl acetate copolymer (Zanetti *et al.*, 2001), polypropylene (Kato *et al.*, 1997), polycaprolactone (Zhenyang *et al.*, 2007), ethylene propylene diene methylene (Usuki *et al.*, 2002), polymethyl methacrylate (Okamoto *et al.*, 2001), polystyrene (Meneghetti and Qutubuddin, 2006) and others.

9. Type of polymer /clay nan-ocomposites

Three main types of composites can be formed when the layered clay is incorporated with a polymer, as shown in Figure 2 (Alexandre and Dubois, 2000). Types of composites formed mostly depend on the nature of the components used (layered silicate, organic cation and polymer matrix) and the method of preparation.

Micro-composites are formed when the polymer chain is unable to intercalate into the silicate layer and therefore phase separated polymer/clay composites are formed. Their properties remain the same as the conventional micro-composites as shown in Figure 2(a). Intercalated nano-composite is obtained when the polymer chain is inserted between clay layers such that the interlayer spacing is expanded, but the layers still bear a well-defined spatial relationship to each other as shown in Figure 2(b). Exfoliated nano-composites are formed when the layers of the clay have been completely separated and the individual layers are distributed throughout the organic matrix as shown in Figure 2(c).

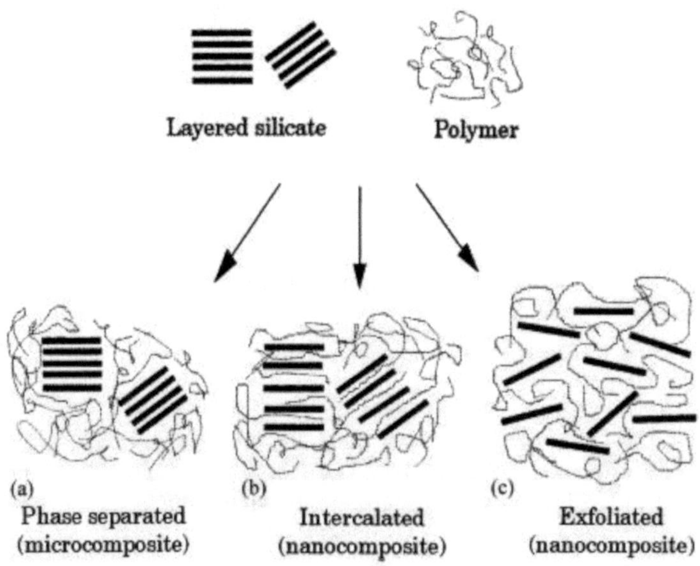

Fig. 2. Three types of composites when layered clays are incorporated with the polymer (Alexandre and Dubois, 2000)

The nano-scale structures in polymer layered-silicate nano-composites can be thoroughly characterized by X-ray diffraction (XRD) and transmission electron microscopy (TEM). XRD is used to identify intercalated structures. XRD allows quantification of changes in layer spacing and the most commonly used to probe the nano-composite structure and occasionally to study the kinetics of the polymer melt intercalation (Ray and Bousmina, 2005). The intercalation of the polymer chains usually increases the spacing of interlayer, in comparison with the spacing of the organo-clay used and can be observed as a shift of the diffraction peak towards lower angles. The values of angle and layer spacing are being related through the Bragg's relation $\lambda = 2d \sin \theta$, where λ corresponds to the wavelength of the X-rays radiation used in the diffraction experiment, d is the spacing between diffraction planes and θ is the measured diffraction angle (Alexandre and Dubois, 2000).

On the other hand, TEM shows a qualitative understanding of the internal structure, spatial distribution of the various phases, and views of the defect structure through direct visualization (Ray and Bousmina, 2005).

10. PLA modifications

Although PLA is an eco-friendly boiplastic with good biocompatibitity, poor hardiness, slow degradation, hydrophobicity, and lack of reactive side-chain groups limit its application (Rasal *et al.*, 2010). Therefore, the tailoring of its properties to reach end-users demands is required. In addition, mechanical properties being better than or comparable to conventional plastics, controlled surface properties such as hydrophilicity, roughness, and reactive functionalities are the successful implementation of PLA in consumer and biomedical applications.

In recent times, blending was the main approach to improve the mechanical properties of PLA by using different plasticizers and polymers to reach desired mechanical properties. PLA, which is a glassy polymer at room temperature, has a poor elongation at break (Rasal *et al.*, 2008). Various biodegradable and non-biodegradable plasticizers were used to lower the glass transition temperature (Tg), improve processibility, and increase flexibility (Mascia *et al.*, 1992). These aspects were carried out by modification some of the plasticizer properties: polarity, molecular weight, and end groups.

Lactide is a natural choice to plasticize PLA and showed a significant increase in elongation at break (Sinclair, 1996) but undergo stiffening with time due to migration of low molecular mass lactide toward the surface (Jacobsen *et al.*, 1999). Oligomeric plasticizers that would not migrate toward the surface due to their relatively higher molecular mass were also used. Glycerol, citrate ester, polyethylene glycol, and oligomeric lactic acid were used to plasticize PLA and found that oligomeric lactic acid and low molecular mass polyethylene glycol gave finest results while glycerol was found to be the least competent plasticizer (Martin and Avérous, 2001). Citrate esters were found to be miscible with PLA at all compositions. Elongation at break was significantly improved accompanied with considerable loss of tensile yield strength (Labrecque *et al.*, 1997).

Ljungberg and Wesslén (2002) plasticized PLA using triacetin and tributyl citrate, and succeeded in lowering Tg to around 10°C at 25 wt%, after which phase separation occurred. Triacetine- or tributyl-citrate-plasticized PLA films underwent crystallization, and plasticizer molecules migrated toward the surface with storage time due to their low molecular mass (Ljungberg *et al.*, 2003). To overcome the aging problem, tributyl citrate oligomers (Scheme 2a) were synthesized by trans-esterification of tributyl citrate and diethylene glycol. However, these oligomeric tributyl citratre plasticizers also underwent phase separation over time (Ljungberg and Wesslén, 2003). To obtain better stability, these researchers used diethyl bishydroxymethyl malonate (DBM) and its oligomer (Scheme 2b), synthesized through an esterification reaction between DBM and dichloride. When DBM alone was used as a plasticizer, it showed a tendency to phase separate and migrate toward the surface. DBM-oligomer plasticized PLA demonstrated morphological stability with storage time (Ljungberg *et al.*, 2004). Oligomeric polyesters and esteramides (Scheme 2c) have also been used to plasticize PLA, showing better plasticizing properties due to an increased number of polar amide groups (Ljungberg *et al.*, 2005).

Epoxidized oils were also used to modify PLA; Ali *et al.* (2009) reported that its use as a plasticizer to improve flexibility. Thermal and scanning electron microscope analysis revealed that epoxidized soybean oil is partially miscible with PLA. Rheological and mechanical properties of PLA/epoxidized soybean oil blends were studied by Xu and Qu (2009) Epoxidized soybean oil exhibited a positive effect on both the elongation at break and melt rheology. Al-Mulla et al. (2010b) also reported that plasticization of PLA (epoxidized palm oil) was carried out via solution casting process using chloroform as a solvent. The results indicated that improved flexibility could be achieved by incorporation of epoxidized palm oil.

PLA-biodegradable polymer blends were studied and extensively investigated more than PLA-non-biodegradable polymer blends due to their property improvements without compromising biodegradability. For instance, polyhydroxyalkanoates (PHA) produced by bacteria from biodegradable aliphatic homo or copolyesters with over 150 unlike kinds composed of dissimilar monomers (Steinbüchel and Lütke-Eversloh, 2003). Poly(3-hydroxy butyrate) (PHB) and its copolymers with 3-hydroxyvalerate (PHBHV), 3-hydroxyoctanoate (PHBHO), and 3-hydroxyhexanoate (PHBHHx) units are among the largely employed PHAs (Bluhm *et al.*, 1986; Doi *et al.*, 1988; Noda *et al.*, 2004). The high crystallinity of PHB homopolymer results in a hard and brittle material, inconvenient to blend with PLA.

PLA/PHBHV solvent cast blends were incompatible and showed minimal improvement in elongation at break (Iannace *et al.*, 1994). Although Takagi *et al.* (2004) found PLA/ poly(3-hydroxyoctanoate) (PHO) blends to be immiscible, they exhibited better impact toughness. PLA phase in PLA-PHBHHx blends (90%wt PLA) undergo when stored quick physical aging comes out with major toughness loss (Rasal *et al.*, 2008). A similar observation was reported for PLA/starch blends, where blends lost their toughness with physical properties (Drumright *et al.*, 2000).

PLA/polycaprolactone (PCL) is another extensively studied biodegradable PLA blend system. PCL as a rubbery polymer, has a low Tg and degrades by hydrolytic or enzymatic pathways. Broz *et al.* (2003) observed changes in modulus, elongation at break and tensile strength by blending PCl with PLA. Elongation at break increased only above 60 wt% PCLl. However, the elongation at break improvement was not significant and resulted in a large modulus and tensile strength loss. The addition of a small quantity of surfactant (copolymer of ethylene oxide and propylene oxide) did not lead to any significant elongation at break improvement for PLA/PCL blends (Chen *et al.*, 2003). However, addition of a small amount of PLA-PCL-PLA tri-block copolymer (4%wt) to PLA/PCL (70/30, w/w) blends enhanced the dispersion of PCL in PLA and improved the flexibility of the resultant blend.

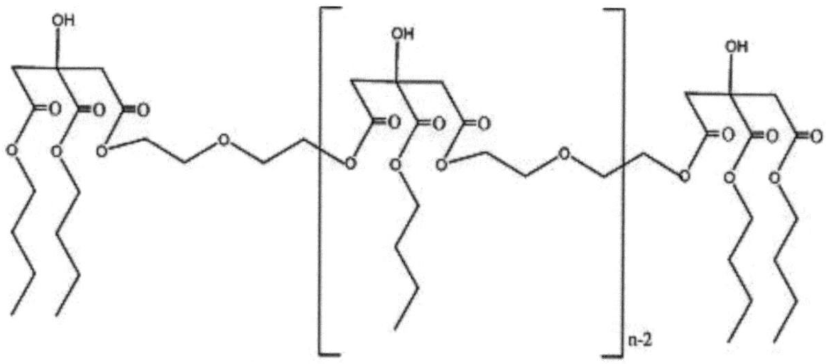

(a) Oligomeric tributyl citrate (Ljungberg *et al.*, 2003)

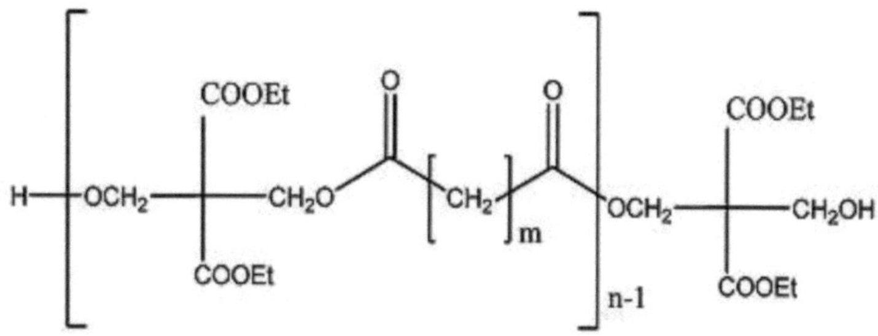

(b) Oligomeric diethyl bishydroxymethyl malonate (DBM) (Ljungberg *et al.*, 2004)

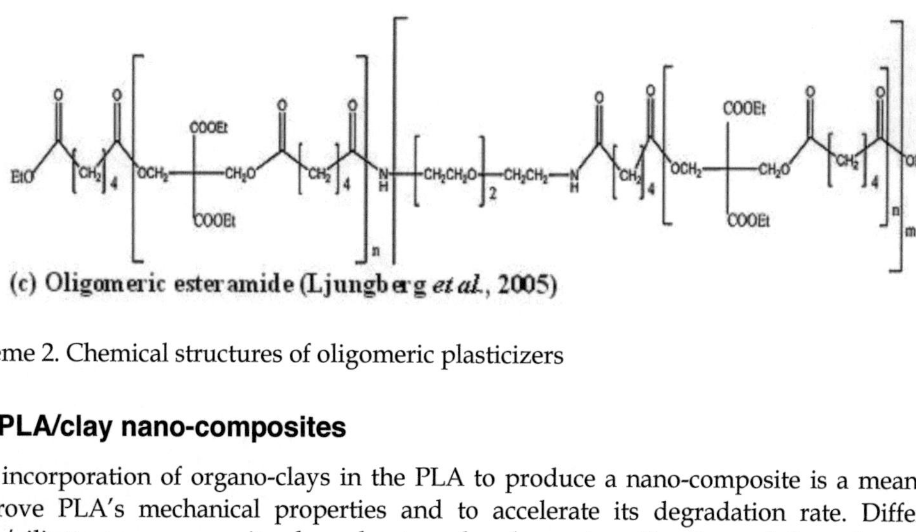

(c) Oligomeric esteramide (Ljungberg *et al.*, 2005)

Scheme 2. Chemical structures of oligomeric plasticizers

11. PLA/clay nano-composites

The incorporation of organo-clays in the PLA to produce a nano-composite is a means to improve PLA's mechanical properties and to accelerate its degradation rate. Different PLA/silicate nano-composites have been explored: montmorillonites and fluorohectorites clays or organo-clays, were blended with the PLA (Oliva *et al.*, 2007; Aguzzi *et al.*, 2007). Properties and the behaviour of the PLA/clays composites in aqueous environment were studied. The combination of PLA and clays, at the nano-scale, often results in remarkably improved mechanical and functional properties compared with pure PLA or conventional composites (Okamoto et al., 2001). Higher modulus, increased strength and increased degradation rate in the case of biodegradable PLA have been reported.

Ogata *et al.* (1997) first prepared PLA/organoclay (OMMT) blends by dissolving the polymer in hot chloroform in the presence of dimethyl distearyl ammonium modified MMT ($2C_{18}MMT$). XRD results show that the silicate layers forming the clay could not be intercalated in the PLA/MMT blends, prepared by the solvent-cast method. Thus, the clay existed in the form of tactoids, consisting several stacked silicate monolayers.

These tactoids are responsible for the particular geometrical structures formation in the blends, which leads to the formation of superstructures in the thickness of the blended film. The Young's modulus of the hybrid is increased by this kind of structural feature. After that, the preparation of intercalated PLA/OMMT nano-composites with much improved mechanical and thermal properties was reported by Bandyopadhyay *et al.* (1999).

Sinha *et al.* (2002) used melt intercalation technique for the preparation of intercalated PLA/layered silicate nano-composites using octadecyl ammonium modified MMT ($C_{18}MMT$). Nano-composites loaded with a very small amount of PCL as a compatibilizer were also prepared in order to understand the effect of PCL on the morphology and properties of PLACNs. XRD patterns and TEM results clearly indicated that the silicate layers of the clay were intercalated, and randomly distributed in the PLA matrix. Incorporation of a very small amount of PCL as a compatibilizer in the nano-composites led to better parallel stacking of the silicate layers, and also to much stronger flocculation due to the hydroxylated edge–edge interaction of the silicate layers. Owing to the interaction between the PLA matrix and clay platelets in the presence of a very small amount of PCL, the strength of the disk–disk interaction plays an important role in determining the stability of the clay particles, and hence the enhancement of mechanical properties of such nano-composites.

In the matrix of PLA/polycaprilactone (PCL)/OMMT nano-composites, the silicate layers of the organoclay were intercalated and randomly distributed (Zhenyang *et al.*, 2007). The PLA/PCL blend significantly improved the tensile and other mechanical properties by addition of OMMT. Thermal stability of PLA/PCL blends was also explicitly improved when the OMMT content is less than 5%wt. Preparation of PLA/thermoplastic starch/MMT nano-composites have been investigated and the products have been characterized using X-Ray diffraction, transmission electron microscopy and tensile measurements. The results show improvement in the tensile and modulus, and reduction in fracture toughness (Arroyo *et al.*, 2010).

PLA/PCL-OMMT nano-composites were prepared effectively using fatty amides as clay modifier. The nano-composites shows increasing mechanical properties and thermal stability (Hoidy et al, 2010c). New biopolymer nano-composites were prepared by treatment of epoxidized soybean oil and palm oil, respectively plasticized PLA modified MMT with fatty nitrogen compounds. The XRD and TEM results confirmed the production of nano-composites. The novelty of these studies is use of fatty nitrogen compounds which reduces the dependence on petroleum-based surfactants (Al-Mulla *et al.*, 2011; Al-Mulla *et al.*, 2011; Al- Mulla *et al.*, 2010c).

Plasticized PLA-based nano-composites were prepared and characterized with polyethylene glycol and MMT. It is reported that the organo-modified MMT-based composites show the possible competition between the polymer matrix and the plasticizer for the intercalation between the alumino-silicate layers (Paul *et al.*, 2002).

12. References

Aguzzi, C., Cerezo, P., Viseras, C. and Caramella, C. 2007. Use of clays as drug delivery systems: Possibilities and limitations. *Applied Clay Science* 36: 22–36.

Alexandre, M. and Dubois, P. 2000. Polymer-layered silicate nano-composites: preparation, properties and uses of a new class of materials. *Materials Science and* Engineering 28: 1- 63.

Ali, F., Chang, Y., Kang, S.C., Yoon, J.Y. 2009. Thermal, mechanical and rheological properties of poly (lactic acid)/epoxidized soybean oil blends. *Polymer Bulletin.* 62: 91-98.

Al-Mulla, E.A.J., Yunus, W.M.Z., Ibrahim, N.A. and Abdul Rahman, M.Z. 2009. Synthesis and characterization of *N,N'*-carbonyl difatty amides from palm oil. *Journal of Oleo Science* 58: 467-471.

Al-Mulla, E.A.J., Yunus, W.M.Z., Ibrahim, N.A., Abdul Rahman, M.Z. 2010a. Enzymatic synthesis of fatty amides from palm olein. *Journal of Oleo Science* 59: 59-64.

Al-Mulla, E.A.J., Yunus, W.M.Z., Ibrahim, N.A. and Abdul Rahman M.Z. 2010b. Properties of epoxidized palm oil plasticized polylactic acid. *Journal of Materisls Science* 45: 1942-1946.

Al-Mulla, E.A.J., Yunus, W.M.Z., Ibrahim, N.A. and Abdul Rahman M.Z. 2010c. Epoxidized palm oil plasticized polylactic acid /fatty nitrogen compounds modified clay nanocomposites: preparation and characterizations. *Polymers and Polymer Composites* 18, 451-459 .

Al-Mulla, E.A.J., Suhail, A. H. and Aowda, S. 2011. A. New biopolymer nanocomposites based on epoxidized soybean oil plasticized poly(lactic acid)/fatty nitrogen compounds modified clay: Preparation and characterization. *Industrial Crops and Products* 33. 23-29.

Al-Mulla, E.A.J. 2011. Polylactic acid/epoxidized palm oil/fatty nitrogen compounds modified clay nanocomposites: Preparation and characterization. *Korean Journal of Chemical Engineering* 28, 620-626.

Bandyopadhyay, S., Chen, R. and Giannelis, E.P. 1999. Biodegradable organic-inorganic hybrids based on poly(L-lactide). *polymer material science engineering* 81: 159–160.

Bluhm, T.L., Hamer, G.K., Machessault, R.H., Fyfe, C.A. and Veregin, R.P. 1986. Isodimorphism in bacterial poly(_-hydroxybutyrate-co-hydroxyvalerate). *Macromolecules* 19: 2871–6.

Broz, M.E., VanderHart, D.L. and Washburn, N.R. 2003. Structure and mechanical properties of poly(d,l-lactic acid)/poly(e-caprolactone) blends. *Biomaterials* 24: 4181–4190.

Chen, C.C., Chueh, J.Y., Tseng, H., Huang, H.M. and Lee, S.Y. 2003. Preparation and characterization of biodegradable PLA polymeric blends. *Biomaterials* 24: 1167–1173.

Donald, V.R. and Dominick, V. 1994. *Reinforced plastics handbook,* 3rd Ed. Elsevier Advanced Technology, Oxford :pp 16-18.

Drumright, R.E., Gruber, P.R. and Henton, D.E. 2000. Polylactic acid technology. *Advanced Materials* 12: 1841–6.

Giannelis, E.P. 1996. Polymer layered silicate nanocomposites. *Advanced Materials* 8: 29–35.

Giannelis, E.P. 1998. Polymer- layered silicate nanocomposites: synthesis, properties and applications. *Applied Organometallic Chemistry* 12: 675–680.

Gilman, J.W., Jackson, C.L., Morgan, A.B., Harris, J.R., Manias, E., Giannelis, E.P., Wuthenow, M., Hilton, D. and Phillips, S.H. 2000. Flammability properties of polymer- layered silicate nanocomposites, polypropylene and polystyrene nanocomposites. *Journal of Materials Chemistry* 12: 1866–1873.

Hoidy, W.H., Ahmad, M.B., Al-Mulla, E.A.J., Yunus, W.M.Z. and Ibrahim, N.A. 2010a. Synthesis and characterization of fatty hydroxamic acids from triacylglycerides. *Journal of Oleo Science* 59, 15-19.

Hoidy, W.H., Ahmad, M.B., Al-Mulla, E.A.J., Yunus, W.M.Z. and Ibrahim, N.A. 2010b. Chemical synthesis and characterization of palm oil-based difatty acyl thiourea. Journal of Oleo Science 59, 229-233.

Hoidy, W.H., Al-Mulla, E.A.J., and Al-Janabi, K.W. 2010c. Mechanical and thermal properties of PLLA/PCL modified clay nanocomposites. *Journal of Polymers and the Environment*. 18, 608-616.

Homminga, D., Goderis, B., Hoffman, S., Reynaers, H. and Groeninckx, G. 2005. Influence of shear flow on the preparation of polymer layered silicate nanocomposites. *Polymer* 46: 9941-9954.

Hussain, F., Chan, J., and Hojjati, M. 2007. Epoxy-silicate nanocomposites: Cure monitoring and Characterization. *Material Science and Engineering A* 445-446: 467-476.

Jacobsen, S. and Fritz, H.G. 1999. Plasticizing polylactide, the effect of different plasticizers on the mechanical properties. *Polymer Engineering and Science* 39: 1303–1310.

Krishnamoorti, R., Viva, R.A. and Giannelis, E.P. 1996. Structure and dynamics of polymer-layered silicate nanocomposite. *Chemistry of Materials* 8: 1728-1734.

Kulinski, Z. and Piorkowska, E. 2005. Crystallization, structure and properties of plasticized poly(L-lactide). *Polymer* 46: 10290-10300.

Labrecque, L.V., Kumar, R.A., Dave, V., Gross, R.A. and McCarthy, S.P. 1997. Citrate esters as plasticizers for poly(lactic acid). *Journal of Applied Polymer Science* 66:1507–13.

Lannace, S., Ambrosio, L., Huang, S.J. and Nicolais, L. 1994. Poly(3- hydroxybutyrate)-co-(3-hydroxyvalerate)/poly-l-lactide blends: thermal and mechanical properties. *Journal of Applied Polymer Science* 54: 1525–35.

Ljungberg, N. and Wesslén, B. 2002. The effects of plasticizers on the dynamic mechanical and thermal properties of poly(lactic acid). *Journal of Applied Polymer Science* 86: 1227–1234.

Ljungberg, N., Andersson, T. and Wesslén, B. 2003. Film extrusion and film weldability of poly(lactic acid) plasticized with triacetine and tributyl citrate. *Journal of Applied Polymer Science* 88: 3239–3247.

Ljungberg, N. and Wesslén, B. 2003. Tributyl citrate oligomers as plasticizers for poly(lactic acid): Thermo-mechanical film properties and aging. *Polymer* 44: 7679–7688.

Ljungberg, N. and Wesslén, B. 2004. Thermomechanical film properties and aging of blends of poly(lactic acid) and malonate oligomers. *Journal of Applied Polymer Science* 94: 2140–2149.

Ljungberg, N., Colombini, D. and Wesslén, B. 2005. Plasticization of poly(lactic acid) with oligomeric malonate esteramides: dynamic mechanical and thermal film properties. *Journal of Applied Polymer Science* 96: 992–1002.

Martin, O. and Avérous, L. 2001. Poly(lactic acid): Plasticization and properties of biodegradable multiphase system. *Polymer* 42: 6209–6219.

Mascia, L. and Xanthos, M.A. 1992. Overview of additives and modifiers for polymer blends: facts, deductions, and uncertainties. *Advances in Polymer Technology* 11: 237–248.

Meneghetti, P. and Qutubuddin, S. 2006. Synthesis, thermal properties and applications of polymer-clay nanocomposites. *Thermochimica Acta* 442: 74–77.

Murariu, M., Ferreira, A.S., Pluta, M., Bonnaud, L., Alexandre, M. and Duboi, P. 2008. Polylactide (PLA)-CaSO4 composites toughened with low molecular weight and polymeric ester-like plasticizers and related performances. *European Polymer Journal* 44: 3842-3852.

Noda, I., Satkowski, M.M., Dowrey, A.E. and Marcott, C. 2004. Polymer alloys of nodax copolymers and poly(lactic acid). *Macromolecular Bioscience* 4: 269-275.

Ogata, N., Jimenez, G., Kawai, H. and Ogihara, T. 1997. Structure and thermal/mechanical properties of poly(Llactide)- clay blend. *Journal of Polymer Science, Part B: Polymer Physics* 35: 389-396.

Okada, A. and Usuki, A. 2006. Twenty years of polymer-clay nanocomposites. *Macromolecular Materials and Engineering* 291: 1449-1476.

Okada, A., Kawasumi, M., Usuki, A., Kojima, Y., Kurauchi, T. and Kamigaito, O. 1990. Synthesis and properties of nylon-6/clay hybrids. In: Schaefer DW, Mark JE. eds. Polymer based molecular composites. Pittsburgh. *MRS Symposium Proceedings* 171: 45-50.

Okamoto, M., Morita, S., Kim, H.Y., Kotaka, T. and Tateyama, H. 2001. Dispersed structure change of smectic clay/poly(methyl methacrylate) nanocomposites by copolymerization with polar comonomer. *Polymer* 42: 1201-1206.

Oliva, J., Paya, P., Camara, M.A. and Barba, A. 2007. Removal of famoxadone, fluquinconazole and trifloxystrobin residues in red wines: Effects of clarification and filtration processes. *Journal of Environmental Science and Health, Part B* 42: 775-781.

Paul, M., Alexandre, M., Degée, P., Henrist, C., Rulmont , A., and Dubois, P. 2002. New nanocomposite materials based on plasticized poly(L -lactide) and organo-modified montmorillonites: thermal and morphological study. *Polymer* 44: 443-450.

Rasal, R.M. and Hirt, D.E. 2008. Toughness decrease of PLA-PHBHHx blend films upon surface-confined photopolymerization. *Journal of Biomedical Materials Research Part A* 88: 1079-1086.

Rasal, R.M., Janorkar, A.V. and Hirt, D.E. 2010. Poly(lactic acide) modifications. *Progress in Polymer Science* 33: 338-356.

Ray, S.S. and Bousmina, M. 2005. Biodegradable polymers and their layered silicate nanocomposites: In greening the 21st century materials world. *Progress in Materials Science* 50: 962-1079.

Ray, S.S. and Okamoto, M. 2003. Polymer/layered silicate nanocomposite: a review from preparation to processing. *Progress in Polymer Science* 28: 1539-1641.

Sinclair, R.G. 1996. The case for polylactic acid as a commodity packaging plastic. *Journal of Macromolecular Science: Pure and Applied Chemistry* 33: 585-97.

Sinha, R.S., Yamada, K., Okamoto, M. and Ueda, K. 2002. New polylactide/layered silicate nanocomposite: A novel biodegradable material. *Nano Letters* 2: 1093-1096.

Steinbüchel, A. and Lütke-Eversloh, T. 2003. Metabolic engineering and pathway construction for biotechnological production of relevant polyhydroxyalkanoates in microorganisms. *Biochemical Engineering Journal* 16: 81-96.

Takagi, Y., Yasuda, R., Yamaoka, M. and Yamane, T. 2004. Morphologies and mechanical properties of polylactide blends with medium chain length poly(3-hydroxyalkanoate) and chemically modified poly(3-hydroxyalkanoate). *Journal of Applied Polymer Science* 93: 2363-2369.

Theng, B.K.G. 1979. Formation and properties of clay-polymer complexes. *Amsterdam. Elsevier* pp. 362.

Usuki, A., Kawasumi, M., Kojima, Y., Okada, A., Kurauchi, T., and Kamigato, O. 1993. Preparation and properties of polypropylene/clay nanocomposites. *Journal of Materials Research* 8: 11-47.

Usuki, A., Tukigase, A. and Kato, M. 2002. Preparation and properties of EPDM- clay hybrids. *Polymer* 43: 2185–2189.

Vaia, R.A., Ishii, H. and Giannelis, E.P. 1993. Synthesis and properties of two dimensional nanostructures by direct intercalation of polymer melts in layered silicates. *Chemistry of Material* 5: 1694-1696.

Xu, R., Manias, E., Snyder, A.J. and Runt, J. 2001. New biomedical poly(urethane uera)-layered silicate nanocomposites. *Macromolecules* 34: 337–339.

Xu, Y. and Qu, J. 2009. Mechanical and rheological properties of epoxidized soybean oil plasticized poly(lactic acid). *Journal of Applied Polymer Science* 112: 3185 - 3191.

Zanetti, M., Camino, G., Thomann, R. and Mulhaupt, R. 2001. Synthesis and thermal behaviour of layered silicate-EVA nanocomposites. *Polymer* 42: 4501– 4507.

Zhenyang, Y., Jingbo, Y., Shifeng, Y., Yongtao, X., Jia, M. and Xuesi, C. 2007. Biodegradable poly(L-lactide)/poly(3-caprolactone)-modified montmorillonite nanocomposites: Preparation and characterization. *Polymer* 48: 6439-6447.

3

Fermentation of Sweet Sorghum into Added Value Biopolymer of Polyhydroxyalkanoates (PHAs)

Pakawadee Kaewkannetra
Department of Biotechnology, Faculty of Technology
Khon Kaen University, Khon Kaen
Thailand

1. Introduction

The world today cannot deny the prospect of 'peak oil', higher prices and depletion of petrochemical feed stocks. At the same time, there is an environmental concern of the widely used synthesized plastic from petroleum industry because of its non- degradable nature. Plastics, solid wastes, and pollutants of all kinds not only accumulate as carbon footprint, but also pose a threat to the global warming issues. If they are disposed by open air burning. The United Nation's International Center for Science and High Technology (ICS) thus launched a program focusing on the development of degradable biopolymeric materials and plastic waste disposal in developing countries. Recycling, reuse, incineration, composting, and new technology for development of environmental friendly degradable plastics are making a highly efficient contribution to the mitigation of environmental problems. In addition, concerned researchers and the industrial sector have seen the importance of producing bio-based plastics and biopolymers from agricultural crops based on locally available biomass resources.

2. Polyhydroxyalkanoates (PHAs)

2.1 Definition

Polyhydroxyalkanoates (PHAs) or polyhydroxyalcanoic acid, the main kind of biodegradable and biocompatible biopolymer, are classified as linear polyesters. Typically, PHAs can be produced in nature by several microorganisms such as yeast, fungi and mostly by various bacterial strains. The bacteria could accumulate as intracellular carbon and energy storage under imbalanced growth conditions such as excess in carbon sources but limiting in nutrients of nitrogen, phosphorous and potassium etc (Yu et al., 2005). In addition, fermentation processes from renewable resources such as sugar, starch and even lipid based materials are also affected on the production of PHAs (Kaewkannetra et al, 2008).

2.2 Classification

Since 1925, PHAs were the first biomaterial, discovered by the French microbiologist, M. Lemoigne, accumulated as intracellular substance in a bacterial strain of *Bacillus megaterium* (Lemoigne, 1926; Anderson & Dawes, 1990; Jacquel et al., 2008). Nowadays, biopolymers have been synthesised or are formed in nature during the growth curves of all microorganisms. Depending on the evolution of the synthesis process, different classifications of the different biopolymers have been proposed. In this case, they are classified into the following 4 categories. However, it should be noted that only 3 categories (from 2.2.1 to 2.2.3) are obtained from renewable resources and the remainder is obtained by chemical synthesis.

- Biopolymers derived from biomass such as from agro-resources (e.g., starch, lingo-cellulosic materials, protein and lipids)
- Biopolymers obtained by microbial production as the PHAs
- Biopolymers conventionally and chemically synthesised and the monomers are obtained from agro-resources, e.g., the poly-lactic acids or PLAs
- Biopolymers whose monomers and polymers are obtained conventionally by chemical synthesis such as aliphatic and aromatic hydrocarbon.

2.3 PHAs structures

PHAs are produced by the bacteria to store carbon and energy reserves (Keshavarz, Roy, 2010). Previous works stated that an intracellular accumulation of PHAs improves the survival of general bacteria under environmental stress conditions (Kadouri et al., 2005; Zhao et al., 2007). Various microorganisms are produced in different properties of biopolymer depending on the types of microorganisms and carbon sources used. More than 150 different monomers can be combined within this family to give materials with extremely different properties (Chen & Wu, 2005).

PHAs structure is composed of a monomer of 3-hydroxyalkanoic (HA) acid. The general formulae of the monomer unit is -[O-CH(R)-CH_2-CO]- as seen in Fig.1 (Lee, 1995). The (R)-3HA monomer units are all in the R configuration due to sterospecificity of polymerizing enzyme PHAs synthase. According to the size of the alkyl substituent (R), the mechanical properties of PHAs can typically be divided into three groups by number of carbon atoms in their side chain. Short chain length (*scl*) PHAs are composed of 3-5 carbon atoms, while medium chain length (*mcl*) PHAs consist of 6-15 carbon atoms and long chain length (*lcl*) ones comprise 15 and above carbon atoms. The structure of PHAs depends on supplying carbon sources and microbial types. The composition of PHAs depends on the microorganisms and nature of the carbon sources allowing the formulation of new polymers with different physicochemical properties, i.e., short or mid-chain and long-chain fatty acids.

The most common PHAs' forms found in microorganism cells are polyhydroxybutyrate (PHB) and polyhydroxyvalerate (PHV). However, the majority of the published research on PHAs rather than PHB has concentrated on two bacterial strains, i.e., *Alcaligenes eutrophus* and *A. latus* (Slater, et al, 1988; Kim, et al, 1994; Yamane, et al, 1996, Shi, et al, 1997; Wang & Lee, 1997; Tsuge, et al, 1999; Yu, et al, 2005; Yezza, et al, 2007).

These monomers are biodegradeable and used for the production of bioplastics. PHAs produced from the process are usually composed of 100-30,000 monomers and exist in a short chain. Naturally, the properties of PHAs are similar to thermoplastics that are obtained from petrochemical industry such as polypropylene (PP) and polyethylene (PE) as shown in Table 1 (Evan and Sikdar, 1990).

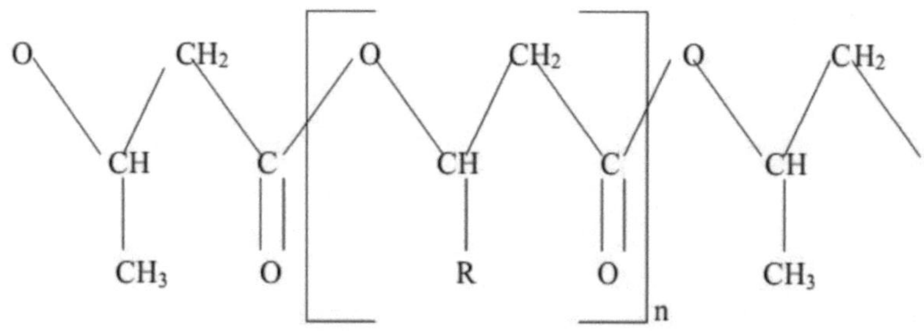

n varies from 600 to 35000

R= hydrogen	Poly(3-hydroxypropionate)
R=methyl	Poly(3-Hydroxybutyrate)
R=ethyl	Poly(3-hydroxyvalerate)
R=propyl	Poly(3-hydroxyhexanoate)
R=pentyl	Poly(3-hydroxyoctanoate)
R=nonyl	Poly(3-hydroxydodecanoate)

Fig. 1. PHAs structure (Lee, 1995)

Property	PHB	Polypropylene
Melting point (T_m), °C	175	176
Crystallinity, %	80	70
Molecular weight, Daltons	5×10^5	20×10^5
Glass transition (T_g), °C	15	-10
Density, g/cm3	1.25	0.905
Tensile strength, Mpa	40	38

(Evan and Sikder, 1990)

Table 1. Properties of polyhydroxybutyrate (PHB) and polypropylene (PP)

They can be either thermoplastic or elastomeric materials with melting points ranging from 40 to 180°C and the percentage of crystallinity (up to 70-80) is similar (Blumm & Owen, 1995). Thus, they can tolerate organic solvents and even lipid and oil. The mechanical and biocompatibility of PHAs can also be easily changed by blending, forming, modifying the surface or combining PHAs with other polymers, enzymes and inorganic materials, making it possible for a wider range of applications such as bottles, bags and wrap films and even in pharmaceutical and medical areas such as drug coating and drug delivery (Steinbuchel & Fuchtenbusch, 1998, Jacquel, et al., 2008).

3. Sweet sorghum

Sweet sorghum (*Sorghum bicolor* L. Moench) is a 3 annual crop and a 4 carbons (C4) containing plant with high biomass productivity. As a high photosynthetic efficient crop, it does not only produce grains, but also yields large amount of sugars in the stems. Typically, it mainly consists of sucrose (up to 55% of dry weight biomass), fructose and glucose which are ideal for preparing fermentation media (Kaewkannetra et al., 2008; Laopaiboon et al., 2009; Gao et al., 2010).

Fig. 2. Sweet sorghum (strain Khon Kean University 40 : KKU 40) from agricultural plantation area of Khon Kaen University, Khon Kaen, Thailand

In the age of petroleum crisis, sweet sorghum (see Fig. 2), especially grains and stems, has already proven advantages over other crops such as sugar cane, cassava, palm, in terms of residue or agricultural wastewater. These crops are feedstock for producing bio-fuels by squeezing the juice and then fermenting into bio-ethanol (Laopaiboon et al (2009), bio-hydrogen (Antonopoulou, et al., 2008). They have recently been used as carbon sources for algae during bio-diesel production (Gao et al., 2010) owing to their sustainability, processing efficiency and superior byproducts, such as bagasse, which serves as high-quality cattle feed. The crops can be cultivated under dry, non-arable land, or warm conditions and are inexpensive to grow. Thus, they are more typically grown for forage, silage, and sugar production than many other crops. The crops are also competitive on the cost aspect of ethanol production.

Presently, the cost of PHAs production is a main limiting factor for extensive production in an industrial scale. The carbon source contributes most significantly (up to 50%) to the overall cost in PHAs production. An attempt to produce PHAs by applying cheap carbon sources could reduce the total cost of the production. Previously, production of PHAs by using cheap carbon sources such as molasses, maple sap and cassava were studied (Grothe et al 1999; Yezza, et al., 2007; Koller et al., 2008). However, sweet sorghum had not been explored so far as a carbon source for PHAs production until in 2008 (Kaewkannetra, et al., 2008). Currently, it is planted in more than 90 countries around the world including Thailand, which is one of the agro-industrial countries in the Southeast Asia with plentiful cheap carbon sources. A number of crops such as cassava, sugar cane, corn, potato, can be extensively used as raw materials to produce degradable polymer and biomaterials via fermentation process by microorganisms. Sweet sorghum, among all the crops mentioned, is of high potentiality to help mitigate environment problems if it can be produced as biodegradable plastics.

3.1 Classification

Sorghum can be classified into 5 categories: grain, grass, broom, pop and sorgo or sweet sorghum. Typically, it is used for animal feed and as sweeteners, primarily in the form of sweet sorghum syrup similar to sugar cane. Recently researchers found rich sugar content in the stem of sweet sorghum, commonly expressed with juice brix degree, but the relation between sugar content and brix degree has not been very clear due to different varieties during their growth. The results revealed a scientific basis for the arrangement in their sowing dates.

3.2 Characterization

Since the duration of sweet sorghum for maturation is approximately 3-5 months, 2-3 crops could be harvested annually in Thailand. Therefore, the production yields of sugar would be double or triple. We can see the possibility of reducing the world energy crisis if sweet sorghum can be converted to energy efficiently. Typically, after harvest, the leaves of the fresh crop are stripped and the stems are squeezed in a roller mill (Fig.3) to obtain sweet sorghum juice. It should be noted that the stems are stored at 4° C while the juice is kept at -20° C prior to use.

Fig. 3. Stems of sweet sorghum are squeezed by a roller mill for preparing sweet sorghum juice (SSJ)

Table 1 shows variations of types and sugar content in sweet sorghum collected from several areas, which depend on the strain, planting seasons, areal conditions, etc. For examples, the total sugar, analyzed by phenol-sulfuric method, was 207.43 g/l. Analysis by means of High Performance Liquid Chromatography (HPLC) has proved contents of 175.97 g/l sucrose, 12.32 g/l glucose, 5.75 g/l fructose in sweet sorghum juice. Normally, the initial sugar concentration of 30-40 g/l is suitable and sufficient for using as carbon source for microbial growth. Therefore, the juice for 1 liter can be diluted as fermentation medium for 4-5 liters (Kaewkannetra et al., 2008).

cultivation area	Harvest month	Sugar types (g/l)			Total Sugar (g/l)	Sources
		sucrose	glucose	fructose		
Thailand	July	175.97	12.32	5.75	194.04	Kaewkannetra et al (2008)
Thailand	*	124.1	20.9	16.8	161.8	Laopaiboon et al (2009)
USA	October	143.3	39.3	61.0	242.6	Liang et al (2010)
Hungary	September	75.1	25.0	18.1	118.2	Sipos et al (2009)
Greece	November	211.9	20.1	-	232.0	Mamma et al (1995)

*Not specific

Table 1. Type and sugar contents in sweet sorghum collected from several areas

4. Fermentation

4.1 Definition

Fermentation implies an intracellular biochemistry process. It is believed to have been the primary means of energy production in earlier organisms before oxygen was at high concentration in the atmosphere. Therefore, it would generate Adenosine Triphosphate (ATP) of the energy molecule of cells even in the presence of oxygen and is synthesized mainly in mitochondria and chloroplasts. In other words, it means the anaerobic enzymatic conversion of organic compounds to simple compounds producing energy in the form of ATP.

Fig. 4. Fermentation of sweet sorghum in fermentor (Kaewkannetra et al., 2008)

Fermentation occurs naturally in various microorganisms such as bacteria, yeasts, fungi and in mammalian muscle. Yeasts were discovered to have connection with fermentation as observed by Louis Pasteur and originally defined as *respiration without air*. However, it does not have to always occur in anaerobic condition. For example, starch when fermented under anaerobic conditions gives alcohols or acids. Yeasts, in ethanol fermentation, use an anaerobic respiration primarily when oxygen is not present in sufficient quantity for normal cellular respiration. However, in large-scale fermentation, the breakdown and re-assembly of biochemicals for industry often carry on in aerobic growth conditions.

4.2 Types of fermentation

Normally, fermentation processes can be classified depending on the objective of study. For example, in terms of products fermentation is divided into 4 types, namely, microbial cell, microbial enzyme, microbial metabolite and transformation process. If considering due to its contaminating conditions, it will be classified into 3 types: septic, semi-septic and aseptic fermentation. However, in general, the fermentation processed are classified into 3 types as follows.

4.2.1 Batch fermentation

Batch fermentation means the cultivation of microorganisms, where the sterile growth medium in desired volume is inoculated with the microorganisms into the bioreactor and no additional growth medium is added during the fermentation. The product will be harvested at the end of the process. Typically, PHA's production is performed using batch fermentation because of low cost for investment and no special control. In addition, sterilization of the feed stock is easier than other fermentation processes, and operation is flexible.

Previous studies have investigated batch fermentation of both carbon sources and microorganisms. The microorganisms will accumulate PHAs after the cell reached the maximum growth coupled with the depletion of nitrogen or phosphorus (Braunegg et al 1998; Wang & Bakken, 1998; Chien et al, 2007).

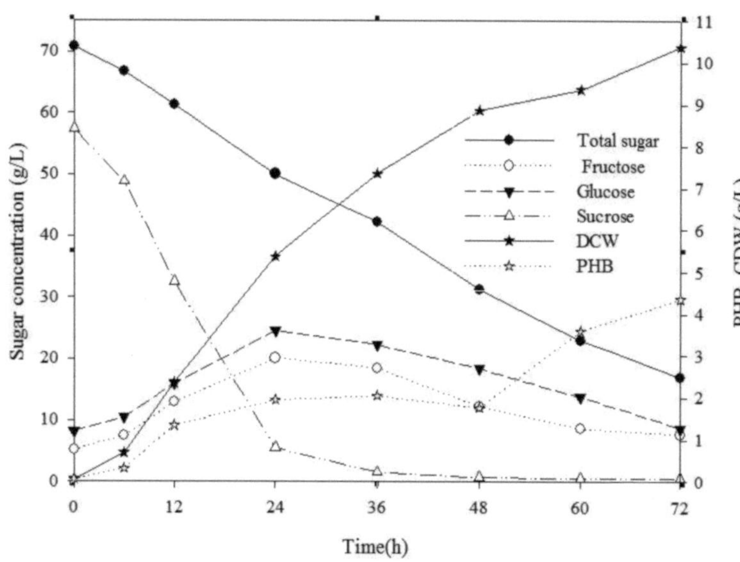

Fig. 5. Growth curve and PHAs production during batch fermentation of sweet sorghum by Bacillus arybhattai in 3 L fermentor (Tanamool et al., 2011)

Fig. 5 shows batch fermentation of sweet sorghum juice (SSJ) by *Bacillus aryabhattai* in 3 L fermentor under cultivating condition with agitation rate at 200 rpm, air rate at 1.5 l/min and temperature at 30° C. Growth monitoring and PHAs production were investigated including total sugar, dry cell weight (DCW) and variations of fructose, glucose and sucrose as functions of time. It was found that the maximum cell and PHAs product reached at about 10.38 g/l and 4.36 g/l. Both slightly increased after 72 hr. The sugar trend then changed. Sucrose was almost depleted within 36 hr while glucose and fructose were slightly increased at the beginning and reached maximum at 24 hr. Then the cells started to use both types of sugar (Tanamool et al., 2011).

Bacterium	PHAs	carbon source	Culture time (h)	Cell conc. (gl^{-1})	PHAs conc. (gl^{-1})	PHAs content (%)	Productivity ($gl^{-1}h^{-1}$)
Alcaligenes eutrophus	P(3HB)	Glucose	50	164	121	76	2.42
A. eutrophus	P(3HB)	CO_2	40	91.3	61.9	67.8	1.55
A. eutrophus	P(3HB)	Tapioca hydrolysate	59	106	61.9	57.5	1.03
A. eutrophus	P(3HB-co-3HV)	Glucose+ propionic acid	46	158	117	74	2.55
A. latus	P(3HB)	Sucrose	18	143	71.4	50	3.97
Azotobacter vinelandii	P(3HB)	Glucose	47	40.1	32	79.8	0.68
Chromobacterium violaceum	P(3HV)	Valeric acid	-	39.5	24.5	62	-
Methylobacterium organophilum	P(3HB)	Methanol	70	250	130	52	1.68
Protomonas extoquens	P(3HB)	Methanol	170	233	149	64	0.88
Pseudomonas olevorans	P(3HBx-co-3HO)	n-octane	$D=0.2h^{-3}$	11.6	2.9	25	0.58
P. oleovorans	P(3HBx-co-3HO)	n-octane	38	37.1	12.1	33	0.32
Recombinant *Escherichia coli*	P(3HB)	Glucose	39	101.4	81.2	80.1	2.08
Rec. *Klebsiella aerogenes*	P(3HB)	Molasses	32	37	24	65	0.75

Table 2. Production of PHAs by various bacteria and carbon sources (Yamane, 1996)

In Table 2, different carbon sources were evaluated and it was found that each strain produced different amounts of PHAs. For example, *Alcaligenes eutrophus* or *Rastonia eutropha* prefers to use fructose (Khanna & Srivastava, 2005). *A. latus* is in favour of sucrose (Grothe et al, 1999) while *Azotobacter valandii* can use glucose (Lin & Sadoff, 1968). Currently, genetic engineering techniques can modify strains such as rec. *E. coli* and rec. *Klebsiella aerogenes* are used to produce PHAs in cheap carbon sources, such as molasses (Slater et al., 1988; Ramachander et al., 2002).

4.2.2 Fed-batch fermentation

Fed-batch fermentation process is a production technique between batch and continuous fermentation. A proper medium feed rate is required to add sequentially into the fermentor during the process and the product is harvested at the end of fermentation just like a batch type.

Fed-batch contains many advantages compared to other cultures. It can be easily concluded that under controllable conditions and with the required knowledge of the microorganism involved in the process, the feed of the required components for growth and/or other substrates are also important due to catabolic repression. The product will never be depleted, the nutritional environment can be maintained approximately constant and the extension of the operating time, high cell concentrations can be achieved and thereby, improve productivity. This is greatly favored in the production of growth-associated products.

Fig. 6 shows a fed batch fermentation of sweet sorghum juice (SSJ) by *Bacillus aryabhattai* in 3 L fermentor under cultivating condition with agitation rate at 200 rpm, air rate of 1.5 l/min, at 30° C and feeding time at 18 and 24 hr during log phase of the culture. It was found that the cell could continuously produce both biomass and PHAs. Maximum cells were obtained at about 14.20 g/l at 54 hr when PHAs content reached 4.84 g/l after 66 hr (Tanamool et al., 2011). In addition, in Table 2, fed batch fermentation by *A. latus* was used for the production of PHAs (Yamane et al, 1996; Wang & Lee, 1997). It could yield high productivity with the use of cheap carbon sources.

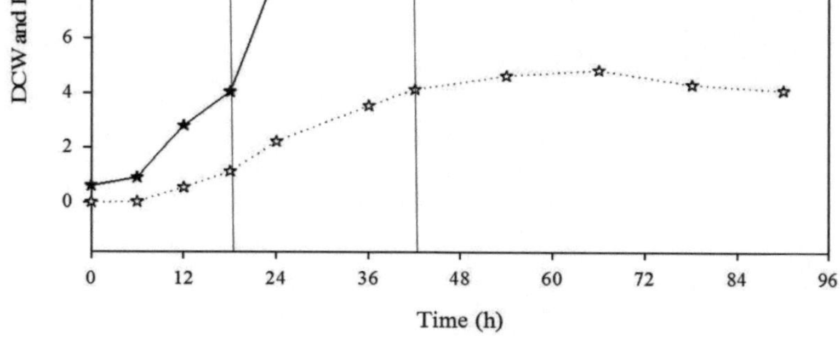

Fig. 6. Growth monitoring and PHAs production during fed-batch fermentation of sweet sorghum by Bacillus arybhattai in 3 L fermentor (Tanamool et al., 2011)

4.2.3 Continuous fermentation

Continuous fermentation is process in which feeds, containing substrate, culture medium and other requirements, are pumped continuously into the agitated bioreactor where the microorganisms are active. Final product is withdrawn from the upper part of the tank. Typically the process is performed by eliminating the inherent down time for cleaning and sterilization and the long lags before the organisms enter a brief period of high productivity.

However, it should be noted that there is a small percentage of the total time in which productivity rate is near its maximum. It is sometimes possible to maintain very high rates of products for a long time with continuous fermentation. Although it can get much more productivity from the fermentor, enhancement over batch fermentation in terms of the total volume of fermentor is not high because equipment needs to be sterilized to support the continuous tank.

Previously, Yu et al (2005) studied the increase of a co-biopolymer of PHBV by *Ralstonia eutropha* in a continuous stirred tank reactor. It was found that the productivity rate increased when sodium propionate was used as the carbon source. Later, Yezza et al (2007) investigated the use of maple sap as a carbon for PHB production by *A. latus*. The productivity of PHB reached 2.6 $gl^{-1} h^{-1}$.

5. Microorganisms

The majority of PHAs biosynthesis is performed by various microorganisms, especially bacteria. They can produce PHAs from a number of substrates and accumulate in their cells as carbon source and energy reserve under imbalanced growth conditions such as nutrient limitation. Fig.7 shows PHA accumulated in their cells that are characterized by transmission electron microscopic (TEM) technique.

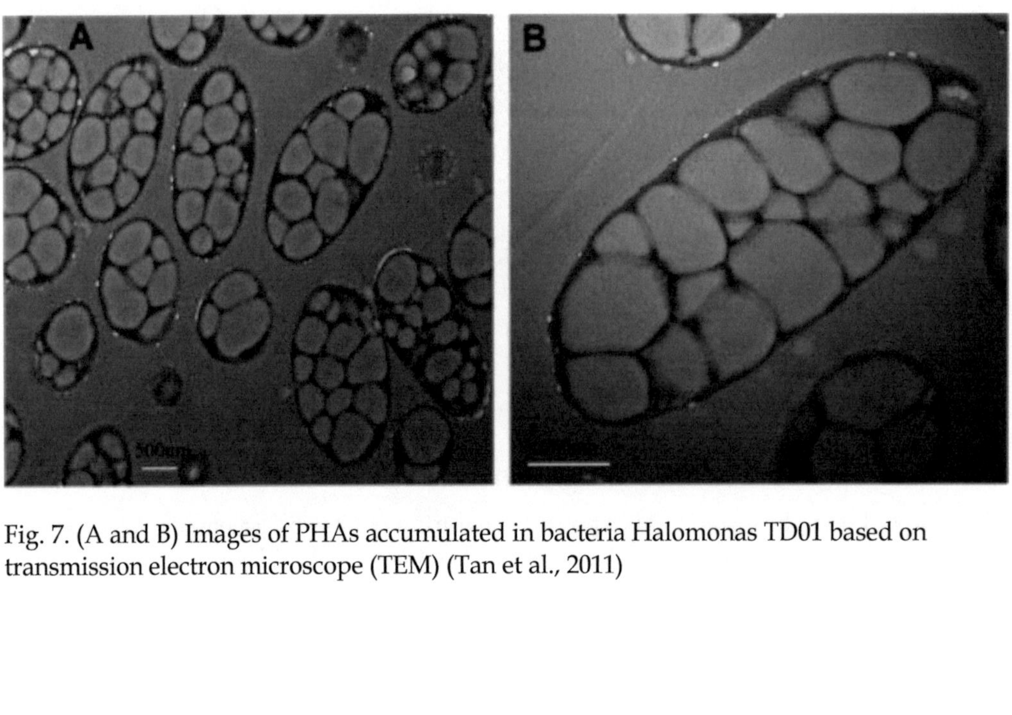

Fig. 7. (A and B) Images of PHAs accumulated in bacteria Halomonas TD01 based on transmission electron microscope (TEM) (Tan et al., 2011)

Depending on PHAs accumulation behaviour, microorganisms can be categorized into two groups. The first group requires limitation of some nutrients such as nitrogen or phosphate while there is an excess in carbon. The members of this group belong to *Alcaligenes eutrophus,* (Kaewkannetra et al, 2008), *Cupiavidus necator* (formerly known as *Ralstronia eutropha*) (Kim et al, 1994; Yu et al, 2005) and *Hydrogenophaga* sp. (Yoon & Choi, 1999; Mahmoudi et al, 2010). The second group does not depend on nutritional limitation but can accumulate PHAs during the growth. Examples are *Alcaligenes latus* (Yamane et al, 1996; Yezza, et al, 2007), *Azotobacter vinelandii, P. aeruginosa,* (Fernandez et al, 2005), *Bacillus mycoides* (Thakur et al, 2001) and *Escherichia coli*, etc.

5.1 Screening, isolation and identification

Although PHAs obtain interest and are widely studied by many researchers, PHAs production is limited by production cost. A major problem to the commercialization of PHAs is much higher production cost than petrochemical-based synthetic plastic (Luengo, 2003).

Many attempts have been devoted to investigate for reducing the cost of PHAs by the isolation of better bacterial strains from various sources such as sludge from wastewater (Kasemsap & Wantawin, 2007) or marine (Chien et al, 2007) and soil environment (Tanamool et al, 2011) using a simple method that can monitor the accumulation of PHAs. For example, efficient conditions for bacterial PHB production can be found in soil due to its heterogeneous nature. Nitrogen availability in soil varies with microsites. It may become a limiting factor for bacterial growth, especially in some nitrogen-poor (carbon-rich) sites (Thakur, 2001). Accordingly, some PHAs-producing bacterial strains such as *Bacillus, Arthrobacter, Corynebacterium, Curtobacterium, Pseudomonas, Micro- coccus,* and *Acinetobacter,* etc. can be isolated from soil environments.

Several simple methods used for detecting intracellular PHAs granules are applied to the screening of PHA producers such as Sudan Black staining (Anderson & Dawes, 1990) and Nile blue A staining (Wang & Bakken, 1998). The positive result shows a black color or fluorescent granules under microscope. Although these methods are feasible and easy, they are rather time- and labor-consuming due to requirement of environmental isolation. Alternative staining method has recently been developed for directly staining colonies or growing bacteria on plates containing Nile blue A or Nile red (Ostle & Holt, 1982). The dye can be directly diffused into microbial cytoplasm, resulting in fluorescent colonies that can be observed under UV illumination without microscopic observation. It is believed that the colony-staining is a suitable method for screening a large number of microbial strains.

After the samples have been collected from the environment, they are preserved at 4°C to protect against contamination prior to isolation. Recently, Tanamool et al, 2011 explained the primary isolation of soil microbes by serial dilution technique. In short, the soil sample was transferred to nutrient broth (NB) and incubated. Then the culture was diluted in normal saline and spread on nutrient agar (NA). Finally, several single colonies were picked and transferred to mineral salt agar. After incubation, PHAs producing capabilities of the microbes were confirmed by Sudan Black B and Nile blue A staining methods as described by Ostle & Holt, 1982; Wang & Bakken, 1998. Viable colonies were directly observed under UV light and fluorescence to detect the accumulation of PHAs (Spiekermann et al, 1999).

The morphology and biochemical characteristics of the isolate then are observed. The PHAs-producing strains can be identified by sequencing partial sequences of their 16S rDNA as described by Edwards et al, (1989). Fig. 8 shows phylogenetic analysis of 16S rDNA sequence of the isolate V 33. It was closely related to bacterial strain of *Hydrogenophaga* sp. (99% identity) (Tanamool et al, 2011). Previous studies have reported that *Hydrogenophaga* sp. can be isolated from soil, mud, and water. The strains were gram-negative, rod-shaped cell and yellow pigmented hydrogen-oxidizing bacteria (Willem et al, 1989). They can grow and produce PHAs from various substrates such as glucose, galactose, xylose, arabinose (Choi et al, 1995) whey lactose (Koller et al, 2008) and in fructose (Mahmudi et al, 2010).

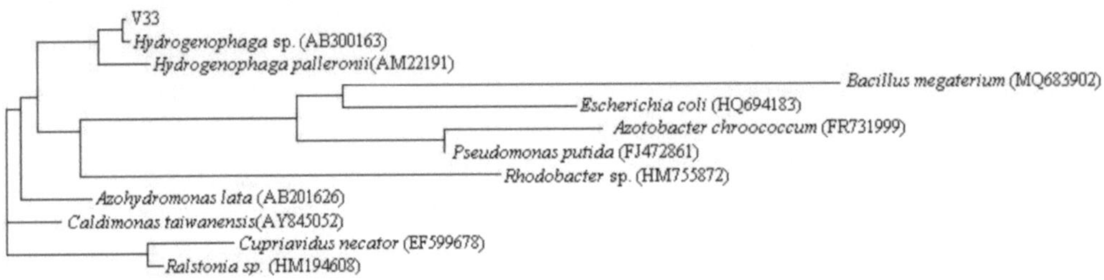

Fig. 8. Phylogenetic analysis of 16S rDNA sequence of the isolate V 33 (Tanamool et al, 2011)

6. Factors affecting fermentation process

6.1 Microbial strain

Tim & Steinbuchel (1990) studied the mechanisms of *Alcaligenes* spp. and *Rhodospirillum rubrum* for PHB synthesis by using butyric acid. Later, Steinbuchel (1991) confirmed the results for the potential of PHAs production and the compositions of PHAs. Under various microbial strains used, different contents of PHAs were obtained.

Table 3 shows PHAs contents in different weights compared to dry cell mass accumulated in various microorganisms. It was found that *Alcaligenes* spp. contains a maximum of 96% PHAs in its cell.

Microbial strain	Weight/ dry cell mass (%)
Ralstonia eutropha (reclassified Alcaligenes eutrophus)	96
Azospirillum	75
Azotobacter	73
Baggiatoa	57
Leptothrix	67
Methylocystis	70
Pseudomonas	67
Rhizobium	57
Rhodobacter	80

(Available from *www: http://members.rediff.com/jogsn/bp6.htm)*

Table 3. Variations of PHAs contents in various microbial strains

6.2 Nitrogen sources

Nitrogen sources are a nutrient effect on structure and the accumulation of PHAs in microbial cells. Nitrogen sources compose one limiting factor for microorganisms during cultivation. In 1982, Rhee et al studied the production of PHAs from *A. eutrophus*. Variation of nitrogen sources was also studied and the results were summarized in Table 4. Although the microorganisms can be used in the same carbon source, the cell growth and their PHAs products are different when nitrogen sources vary.

Carbon source	Nitrogen source	Cell dry Weight (g/l)	PHAs content (wt%)	PHAs	
				3HB	3HV
Glucose	Yeast extract	3.2	45.2	98.4	1.6
	Urea	1.4	14.8	85.6	14.4
	$(NH_4)_2SO_4$	1.8	29.3	93.3	6.7
Sucrose	Yeast extract	1.5	18.9	98.3	1.7
	Urea	1.3	4.0	93.5	6.5
	$(NH_4)_2SO_4$	1.2	15.1	92.0	8.0
Sorbitol	Yeast extract	3.1	44.8	93.4	6.9
	Urea	1.8	37.2	85.7	14.3
	$(NH_4)_2SO_4$	1.7	28.1	93.5	6.5
Mannitol	Yeast extract	3.4	58.7	94.1	5.9
	Urea	2.2	18.2	92.5	7.5
	$(NH_4)_2SO_4$	1.8	29.0	93.3	6.7
Na-gluconate	Yeast extract	2.3	34.5	91.9	8.1
	Urea	1.6	5.3	78.1	21.9
	$(NH_4)_2SO_4$	2.7	41.1	66.7	13.3

Table 4. Effect of nitrogen sources on PHAs production (Rhee et al, 1992)

6.3 Fermentation time

Timm & Steinbuchel (1990) revealed the production of PHAs by *Pseudomonas aeruginosa*. It was found that when the cell reached to stationary phase, the intracellular PHAs were decreased. Lageveen et al (1998) produced PHAs from n-octane by *Pseudomonas oleovorans*.

It was found that the microorganism has a specific duration for PHAs recovery. However, only some strains can accumulate PHAs along their growth or growth-associated strain.

6.4 Mineral substances

Mineral substances such as phosphorus, sulphur, magnesium, etc. could affect microbial growths in several functions. Phosphorous is assimilated only in form of orthophosphate ($H_2PO_4^-$) and it is essential for all microorganisms. Sulphur is incorporated into S-containing amino acid in protein and magnesium functions as a cofactor or effectors for many enzymes in microbial metabolism. Each microbial strain needs nutrients and minerals in different contents due to suitable limitation of nutrients for PHAs-producing microorganism.

6.5 Temperature and pH

Normally, the production of PHAs usually carries on under moderate temperature range of 30-35°C and at pH 6.5-7.0. This condition may be suitable for most PHA3s-producing microbes.

7. PHAs recovery and their properties

After fermentation, subsequent midstream to downstream processes such as cell disruption, centrifugation, extraction and drying will be carried on for product recovery. Fig. 9 shows a white sheet of PHB obtained from fermentation of sweet sorghum juice (SSJ) by *Bacillus aryabhattai*.

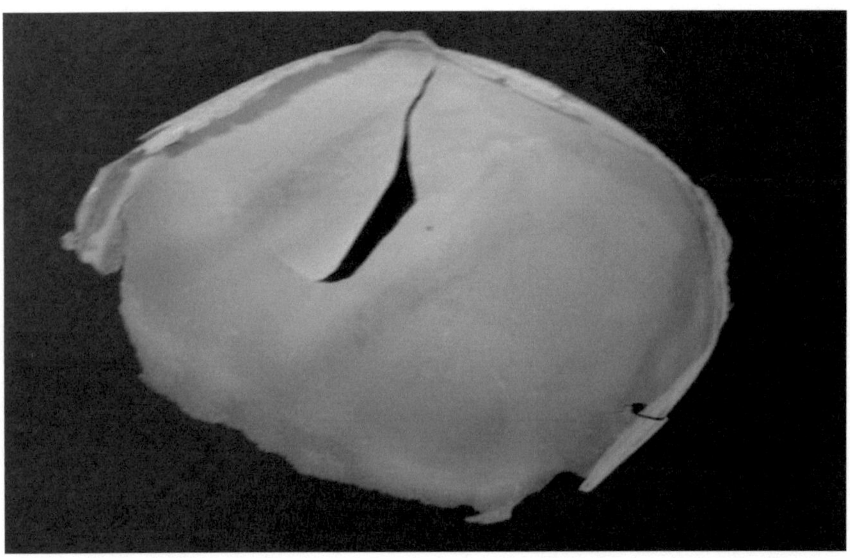

Fig. 9. Biopolymer of PHB obtained from *Bacillus aryabhattai*

In Fig. 10 the sheet was then characterized for the properties such as thermal properties of glass temperature (Tg) and melting temperature (Tm) analyzed by Differential Scanning Calorimetric (DSC) analysis. The Tg and Tm of the PHB sheet reached 1.11°C and 167.3°C when compared to the standard PHB (99.5%) where Tg and Tm are at 2.81°C and 176.29°C. Thermal properties of PHB sheet obtained were lower than the standard values of PHB. This implies that the PHB sheet can be easily further used to blend and form with other cheap biopolymers or raw materials by heat treatment during bioplastic production process. In Fig.11 shows the degrading temperatures analyzed by Thermo gravimetric analysis (TGA) in both standard PHB and the PHB sheet. The degradation will be completed at temperature of 450°C for the standard PHB, while the PHB sheet is completely degraded at 300°C. It means that the PHB sheet obtained containing of improved quality that was better than the standard PHB.

In addition, physico-chemical characterization, blending and forming steps need to be fulfilled to get novel biomaterials for replacing conventional plastic in a wide range of further applications. These include packing containers, bottles, wrappings, bags, thin films and disposable items (diapers or feminine hygiene products).

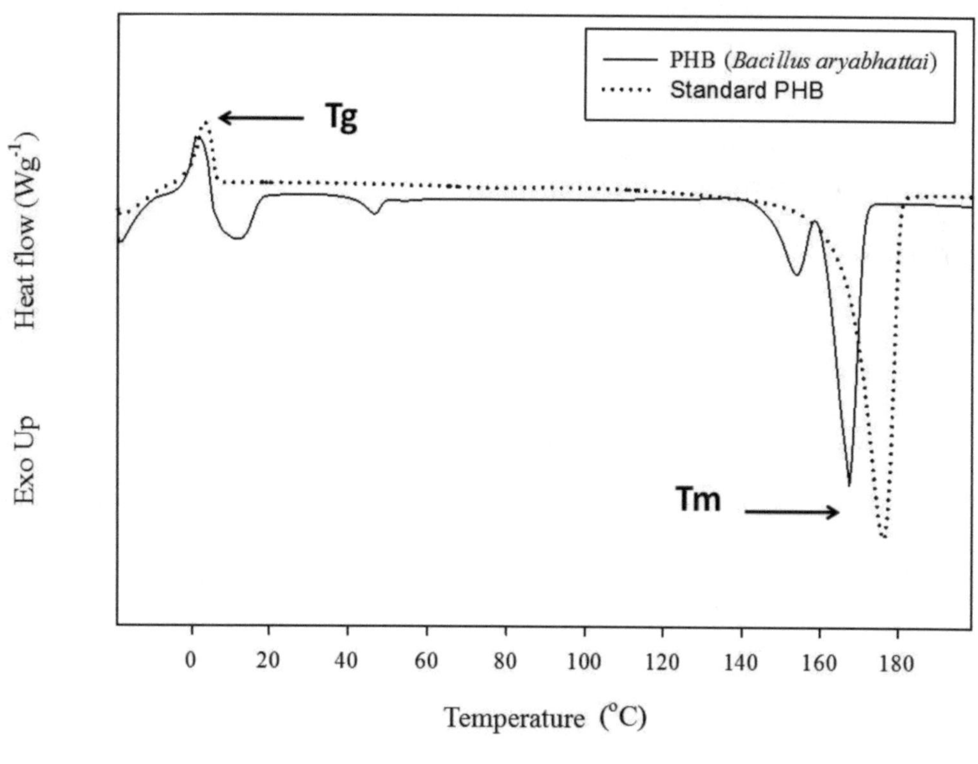

Fig. 10. Tg and Tm of standard PHB and PHB obtained from *Bacillus aryabhattai*

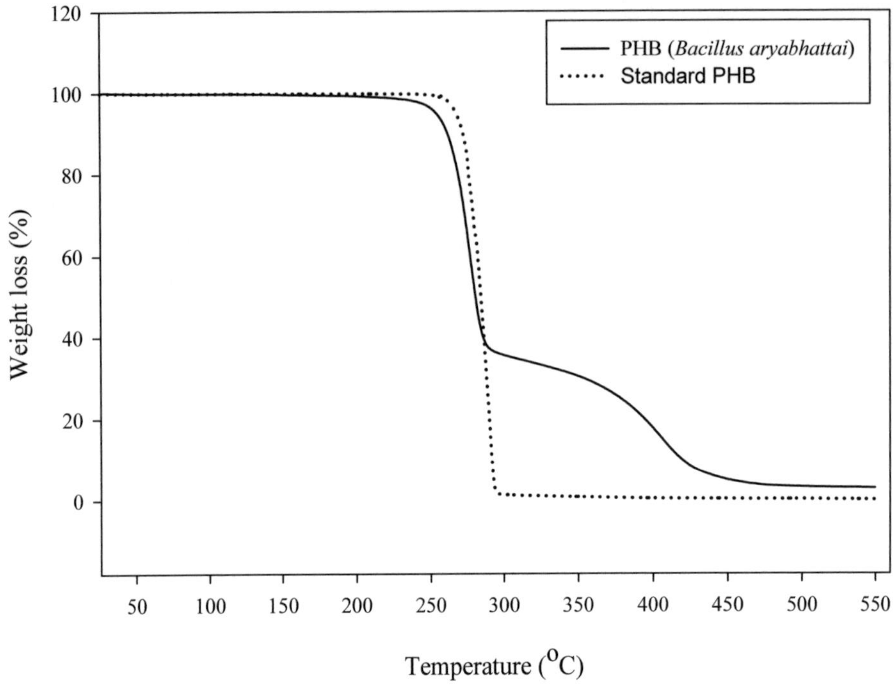

Fig. 11. Thermal degradation temperature of standard PHB and PHB obtained from *Bacillus aryabhattai* by Thermo-gravimetric analysis (TGA)

8. Summary

Although PHAs are being interested and broadly studied by many researchers for a long time, the most important obstacle to commercial application of PHAs is their price. PHAs' production cost is roughly 10 times or more compared to petrochemical-based synthetic plastic materials such as PE and PP. Much attempt has been devoted to reduce the production cost of PHAs in different ways, for example, use of isolated bacterial strains, development of improving strains by genetic techniques such as recombinant DNA of *E. coli* and *Streptomyces aureofacienc* NRRL, controlling a culture condition via various fermentation processes such as batch, fed-batch, repeated-batch/repeated fed-batch, enhancing production via optimization of fermentation process using response surface methodology (RSM), more economical recovery process, and most importantly, the use of cheaper carbon sources.

A novel non-petroleum based biodegradable plastic produced from sugar based agricultural raw materials as sweet sorghum, sugarcane and molasses, having potential properties comparable with conventional or synthetic plastics, is under development and could lower the contribution of plastic wastes to municipal landfills at about 20% of the total waste by volume and 10% by weight and can achieve a satisfactory for the environmental imperative.

The gradual transition towards the biobased economy brings opportunities for 'developing' countries to leapfrog beyond the petroleum era and into a cleaner, greener and more renewable future based on biotechnology knowledge.

9. References

[1] Alias, Z., Tan, I., 2005. Isolation of palm oil utilizing, polyhydroxyalkanoates (PHA)-bacteria by an enrichment technique. Bioresour. Technol. 96, 2729-2734.

[2] Anderson, A.J., Dawes, E.A., 1990. Occurrence, metabolism, metabolic role, and industrial uses of bacterial polyhydroxyalkanoates. Microbiol. Rev. 54, 450–472.

[3] Antonopoulou, G., Gavala, N.H., Skiadas, V.I, Angelopoulos, K., Lyberatos, G., 2008. Biofuel generation from sweet sorghum: Fermentative hydrogen production and anaerobic digestion of the remaining biomass. Biores. Technol. 99 (1), 110-119.

[4] Atlas, R.M., 2006. Handbook of microbiological media. 3rd Ed. CRC Press. 2051.

[5] Blumm, E., Owen, A. J., 1995. Miscibility, crystallization and melting of poly (3-hydroxy butyrate) / poly (L-lactide) blends. Polymer. 36, 4077-4081.

[6] Braunegg, G., Lefebvre, G., Genser, K. L., 1998. Polyhydroxyalkanoates, biopolyesters from renewable resources: Physio-engineering aspects. J. Biotechnol. 65, 127–161.

[7] Byrom, D. 1987. Polymer synthesis by microorganisms: technology and economics. Trends Biotechnol. 5, 246-250.

[8] Chien, C. C., Chen, C. C., Choi, M. H., Kung, S. S., Wei, Y. H., 2007. Production of poly-β-hydroxybutyrate (PHB) by *Vibrio* spp. isolated from marine environment. J. Biotechnol. 132, 259-263.

[9] Chen, G.Q. and Wu, Q. 2005. The application of polyhydroxyalkanoates as tissue engineering materials. Biomaterials. 26: 6565-6578.

[10] Choi, M. H., Song, J. J., Yoon, S. C., 1995. Biosynthesis of copolyesters by *Hydrogenophaga pseudoflava* from various lactones. Can. J. Microbiol. 41, 60-67.

[11] Edwards, U., Rogall, T., Bocker, H., Emde, M., Bottger, E., 1989. Isolation and direct complete nucleotide determination of entire genes. Characterization of a gene coding for 16S ribosomal DNA. Nucleic Acids Res.17:7843-53. DOI: 10.1093/ nat/17.19.7843.

[12] Evan, J.D., Sikdar, S.K., 1990. Biodegradable plastic; An idea whose time has come. Chemtech. 38-42.

[13] Fernandez, D., Rodriguez, E., Bassas, M., Vinas, M., 2005. Agro-industrial oily wastes as substrates for PHA production by the new strain *Pseudomonas aeruginosa* NCIB 40045: Effect of culture conditions. Biochem. Eng. J. 26, 159-167.

[14] Gao, C., Zhai, Y., Ding, Y., Wu, Q., 2010. Application of sweet sorghum for biodiesel production by heterotrophic microalga *Chlorella protothecoides*. App. Ener. 87, 756-761.

[15] Grothe, E., Moo-Young, M., Chisti, Y., 1999. Fermentation optimization for the production of PHB microbial thermoplastic. Enz. Microb. Tech. 25, 132-141.

[16] Ha, C.S., Cho, W.J., 2002. Miscibility, properties, and biodegradability of microbial polyester containing blends. Prog. Polym. Sci. 27, 759-809.

[17] Jacquel, N., Lo, C.-W. (2008) Isolation and purification of bacterial poly (3-hydroxy alkanoates) Biochemical Engineering Journal. 39 (1),15-27.

[18] Kadouri, A., Derenne, S., Largeau, C., Casadevall, E., Berkaloff, E., 1998. Resistant biopolymer in the outer walls of *Botryococcus braunii,* Brace. Phytochemistry, 27(2): 551-557.

[19] Kaewkannetra, P., Tanonkeo, P., Tanamool, V., Imai, I., 2008. Biorefinery of sweet sorghum juice into value added product of biopolymer, J. Biotechnol. 136, S412.

[20] Kasemsap, C., Wantawin, C., 2007. Batch production of PHA by low-poly phosphate-content activated sludge at varying pH. Biores. Technol. 98, 1020-1027.

[21] Keshavarz, T., Roy, I., 2010. Polyhydroxyalkanoates: bioplastics with a green agenda. Curr. Opin. Microbiol. 13, 321-326.

[22] Khanna, S., Srivastrava, A. K. 2005. Statistical media optimization studies for growth and PHB production *by Ralstonia eutropha.* Process Biochemistry. 40, 2173-2182.

[23] Kim, B.S., Lee, S.C., Lee, S.Y., Chang, H.N., Chang, Y.K and Woo, S.I., 1994. Production of polyhydroxybutyrate by fed batch with glucose concentration control in *Ralstonia eutropha.* Biotechnol. Bioeng. 42, 892-898.

[24] Koller, M., Bona, R., Chiellini, E., Fernandes, E.G., Horvat, P., Kutschera, C., Hesse, P., Braunegg, G., 2008. Polyhydroxyalkanoate production from whey by *Pseudomonas hydrogenovora.* Bioresour. Technol. 99, 4854-4863.

[25] Laopaiboon, L., Nuanpeng, S., Srinophakun, P., Klanrit, P., Laopaiboon, P., 2009. Ethanol production from sweet sorghum juice using very high gravity technology: Effects of carbon and nitrogen supplementations. Biores. Technol. 100(18), 4176-4182.

[26] Law, J.H., Slepecky, R.A., 1961. Assay of poly-β-hydroxybutytic acid. Bacteriol. 82, 33-36.

[27] Lageveen, R.G., Huisman, G.W., Preusting, H., Ketelaar, P., Eggink, G., Witholt, B. 1998. Formation of Polyesters by *Pseudomonas oleovorans:* Effect of Substrates on Formation and Composition of Poly-(R)-3-Hydroxyalkanoates and Poly-(R)-3-Hydroxyalkenoates. Applied Environmental and Microbiology, 54, 2924-2932.

[28] Lemoigne, M., 1926. Produits de deshydration et de polymerization de l'acide - oxybutyric. Bull. Soc. Chim. Biol. 8, 770–782.

[29] Liang, Y., Sarkany, N., Cui, Y., Yesuf, J., Trushenski, J., Blackburn, J.W., 2010. Use of sweet sorghum juice for lipid production by *Schizochytrium limacinum* SR21. Bioresource Technology, 101, 3623-3627.

[30] Lin, L.P., Sadoff, H.L. 1968. Encystment and polymer production by *Azotobacter vinelandii* in the presence of beta-hydroxybutyrate. J. Bacterio. 95 (6), 2336-2343.

[31] Lee, S.Y., 1996. Bacterial polyhydroxyalkanoates. Biotechnol. Bioeng. 49, 1-14.

[32] Luengo, J.M., Garcia, B., Sandoval, A., Naharro, G., Olivera, E.R., 2003. Bioplastics from microorganisms. Curr. Opin. Microbiol. 6, 251–260.

[33] Mahmoudi, M., Sharifzadeh-Baei, M., Najafpour, G. D., Tabandeh, F., Eisazadeh, H.,2010. Kinetic model for polyhydroxybutyrate (PHB) production by *Hydrogenophaga pseudoflava* and verification of growth conditions. Afr. J. Biot. 9, 3151-3157.

[34] Mamma, D., Christakopoulos, P., Koullas, D., Kekos, D., Macris, B.J., Koukios, E., 1995. An alternative approach to the bioconversion of sweet sorghum carbohydrates to ethanol. Biomass and Bioenergy, 8, 99-103.

[35] Mizuno, K., Ohta, A., Hyakutake, M., Ichinomiya, Y., Tsuge, T., 2010. Isolation of polyhydroxyalkanoate-producing bacteria from a polluted soil and characterization of the isolated strain *Bacillus cereus* YB-4. Polym. Degrad. Stab. 95, 1335-1339.

[36] Ostle, A., Holt, J.G., 1982. Nile blue A as a fluorescent stain for poly-b-hydroxybutyrate. Appl. Envi. Microb. 44, 238–241.

[37] Ramachander, T.V.N., Rohini, D., Belhekar, A., Rawal, S.K. 2002. Synthesis of PHB by recombinant *E. coli* harboring an approximately 5 kb genomic DNA fragment from *Streptomyces aureofaciens* NRRL 2209. International Journal of Biological Macromolecules, 31 (1-3): 63-69.

[38] Rhee Y.H., Jang J.H., Roger P.L., 1992. Biopolymer production by an *Alcaligenes* sp. for biodegradation plastics. Austral. Biotechnol. 2, 230–232.

[39] Shi, H., Shiraishi, M. and Shimizu, K., 1997. Metabolic flux analysis for biosynthesis of poly (β-hydroxybutytic acid) in *Alcaligenes eutrophus* from various carbon sources. J Ferment. Bioeng. 84, 579-587.

[40] Sipos, B., Reczey, J., Somorai, Z., Kadar, Z., Dienes, D., Reczey, K., 2009. Sweet Sorghum as Feedstock for Ethanol Production: Enzymatic Hydrolysis of Steam-Pretreated Bagasse. Applied Biochemistry and Biotechnology, 153, 151-162.

[41] Steinbuchel, A., 1991.Polyhydroxyalkanoic acids. In Biomaterials. Edited by D. Byrom. Macmillan Publishers Ltd., Basingstoke. 123-213.

[42] Slater S. C., Voige W. H., Dennis D. E., 1988. Cloning and expression in *Escherichia coli* of the *Alcaligenes eutrophus* H16 poly-3-hydroxybutyrate biosynthetic Pathway. Bact.170, 4431-4436.

[43] Spiekermann, P., Rehm, B.H., Kalscheuer, R., Baumeister, D., Stein buchel, A., 1999. A sensitive, viable-colony staining method using Nile red for direct screening of bacteria that accumulate polyhydroxyalkanoic acids and other lipid storage compounds. Arch. Microbiol. 171, 73–80.

[44] Steinbuchel, A., Fuchtenbusch, B., 1998. Bacterial and other biological systems for polyester production. Trends Biotechnology, 16, 419-27.

[45] Tanamool, V., Imai, T., Danvirutai, P., Kaewkannetra, P. 2011. Biosynthesis of poly hydroxylalkanoate (PHA) by *Hydrogenophaga* sp. isolated from soil environments during batch fermentation. J. Life Science. *In press.*

[46] Thakur, P., Borah, B., Baruah, S., Nigam, J., 2001. Growth-associated production of poly-3 hydroxybutyrate by *Bacillus mycoides*. Folia Microbiol. 46, 488-494.

[47] Timm, A., Steinbuchel, A. 1990. Formation of polyesters consisting of medium-chain-length 3-hydroxyalkanoic acids from gluconate by *Pseudomonas aeruginosa* and other *fluorescent pseudomonads*. Applied Environmental Microbiology, 56, 3360-3367.

[48] Tsuge, T., Shimoda, M., Ishizaki, A., 1999. Optimization of L-Lactic acid feeding for the production of poly-D-3-hydroxybutyric acid by *Alcaligenes eutrophus* in fed-batch culture. J Biosci. Bioeng. 8, 404-409.

[49] Wang, F., Lee, S. Y., 1997. Poly (3-Hydroxybutyrate) production with high productivity and high polymer content by a fed-batch culture of *Alcaligenes latus* under nitrogen limitation. App. Envi. Microb. 63, 3703-3706.

[50] Wang, J. G., Bakken, L. R., 1998. Screening of soil bacteria for PHB production and its role in the survival of starvation. Microb. Eco. 35, 94-101.

[51] Willem, A., Busse, J., Goor, M., Pot, B., Falsen, E., Janzent, E., Hoste, B., Gillis, M., Kersters, K., Auling, G., De-Ley, J., 1989. *Hydrogenophaga*, a new genus of hydrogen-oxidizing bacteria that includes *H. flava* comb. nov. (formerly *Pseudomonas flava*), *H. palleronii* (formerly *Pseudomonas palleronii*), *H. pseudoflava* (formerly *Pseudomonas pseudoflava* and *P. carboxydoflava*), *H. taeniospiralis* (formerly *P. taeniospiralis*). Sys. Bacteriol. 39, 319-333.

[52] Yamane, T., 1996. Yield of poly-D-(-)-3-hydroxybutyrate from various carbon sources: a theorethical study. Biotechnol Bioeng. 41, 165-170.

[53] Yamane, Y., Fukunage, M. and Lee, Y.W., 1996. Increased PHB productivity by high-cell- density fed-batch culture of *Alcaligenes latus*, a growth-associate PHB producer. Biotechnol. Bioeng. 50, 197-202.

[54] Yezza, A., Halasz, A., Levadoux, W., Hawari, J., 2007. Production of PHB by *Alcaligenes latus* from maple sap. Appl. Microb. Biot. 77, 269-274.

[55] Yoon, S. C., Choi, M. H., 1999. Local sequence dependence of polyhydroxyalkanoic acid degradation in *Hydrogenophaga pseudoflava*. Bio. Chem. 274, 37800-37808.

[56] Yu, S.T., Lin, C.C., Too, J.R., 2005. PHBV production by *Ralstonia eutropha* in a continuous stirred tank reactor. Proc. Biochem. 40, 2729-2734.

[57] Zhao Y.H., Li H.M., Qin L.F., Wang H.H., Chen G.Q., 2007. Disruption of the polyhydroxyalkanoate synthase gene in *Aeromonas hydrophila* reduces its survival ability under stress conditions. FEMS Microbiology Letters. 276(1): 34-41.

4

Xylan, a Promising Hemicellulose for Pharmaceutical Use

Acarília Eduardo da Silva[1], Henrique Rodrigues Marcelino[1], Monique Christine Salgado Gomes[1], Elquio Eleamen Oliveira[2], Toshiyuki Nagashima Jr[3] and Eryvaldo Sócrates Tabosa Egito[1]

[1]Universidade Federal do Rio Grande do Norte
[2]Universidade Estadual da Paraíba
[3]Universidade Federal de Campina Grande
Brazil

1. Introduction

Polymers are versatile materials with wide use in several industry fields, such as engineering, textile, automobile, packaging and biomedical. In the pharmaceutical industry, both natural and synthetic polymers have been largely used with different applications for the development and production of cosmetics and traditional dosage forms and novel drug delivery systems. For instance, a number of polymers are used as fillers, lubricants, disintegrants, binders, glidants, solubilizers, and stabilizers in tablets, capsules, creams, suspensions or solutions. Additionally, biodegradable and bioadhesive polymers may play an important role in the development of novel drug delivery systems, especially for controlled drug release.

Polymeric microparticles have been studied and developed for several years. Their contribution in the pharmacy field is of utmost importance in order to improve the efficiency of oral delivery of drugs. As drug carriers, polymer-based microparticles may avoid the early degradation of active molecules in undesirable sites of the gastrointestinal tract, mask unpleasant taste of drugs, reduce doses and side effects and improve bioavailability. Also, they allow the production of site-specific drug targeting, which consists of a suitable approach for the delivery of active molecules into desired tissues or cells in order to increase their efficiency.

Lately, the concern with environment and sustainability has been rising progressively and renewable sources of materials have been increasingly explored.

The aim of this chapter is to summarize some of the research findings on xylan, a natural polymer extracted from corn cobs, which presents a promising application in the development of colon-specific drug carriers. Physicochemical characterization of the polymer regarding particle size and morphology, composition, rheology, thermal behavior, and crystallinity will be provided. Additionally, research data on its extraction and the development of microparticles based on xylan and prepared by different methods will also be presented and discussed.

2. Xylan

For thousands of years, nature has provided humankind with a large variety of materials for the most diversified applications for its survival, such as food, energy, medicinal products, protection and defense tools, and others. The pharmaceutical industry has benefitted from such diversity of biomaterials and has exploited the use of natural products as sources of both drugs and excipients. One example of a promising biomaterial for pharmaceutical use is xylan, a hemicellulose largely found in nature, being considered the second most abundant polysaccharide after cellulose.

Xylan has drawn considerable interest due to its potential for packaging films and coating food, as well as for its use in biomedical products (Li et al., 2011). Because it is referred to as a corn fiber gum with a sticky behavior, xylan has been used as an adhesive, thickener, and additive to plastics. It increases their stretch and breaking resistance as well as their susceptibility to biodegradation (Ünlu et al., 2009). Xylan has also been studied because of its significant mitogenic and comitogenic properties, which enable it to be compared to the commercial immunomodulator Zymosan (Ebringerova et al., 1995). Another interesting application for xylan may be found in the food industry as an emulsifier and protein foam stabilizer during heating (Ebringerova et al., 1995). Previous papers have investigated the suitable use of xylan in papermaking (Ebringerova et al., 1994) and textile printing (Hromadkova et al., 1999). In the drug delivery field, xylan extracted from birch wood has been used for the production of nanoparticles after structural modification by the addition of different ester moieties, namely those with furoate and pyroglutamate functions (Heinze et al., 2007). On the other hand, the esterification of xylan from beech wood via activation of the carboxylic acid with *N,N"*-carbonyldiimidazole has been carried out in order to produce prodrugs for ibuprofen release (Daus & Heinze, 2010).

Egito and colleagues have been working for over a decade on the extraction of xylan from corn cobs and its use for the development of microparticles as drug carriers for colon-specific delivery of anti-inflammatory and toxic drugs, such as sodium diclofenac (SD), 5-aminosalycilic acid (5-ASA), and usnic acid (UA). Xylan-coated microparticles have also been developed by Egito and co-workers in order to deliver magnetite particles (Silva et al., 2007). Different microencapsulation techniques have been used for the production of xylan-based microparticles. Coacervation, interfacial cross-linking polymerization, and spray-drying have been shown to be the most successful methodologies for that purpose (Garcia et al., 2001; Nagashima et al., 2008).

Xylan degradation occurs by the action of hydrolytic enzymes named xylanases and β-xylosidases. Those enzymes are produced by a number of organisms, such as bacteria, algae, fungi, protozoa, gastropods, and arthropods (Kulkarni et al., 1999). The degradation of xylan in ruminants has been well reported, while some human intestinal bacteria have been investigated for their ability to produce xylan-polymer degrading enzymes. Among those intestinal species able to degrade complex carbohydrates, lactobacilli, bacteroides, and non-pathogenic clostridia have demonstrated that ability (Grootaert et al., 2007). Because of the presence of those bacteria in the human colon whether by induction of prebiotics or not, it is believed that xylan is a promising polymer for the composition of biodegradable drug carriers for colonic delivery. They would be able to undergo the upper gastrointestinal tract mostly intact, being degraded by xylanases when reaching the colon.

Additionally, corn cobs correspond to an abundant and low-cost renewable material in several countries worldwide and their recycling plays a very important role in the reduction of waste products. Consequently, such approach would lead to a relevant increase in the sustainability of agriculture around the world.

2.1 Sources, extraction, and structure

Hemicelluloses are the second most abundant polysaccharides in nature after cellulose. They occur in close association with cellulose and lignin and contribute to the rigidity of plant cell walls in lignified tissues. Hemicelluloses constitute about 20–30% of the total mass of annual and perennial plants and have a heterogeneous composition of various sugar units, depending on the type of plant and extraction process, being classified as xylans (β-1,4-linked D-xylose units), mannans (β-1,4-linked D-mannose units), arabinans (α-1,5-linked L-arabinose units), and galactans (β-1,3-linked D-galactose units) (Figure 1) (Belgacem & Gandini, 2008).

Xylans are the main hemicelluloses in hardwood and they also predominate in annual plants and cereals making up to 30% of the cell wall material and one of the major constituents (25–35%) of lignocellulosic materials. The most potential sources of xylans include many agricultural crops such as straw, sorghum, sugar cane, corn stalks and cobs, and hulls and husks from starch production, as well as forest and pulping waste products from hardwoods and softwoods (Ebringerova & Heinze, 2000; Kayserilioglu et al., 2003).

The structural diversity and complexity of xylans are shown to depend on the botanic source. Various suitable extraction procedures for the isolation of xylans from different plant sources are described and compared in the literature. It is suggested that certain structural types of xylans, such as glucuronoxylan, arabinoglucuronoxylan, and arabinoxylan, can be prepared from certain plant sources with similar chemical and physical properties. Its general structure has a linear backbone consisting of 1,4-linked D-xylopyranose residues, a reducing sugar with five carbon atoms. These may be substituted with branches containing acetyl, arabinosyl, and glucuronosyl residues, depending on the botanic source and method of extraction (Den Haan & Van Zyl, 2003; Habibi & Vignon, 2005).

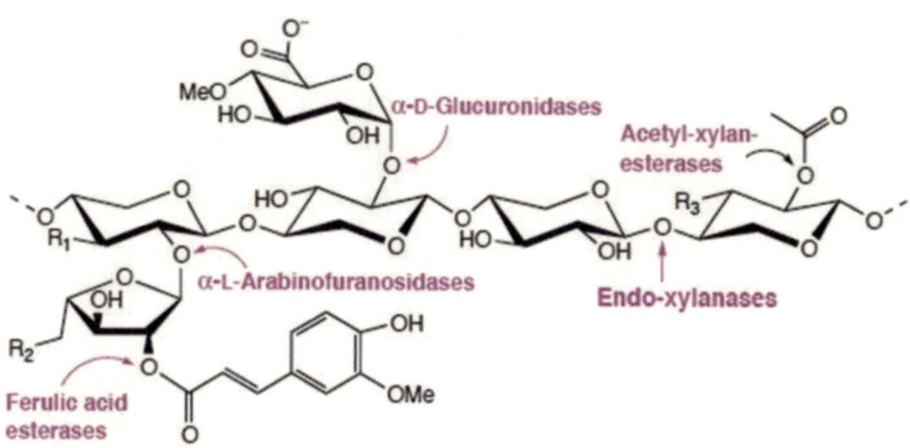

Fig. 1. Chemical structure of xylan (Shallom & Shoham, 2003).

A frequently used classification is based on the degree of substitution and types of side groups for characterization (Ebringerová, 2005; Sedlmeyer, 2011):

a. Homoxylans are linear polysaccharides common in some seaweeds.
b. Glucuronoxylans can be partly acetylated and have units substituted with α-(1→2)-4-O-methyl-D-glucopyranosyl uronic acid (MeGlcUA). They are found in hardwood, depending on the treatment.
c. (Arabino)glucuronoxylans have a substitution with α-(1→3)-L-arabinofuranosyl (ArbF) next to MeGlcUA. They are typical for softwoods.
d. Arabinoxylans with a substitution of the β-(1→4)-D-xylopyranose backbone at position 2 or 3 with ArbF can be esterified partly with phenolic acids. This type is frequently found in the starchy endosperm and the outer layers of cereal grains.
e. (Glucurono)arabinoxylans can be disubstituted with ArbF units, acetylated, and esterified with ferulic acid. This form is typical of lignified tissues of grasses and cereals.
f. Heteroxylans are heavily substituted with various mono- or oligosaccharides and are present in cereal bran, seed, and gum exudates.

Investigation of the xylan structure by various researchers is necessary. The use of xylan as a raw material is directly related to its structure. There is an interest in the application of the xylan polymer in the paper, pharmaceutical, cosmetic, biofuel and food industries. Several medical applications are cited in the literature. The films based on xylan show low oxygen permeability and thus have a potential application in the food packaging and pharmaceutical areas. Numerous studies use the xylan polymer as a specific substrate for xylanases. Besides that, xylan can be hydrolyzed into xylose and subsequently be converted into ethanol (Ebringerova & Heinze, 2000; Ebringerova & Hromadkova, 1999; Ebringerova et al., 1998; Garcia et al., 2000; Kayserilioglu et al., 2003; Oliveira et al., 2010; Sedlmeyer, 2011; Yang et al., 2005).

Previous studies on the corn cob xylan revealed the existence of at least two structurally different components. One is a low-branched arabinoglucuronoxylan, which is mostly water-insoluble (wis-X), and the second is a highly branched, water-soluble heteroxylan (ws-X), which possesses significant mitogenic and comitogenic activities (Ebringerova et al., 1995). The ws-X could be useful also as a food additive because of its emulsifying activity and ability to stabilize protein foam during heating. The wis-X has the ability to remain intact in the physiological stomach environment and small intestine. This property, together with the presence of xylanases (a group of enzymes which degrade the xylan) in the human colon, makes this polymer a suitable raw material for the medical field, especially as a constituent of colon-specific drug carriers (Oliveira et al., 2010; Rubinstein, 1995; Silva et al., 2007).

The most common method to extract xylan is the alkaline extraction. Several pretreatment methods can be used in association in order to break the covalent bonds that exist between xylan and other carbohydrates during the extraction (Wang & Zhang, 2006). A number of articles studied the use of ultrasound on the xylan extraction. Hromadkova and coworkers reported that 36.1% of xylan was extracted from corn cobs with 5% NaOH solution at 60°C for 10 min of ultrasonication in comparison with 31.5% of xylan in the classical extraction. Both extractive methods yielded xylan with immunogenic properties (Hromadkova et al., 1999).

Wang and Zhang also investigated the effects on the xylan extracted from corn cobs enhanced by ultrasound at various lab-scale conditions. Results showed that the optimization conditions of xylan extraction should be carried out using (i) 1.8 M NaOH, (ii) corn cobs to NaOH solution ratio of 1:25 (w/w), (iii) sonication at 200 W ultrasound power for 30 min at 5 min intervals, and (iv) 60 °C (Wang & Zhang, 2006).

The process of the alkaline extraction of xylan from corn cobs was studied by Egito and colleagues (Unpublished data). The methodology applied in this work consisted of milling the corn cobs and separating the powder into different sizes. After that, the dried corn cobs were dispersed in water under stirring for 24h. The sample was treated with 1.3% (v/v) sodium hypochlorite solution in order to remove impurities. Then, an alkaline extraction was carried out by using NaOH solution. The bulk was neutralized with acetic acid, and xylan was extracted by settling down after methanol addition. Afterwards, several washing steps were performed by using methanol and isopropanol. Finally, the sample was filtered and dried at 50°C.

The efficiency of extraction was observed to be inversely proportional to the corn cob particle size. This was expected because the size reduction corresponds to an increase in total particle surface area. An increase in the time of the alkaline extraction and in the NaOH concentration also improves the efficiency of xylan extraction. This happened because when the NaOH concentration was lower, the xylan present in corn cobs could not be fully dissolved in the solution. Thus, it resulted in lower efficiency of xylan extraction. However, when the NaOH concentration was higher than 2 M, the yields decreased with continuously increasing of the NaOH concentration. This is probably due to the alkaline degradation of xylan chains, proceeding at the higher NaOH concentration, which indicated that the ideal NaOH concentration in the extraction was between 1.5 and 1.8 M (Unpublished data).

2.2 Characterization of corn cob xylan

Comprehensive physicochemical characterization of any raw material is a crucial and multi-phased requirement for the selection and validation of that matter as a constituent of a product or part of the product development process (Morris et al., 1998). Such demand is especially important in the pharmaceutical industry because of the presence of several compounds assembled in a formulation, such as active substances and excipients, which highlights the importance of compatibility among them. Besides, variations in raw materials due to different sources, periods of extraction and various environmental factors may lead to failures in production and/or in the dosage form performance (Morris et al., 1998). Additionally, economic issues are also related to the need for investigating the physicochemical characteristics of raw materials since those features may determine the most adequate and low-cost material for specific procedures and dosage forms.

After the extractive process described by Oliveira and colleagues, corn cob xylan appears to be an off-white fine powder with limited flowability. The xylan powder consists of a mixture of aggregated and non-aggregated particles with irregular morphology, a spherical shape, and a rough surface, as could be observed through the scanning electron microscopy (SEM) (Figure 2) (Oliveira et al., 2010).

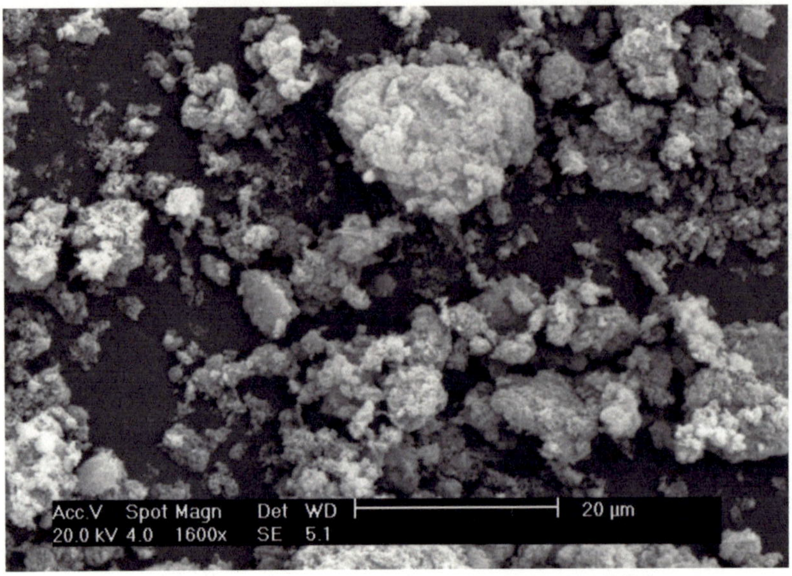

Fig. 2. SEM image of xylan powder after extraction from corn cobs (Oliveira et al., 2010).

The xylan particle size distribution was determined by laser diffraction. It was observed that approximately 90%, 50%, and 10% of the dry extract of xylan was smaller than 65.39 ± 1.76, 23.34 ± 1.2, and 7.68 ± 0.54 µm, respectively, while the mean particle size of xylan was found to be 30.53 ± 1.5 µm (Oliveira et al., 2010) (Figure 3).

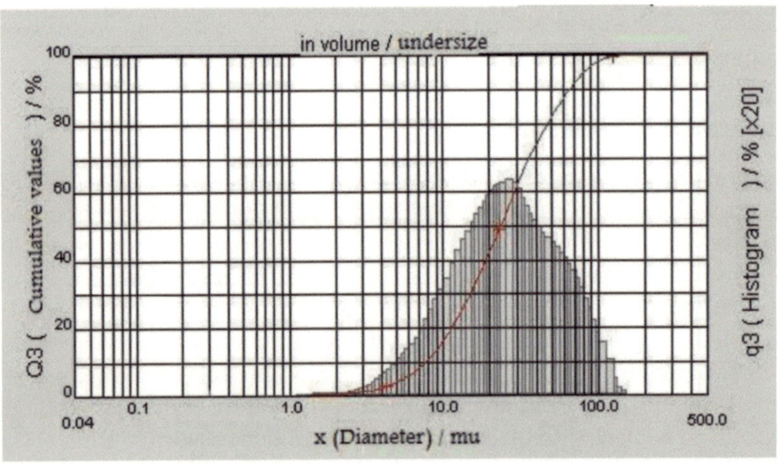

Fig. 3. Particle size distribution of xylan powder after extraction from corn cobs (Oliveira et al., 2010).

As a consequence of the irregular and rough structure of the xylan particles, entanglements between particles are promoted and this fact may explain the poor flow properties of this polymer (Kumar et al., 2002; Nunthanid et al., 2004). Additionally, rheological parameters of xylan powder have also been studied, such as bulk and tapped densities, Hausner ratio, Carr's index, and angle of repose values, and they are summarized in Table 1.

Property	Value (± standard deviation)
Bulk density	0.1336 (± 0.0029) g/ml
Tap density	0.2256 (± 0.0059) g/ml
Compressibility index	40.77 (± 0.0035) %
Hausner ratio	1.68 (± 0.01)
Compactability	32.6 (± 0.1) mL[a]
Angle of repose	40.70 (± 3.2318)°

[a]extrapolating the values to 100 mL

Table 1. Rheological properties of xylan powder extracted from corn cobs

The bulk density of a powder is calculated by dividing its mass by the volume occupied by the powder (Abdullah & Geldart, 1999). Tapped bulk density, or simply tapped density, is the maximum packing density of a powder achieved under the influence of well-defined, externally applied forces (Oliveira et al., 2010). Because the volume includes the spaces between particles as well as the envelope volumes of the particles themselves, the bulk and tapped density of a powder are highly dependent on how the particles are packed. This fact is related to the morphology of its particles and such parameters are able to predict the powder flow properties and its compressibility.

Hausner ratio and the compressibility index measure the interparticle friction and the potential powder arch or bridge strength and stability, respectively (Carr, 1965; Hausner, 1967). They have been widely used to estimate the flow properties of powders. A Hausner ratio value of less than 1.20 is indicative of good flowability of the material, whereas a value of 1.5 or higher suggests a poor flow (Daggupati et al., 2011). The compressibility index is also called the Carr index. According to Carr, a value between 5 and 10, 12 and 16, 18 and 21, and 23 and 28 indicates excellent, good, fair, and poor flow properties of the material, respectively. The Hausner ratio and Carr's index values obtained for xylan are listed in Table 1 and suggest that xylan presents extremely poor flow properties. Although the Hausner ratio and the Carr index correspond to indirect measurements of flowability of materials during preliminary studies, the values obtained for xylan suggest the characterization of this biopolymer as a cohesive powder.

Another parameter of the flow behavior of a powder is the angle of repose, which evaluates the flowability of powders through an orifice onto a flat surface. It is considered a direct measurement. Angles of repose below 30° indicate good flowability, 30°-45° some cohesiveness, 45°-55° true cohesiveness, and > 55° sluggish or very high cohesiveness and very limited flowability (Geldart et al., 2006). The angle of repose for xylan is 40.70°, which confirms its cohesive nature predicted by the aforementioned indirect measurements. This is due to the irregular shape of the xylan particles. Besides, the fine particles of xylan, having high surface-to-mass ratios, are more cohesive than coarser particles; hence, they are more influenced by gravitational force. In addition, it is generally believed that the flowability of powders decreases as the shapes of particles become more irregular (Oliveira et al., 2010).

Regarding the characterization of corn cob xylan by Fourier-transform infrared (FT-IR) spectroscopy, two main absorption bands at 3405 cm^{-1} and 1160 cm^{-1} are revealed. They can be attributed to the OH stretching characteristic of glycosidic groups and to CC and COC stretching in hemicelluloses, respectively (Figure 4).

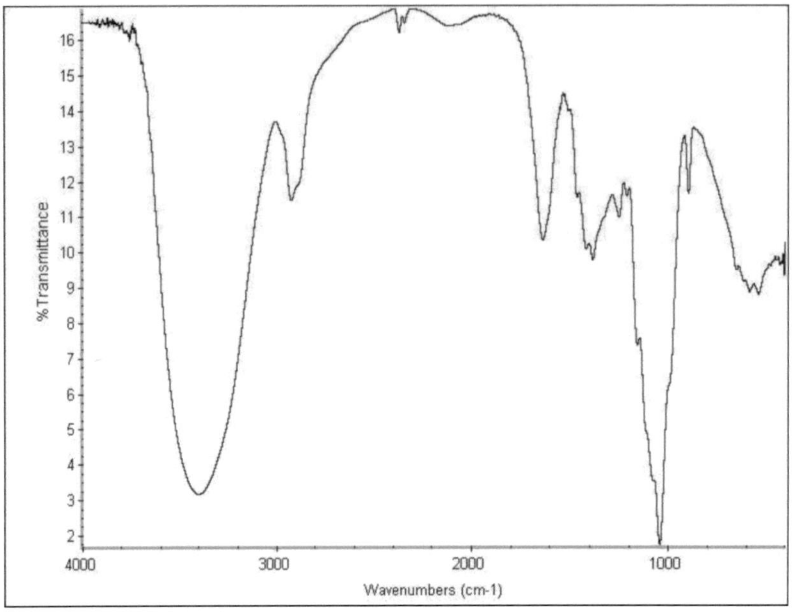

Fig. 4. FT-IR spectrum of xylan powder extracted from corn cobs.

Moreover, an absorption band near 1375 cm^{-1} is detected and it is assigned to the CH bending vibration present in cellulose and hemicellulose chemical structures (Sun et al., 1998). The prominent band at 1044 cm^{-1} is also associated with hemicelluloses and is attributed to the C–OH bending. Finally, a sharp band at 897 cm^{-1}, which is typical of b-glycosidic linkages between the sugar units in hemicelluloses, was detected in the anomeric region (Sun et al., 2005).

A solid-state ^{13}C nuclear magnetic resonance (NMR) experiment was carried out in 4 mm double bearing rotor made from ZrO_2 on a Bruker DSX 200 MHz spectrometer with resonance frequency at 75.468 MHz. The pulse length was 3.5 μs and the contact time of 1H–13C CP was 2–5 ms.

The NMR spectrum of the dry sample showed broad unresolved peaks that correspond to a typical mixture of 4-O-methyl-D-glucuronic acid, L-arabinose and D-xylose, and proteins (Oliveira et al., 2010) (Figure 5).

Concerning the analysis of crystallinity of xylan, the X-ray diffraction detects a few and small peaks, which indicate that xylan presents a low crystallinity (Figure 6).

On the other hand, thermal analysis of xylan by thermogravimetry demonstrates a first event of 8.9% weight loss detected in the range of 62 and 107°C due to dehydration. The second and most relevant event of 49.8% weight loss appears in the range of 250 and 300°C due to the polymer decomposition (Figure 7). The differential scanning calorimetry curve reveals an endothermic peak at 293.04°C, which is attributed to the melting point of the polymer (Figure 7).

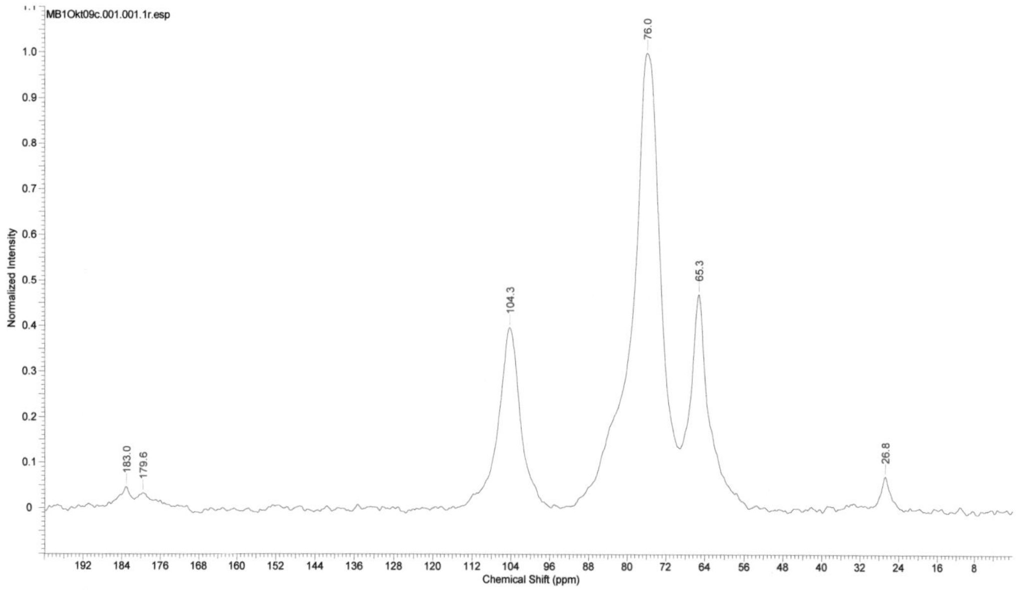

Fig. 5. Solid-state ^{13}C nuclear magnetic resonance spectrum of corn cob xylan.

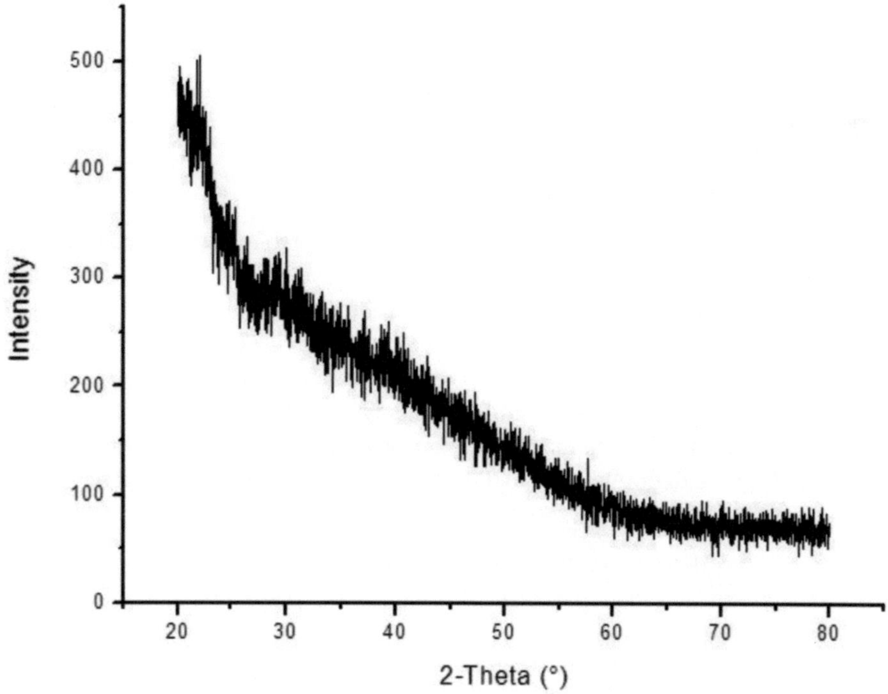

Fig. 6. X-ray diffraction pattern for corn cob xylan (Unpublished data).

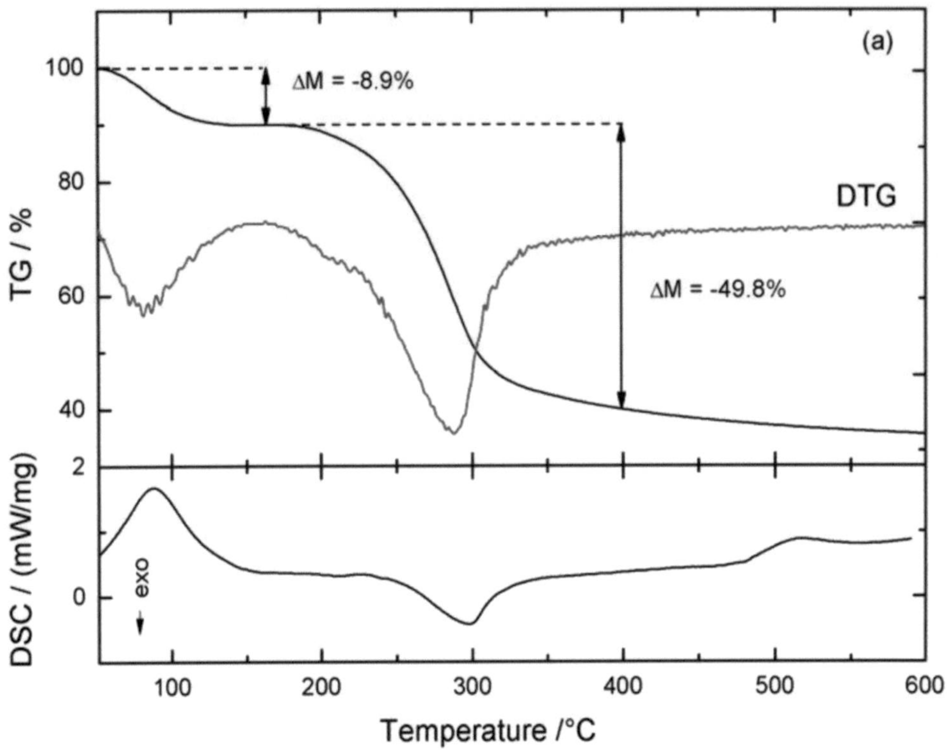

Fig. 7. Thermogravimetry and differential scanning calorimetry curves for corn cob xylan (Unpublished data).

3. Xylan microparticles

As previously described, xylan has been considered as a suitable raw material to produce colonic drug delivery systems due to the ability of enzymes produced by the colonic microflora to degrade the β-glycosidic bonds between the sugar units of the polymer backbone (Kacurakova et al., 2000; Oliveira et al., 2010; Saha, 2000). Regarding the colonic environment, it presents a neutral pH range of the colon and a local blood circulation that prevents the rapid distribution of the drug into the body before circulating into the intestinal blood vessels. As a result, the colonic absorption of drugs is an alternative approach to deliver molecules that are degraded in the stomach medium and are toxic in small quantities in the body (Luo et al., 2011).

A large variety of drug delivery systems are described in the literature, such as liposomes (Torchilin, 2006), micro and nanoparticles (Kumar, 2000), polymeric micelles (Torchilin, 2006), nanocrystals (Muller et al., 2011), among others. Microparticles are usually classified as microcapsules or microspheres (Figure 8). Microspheres are matrix spherical microparticles where the drug may be located on the surface or dissolved into the matrix. Microcapsules are characterized as spherical particles more than 1μm containing a core substance (aqueous or lipid), normally lipid, and are used to deliver poor soluble molecules in hydrophilic medium (Couvreur et al., 2002; Kumar, 2000; Ribeiro et al., 1999). Furthermore, microcapsules may have one or more cores while the microspheres may show a homogenous or heterogeneous aspect with the drug distributed equally or aggregated into the particle.

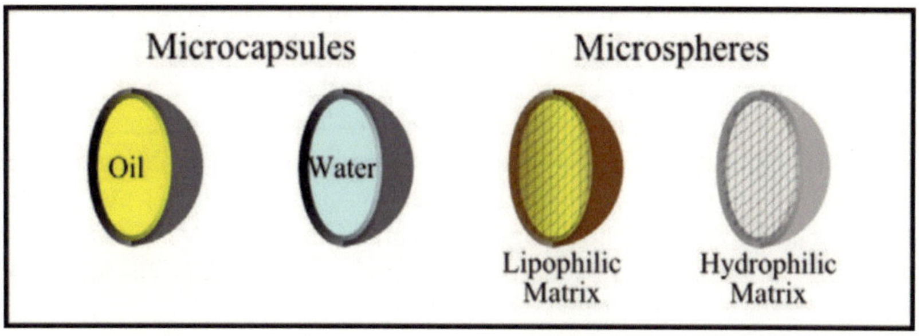

Fig. 8. Structural differences between microcapsules and microspheres.

In the past, microparticles were considered as mere carriers, usually micronized dry material without sophisticated attributes (Vehring, 2008). However, nowadays they have found a number of applications in the pharmaceutical field. For instance, microparticles have been used in order to achieve controlled release of drugs, deliver two or more agents in the same system, improve the bioavailability and the biodistribution of molecules, target drugs to specific cells or issues, or mask the unpleasant taste of some active molecules (Simó et al., 2003; Tran et al., 2011; Vehring, 2008). Xylan microparticles have been successfully produced by the following methods: coacervation (Garcia et al., 2001), interfacial cross-linking (Nagashima et al., 2008) and spray-drying (Unpublished data), all of which are described in the following subsection.

3.1 Methods of production

3.1.1 Coacervation

The coacervation technique is defined as a partial desolvation of a homogeneous polymer solution into a polymer-rich phase (coacervate) and the poor polymer phase (coacervation medium). It was the first process to be scaled-up to an industrial process (Jyothi et al., 2010). However, for the optimization of this method, some changes in the methodology were made and the technique was classified into two types: simple and complex. In simple coacervation the desolvation agent is added to form the coacervate, while the complex coacervation process is guided by the presence of two polymers with different charges, and divided into three steps: (i) formation of three immiscible phases, (ii) deposition of the coating, and (iii) strengthening of the coating (Gouin, 2004; Jyothi et al., 2010; Qv et al., 2011).

After the first step, which includes the formation of three immiscible phases (liquid manufacturing vehicle, core material, and coating material), the core material is dispersed in a solution of the coating polymer. The coating material phase, which corresponds to an immiscible polymer in liquid state, is formed by (i) changing the temperature of the polymer solution, (ii) adding a salt, (iii) adding a non-solvent, (iv) adding an incompatible polymer to the polymer solution, and (v) inducing polymer-polymer interaction. The second step includes deposition of the liquid polymer upon the core material. Finally, the prepared microcapsules are stabilized by cross-linking, desolvation, or thermal treatment (Jyothi et al., 2010; Stuart, 2008).

Xylan-based micro- and nanoparticles have been produced by simple coacervation (Garcia et al., 2001). In the study, sodium hydroxide and chloride acid or acetic acid were used as solvent and non-solvent, respectively. Also, xylan and surfactant concentrations and the molar ratio between sodium hydroxide and chloride acid were observed as parameters for the formation of micro- and nanoparticles by the simple coacervation technique (Garcia et al., 2001). Different xylan concentrations allowed the formation of micro- and nanoparticles. More precisely, microparticles were found for higher concentrations of xylan while nanoparticles were produced for lower concentrations of the polymer solution. When the molar ratio between sodium hydroxide and chloride acid was greater than 1:1, the particles settled more rapidly at pH=7.0. Regarding the surfactant variations, an optimal concentration was found; however, at higher ones a supernatant layer was observed after 30 days (Garcia et al., 2001).

3.1.2 Interfacial cross-linking polymerization

The production of microparticles by this technique involves basically two experimental steps: (i) emulsification and (ii) cross-linking reaction (Figure 9). In fact, the emulsification is the major step of the process to determine the particle size distribution and the aggregation arrangement of the microparticles. Therefore, the chemical reactivity of the cross-linking agent is also important to determine the required time to complete the entire process (Chang, 1964; Jiang et al., 2006; Levy & Andry, 1990; Li et al., 2009).

In the first step of the interfacial cross-linking polymerization, the polymer is dissolved into the solvent, which is the internal phase of the emulsion, and another phase with a non-solvent to the polymer is produced; then the aqueous phase is poured to the organic phase to produce the emulsion. Afterwards, a solution containing the cross-linking agent is added to the emulsion to form a rigid structure of the microparticles (Couvreur et al., 2002; Rao & Geckeler, 2011).

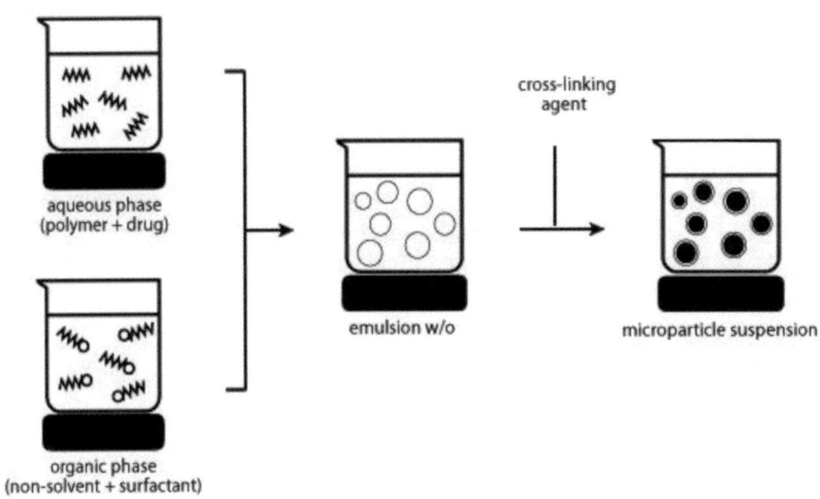

Fig. 9. Scheme for interfacial cross-linking polymerization.

The influence of the lipophilic external phase on the production of xylan-based microparticles by interfacial cross-linking polymerization has been investigated (Nagashima et al., 2008). Three different external phases were investigated: a 1:4 (v/v) chloroform:cyclohexane mixture, soybean oil, and a medium chain triglyceride, with viscosities below 1, 24, and 52 cP, respectively. It was observed that the use of these different lipid phases results in different macroscopic and microscopic aspects of the system (Figure 10).

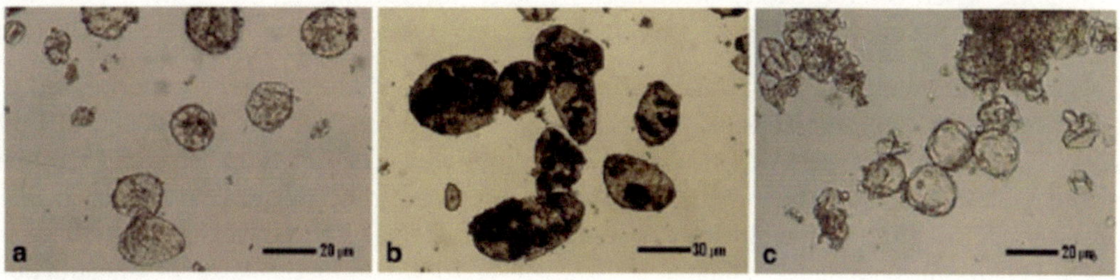

a) 1:4 (v/v) Chloroform: cyclohexane mixture;
b) Soybean oil;
c) Medium chain triglycerides.

Fig. 10. Optical microscopy images of xylan microcapsules produced by interfacial cross-linking polymerization with different lipophilic external phases (Nagashima et al., 2008).

Because emulsions are susceptible to many destabilizing phenomena occurring since the formation of these systems, such as Ostwald ripping (Anton et al., 2008) and coalescence (Li et al., 2009), the formation of the microcapsules may be influenced by those phenomena, which can form aggregates and agglomerates, respectively. Also, the higher viscosity of the lipid phase may support the shaping of microcapsules with a bigger size than the oil phases with a lower viscosity (Nagashima et al., 2008).

The cross-linking agent is present in the interfacial area, where the polymer should be adsorbed due to the poor solubility of the polymer at the external medium. It is known that the chemical reactivity of the cross-linking agent is a limiting parameter to determine the duration and the yield of the process (Li et al., 2009). Terephthaloyl chloride is a cross-linking agent used to produce microcapsules based on polysaccharides, and it was extensively studied by Levy to produce starch derivate microcapsules for pharmaceutical uses. According to Levy, the pH medium, the concentration of the polymer, the stirring speed, and the concentration of terephthaloyl chloride are significant parameters for the formation of the microparticles and their structure (Andry et al., 1996; Andry & Lévy, 1997; Edwards-Lévy et al., 1994; Levy & Andry, 1990).

Cross-linked xylan-based microparticles are produced by the emulsification of an alkaline solution of xylan with a lipophilic phase formed by a mixture of chloroform and cyclohexane by using 5% (w/v) sorbitan triesterate as the surfactant. Subsequently, the cross-linking reaction is carried out for 30 minutes with 5%(w/v) terephthaloyl chloride in order to yield a hard and rigid polymeric shell (Nagashima et al., 2008).

The interfacial cross-linking polymerization has been demonstrated to be a suitable method for the production of xylan microcapsules with high drug encapsulation efficiency. SD-loaded cross-linked xylan microcapsules have been produced with three different amounts of the drug (3.1, 6.2, and 60mg). At the end of the process, yellowish suspensions of spherical polymeric microcapsules were produced. The mean particle size was found to be approximately 12.5 µm (Figure 11). Regarding the encapsulation efficiency, high and inversely concentration-dependent rates were achieved. While the SD concentration of 3.1 mg induced a load ability of 99 ± 2%, 6.2 mg of SD promoted 75.8 ± 1 %, and 60mg of SD yielded a 30.4 ± 6 % load efficiency. Accordingly, the results demonstrated the feasibility of producing xylan microcapsules with and without SD, presenting the same aspect and homogeneity, but concentration-dependent encapsulation rates (Unpublished data).

Regarding the stability of those formulations after storage, studies have been performed in order to evaluate the SD release. As a result of storage for 30 days, it was found that approximately 30 ± 5% of SD had been released to the external medium. This fact may be evidence that some adjustments in the methodology need to be made. One approach that has been shown as a promising strategy to avoid the drug release to the external medium is the spray-drying technique, which will produce a dried product instead of an aqueous suspension of microparticles. It may be used as a complement to the interfacial cross-linking polymerization and is described in the following subsection.

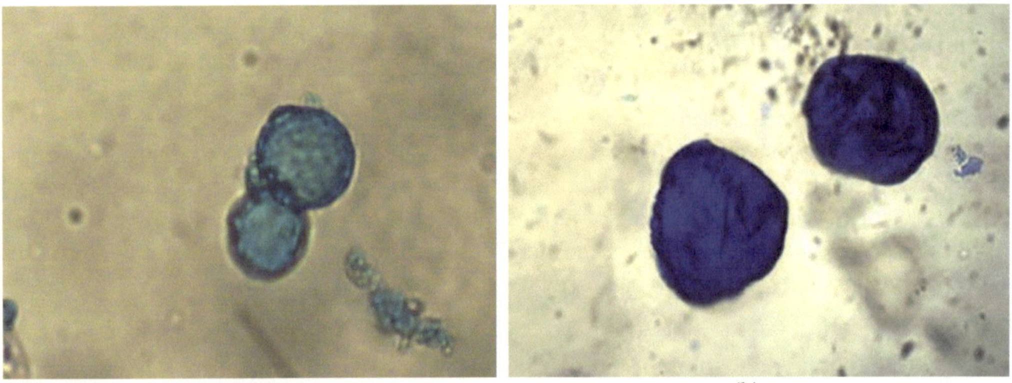

(a) (b)

a) SD-loaded cross-linked xylan microcapsules containing 60 mg of SD
b) SD-loaded cross-linked xylan microcapsules containing 3.1 mg of SD

Fig. 11. Optical microscopy of SD-loaded cross-linked xylan microcapsules at 40x magnification.

Cross-linked xylan microcapsules have also been successfully developed in order to protect superparamagnetic particles from gastric dissolution (Silva et al., 2007). First, magnetic particles were synthesized by coprecipitation using solutions of ferric chloride and ferrous sulphate as a source of iron. Subsequently, xylan was dissolved in 0.6 M NaOH solution and the magnetic suspension was added to the xylan solution after neutralization and sonication. Finally, the emulsification was carried out in chloroform:cyclohexane containing 5% (w/v) sorbitan tristerate followed by the cross-linking reaction with terephthaloyl chloride. As a result, polymeric microparticles with a mean diameter of 25.26 ± 0.42 µm and roughly spherical in shape were produced. They were suggested to involve more than one magnetic particle entity due to their five-fold larger size. Additionally, dissolution studies revealed that only 2.3% of the magnetite content was dissolved in 0.1 M HCl solution at 37 ± 0.1 °C after 120 min. This fact corroborates the feasibility of xylan as a material for colon delivery.

3.1.3 Spray-drying

The spray-drying technique is a one-step continuous operation characterized by the atomization of suspensions or solutions into fine droplets followed by a drying process that leads to the formation of solid particles (Tewa-Tagne et al., 2007). When compared to other approaches for producing and drying systems, this technique exhibits the advantages of low price, rapid process, and the possibility of modulating the physicochemical properties of particles, such as particle size, polydispersity, bulk and tapped densities, and cohesion (Raffin et al., 2006; Tewa-Tagne et al., 2006; Vehring, 2008). Briefly, the main steps of the process are (1) atomization of the feed into a spray, (2) spray-air contact, (3) drying of the spray, and (4) separation of the dried product from the drying gas (Tewa-Tagne et al., 2007; Tewa-Tagne et al., 2006). Because of the dry state of the final product obtained by the spray-drying technique, this method is highly appropriate to improve the stability of microparticulate systems due to the reduction of microbiological contamination, polymer hydrolysis, and physicochemical instability because of the elimination of the water content.

The production of xylan-based microparticles by spray drying has provided useful results. Although some limitation may be observed due to the sticky nature of xylan, which may lead to scarce amounts of final dry product, the use of other materials is very helpful. With that purpose, derivatives of methacrylic acid and methyl-methacrylate, also known as Eudragit®, have been used to prepare suitable xylan-based microparticles. In addition, Eudragit® S-100 (ES100) plays an additional role in the pharmacokinetic properties of the polymeric microparticles. ES100 is a synthetic gastroresistant polymer that has been largely used in the pharmaceutical industry due to its safety and degradation behavior. It is a pH-sensitive copolymer and, because of that, it is able to prevent drug release until the formulation passes through the stomach and reaches some distance down the small intestine (Friend, 2005).

Thus, spray-dried xylan/ES100 microparticles were produced at different polymer weight ratios dissolved in alkaline and neutral solutions, separately. More precisely, xylan and ES100 were dissolved in 1:1 and 1:3 weight ratios in 0.6 N NaOH and phosphate buffer (pH 7.4). Then, the suspensions were spray-dried at the feed rate of 1.2 mL/min (inlet temperature of 120°C) using a Büchi Model 191 laboratory spray-dryer with a 0.7 mm nozzle, separately. Cross-linked xylan microcapsules were also coated by ES100 after spray-drying at the same conditions.

It was observed that this technique was able to produce microparticles with a mean diameter of approximately 10.17 ± 3.02 μm in a reasonable to satisfactory yield depending on the formulation. This value was observed to be higher for the polymer weight ratio of 1:3 (87.00 ± 4.25 %), which indicates that ES100 improves the final result of the spray-drying process. According to the SEM analysis, the polymeric microparticles were shown to be quite similar in shape. Regardless of the formulation, they appeared to be mostly concave and asymmetric (Figure 12).

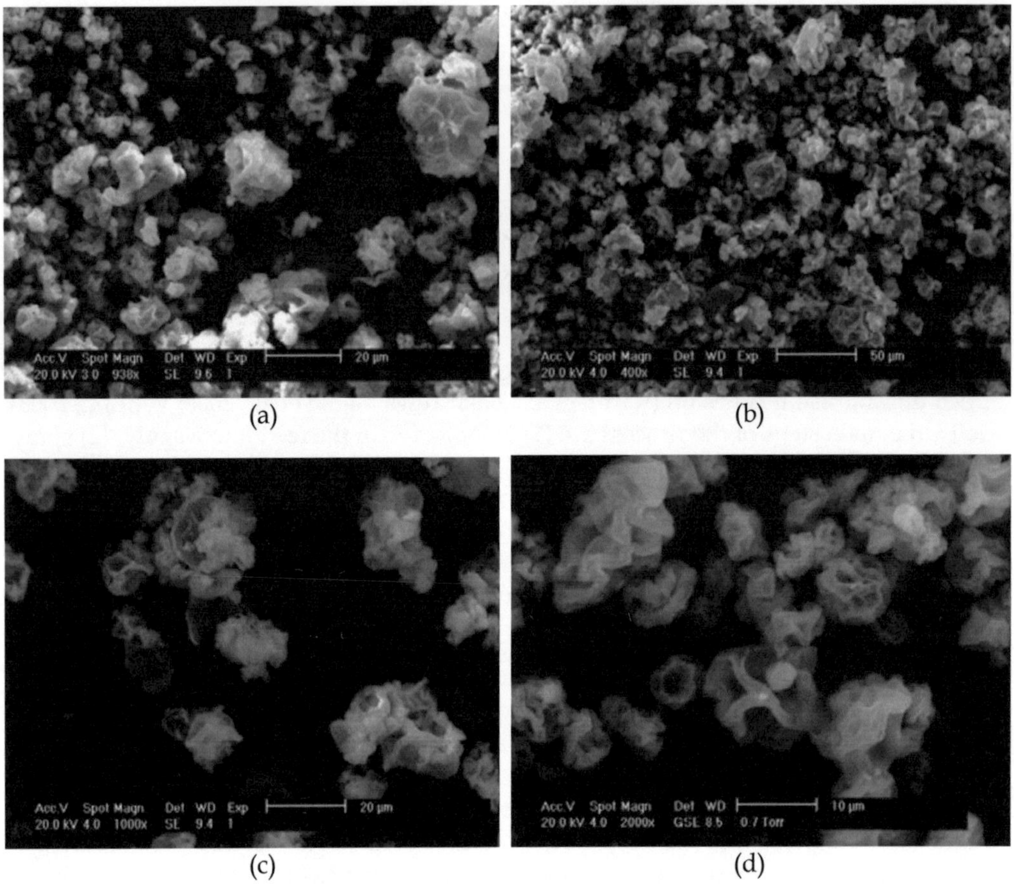

a) 1:1 (w/w) xylan/ES100 microparticles (solvent: NaOH) at 938x magnification.
b) 1:3 (w/w) xylan/ES100 microparticles ((solvent: NaOH) at 400x magnification.
c) 1:3 (w/w) xylan/ES100 microparticles (solvent: phosphate buffer) at 1000x magnification.
d) 1:3 (w/w) xylan/ES100 microparticles (solvent: phosphate buffer) at 2000x magnification.

Fig. 12. SEM images of 5-ASA-loaded spray-dried xylan and ES100 microparticles in different polymer weight ratios (Unpublished data).

4. Biocompatibility of xylan and its products

Among other natural products, biopolymers have been largely studied, due to their numerous applications in which their contact to cells and tissues via their surface is of utmost importance. For instance, micro- and nanocapsules, film coatings, excipients for traditional dosage forms, and novel drug delivery systems have taken much advantage by using biopolymers, especially due to their biocompatibility and biodegradability properties (Drotleff et al., 2004; Villanova et al., 2010). Biopolymers are subject to degradation *in vivo* by hydrolysis or enzymatic attack. The use of these polymers may represent a lower cost compared to other conventional biodegradable polymers (Villanova et al., 2010).

During the development of pharmaceutical products, the toxic effect of biomaterials on cells is considered one of the most important issues to be evaluated. For instance, cell death, cell proliferation, cell morphology, and cell adhesion are features directly correlated with the toxicity *in vitro*. Therefore, loss of viability could be a consequence of a toxic biomaterial (Marques, 2005). Although biopolymers are considered non-toxic and biocompatible, residues from their extraction methodology may cause toxicity issues.

In order to assess the effect of the corn cob xylan on the cell viability and proliferation rate, xylan solutions at concentrations of 0.1, 0.25, 0.50, 0.75, and 1 mg/ml were placed in contact with human cervical adenocarcinoma cells (HeLa cells) for 24 and 72 h. Finally, the cell viability was determined by the MTT assay. It was observed that regardless of the xylan concentration, the samples tested did not affect the viability of HeLa cells after incubation for 24 h (Figure 13) (Unpublished data).

Besides, the statistical analysis of the results obtained confirmed that the xylan samples did not present a significant effect on the cell viability and cell proliferation rate when in direct contact with HeLa cells at the concentrations used in this study and compared to the control.

Similarly, after a longer time of incubation, no significant changes in the cell proliferation rate was detected, as can be seen in the data for 72 h (Figure 13). In fact, this was expected due to the biocompatible nature of xylan. As a natural polyssacharide, this type of biomaterial is considered to be highly stable, non-toxic and hydrophilic (Liu et al., 2008). Accordingly, the alkaline extraction of xylan from corn has proved to be a safe approach for obtaining the polymer with no relevant toxicity (Unpublished data).

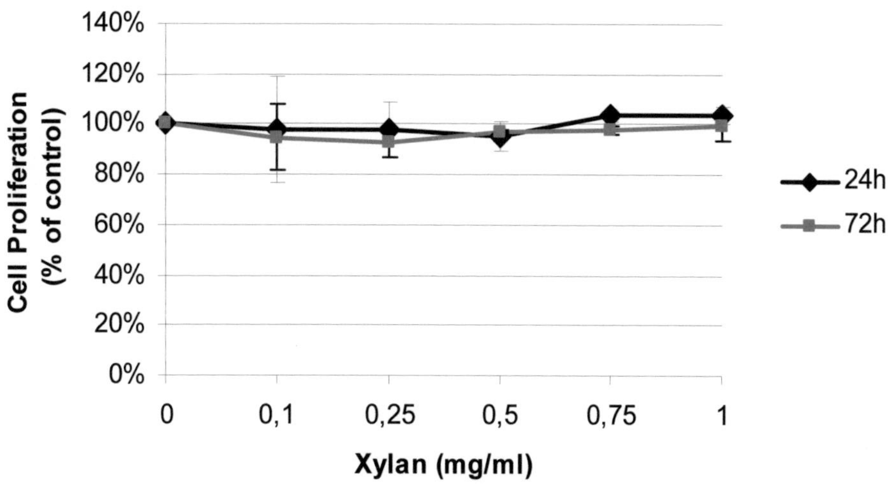

Fig. 13. Viability of HeLa cells after incubation for 24 and 72h with solutions of xylan at different concentrations.

Xylan-based microparticles were also evaluated regarding their *in vitro* toxicity. In fact, cross-liked (CLM) and spray-dried microparticles (SDM) based on xylan and ES100 were produced in order to carry UA and avoid its side effects, namely hepatotoxicity and nephrotoxicity. Additionally, CLM and SDM dispersions at concentrations of 50, 125, 250, and 500 μg/ml were placed in contact with human embryonic lung fibroblasts (MRC-5 cells) for 24 h and the MTT assay was carried out to assess the cell viability. According to the MTT assay results, the cells treated with CLM presented an initial decrease in the cell viability of 56% at the lowest tested concentration (50 µg/mL) while the cell viability rate reached only 12.6% at the highest concentration (500 µg/mL) (Figure 14).

Nevertheless, SDM showed a maximum decrease in the cell survival rate of approximately 12% and 27% at the lowest and highest concentrations of microparticles, respectively (Figure 14). The massive cytotoxicity induced by CLM may be explained by the presence of remaining molecules of terephthaloyl chloride, which plays the role of cross-linking agent during the formation of CLM and is well known as a toxic substance.

In contrast, the MTT assay for SDM did not show high cytotoxicity. This fact confirms the advantage of using spray-drying in order to avoid toxic and hazardous reagents such as terephthaloyl chloride and other cross-linking agents. Additionally, such results indicate a relevant biocompatibility of spray-dried xylan/ES100 microparticles containing UA.

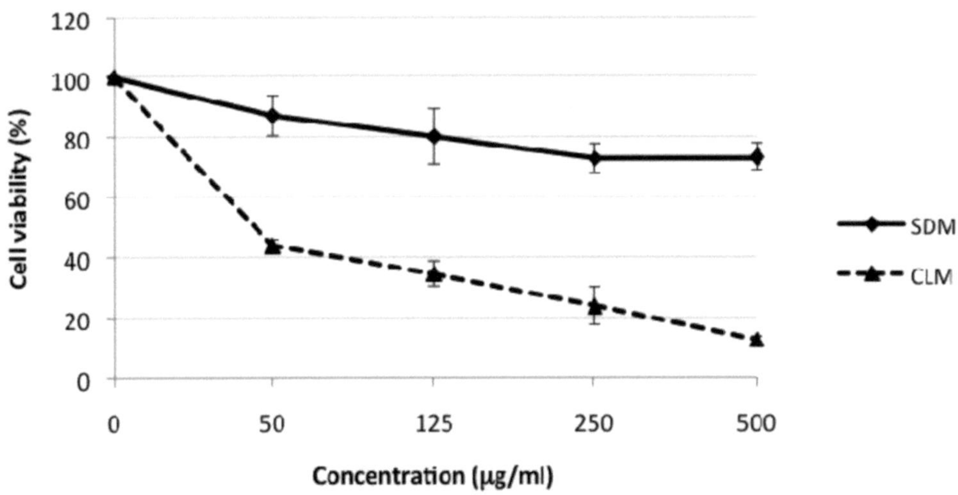

Fig. 14. Viability of MRC-5 cells after incubation for 24h with spray-dried (SDM) and cross-linked xylan microparticles (CLM) containing UA.

5. Conclusions

The need of modern science to achieve a sustainable future development has been shown in many circumstances in society. Finding strategies less harmful to the environment has been a quest for research in several areas, such as pharmaceuticals, biotechnology, and food industries. With that purpose, the increase in research and development of more applications of xylan and its derivatives has shown the versatility of this biopolymer, thus helping the search for sustainable alternatives.

Xylan may be extremely useful in the pharmaceutical field, especially for the production of colon-specific drug carriers, such as micro- and nanoparticles, and film coatings. In addition, because of its abundant sources in nature, its use would bring many benefits, including reducing costs to industry, optimizing the use of natural resources, and reducing environmental damage due to its biodegradability and biocompatibility.

Large amounts of agricultural waste products, such as corn cobs, are continuously provided in several developing countries. Xylan is considered to be a green polymer that may play an essential role in the renewability of waste products due to its biodegradable and biocompatible nature. Furthermore, as shown in this chapter, xylan presents particular properties that allow a wide range of applications.

6. Acknowledgements

The authors are grateful to Professor Dr. Lucymara Fassarela Agnez Lima and Acarízia Eduardo da Silva for the contribution with the cytotoxicity assay.

7. References

Abdullah, E. C. & Geldart, D. (1999). The use of bulk density measurements as flowability indicators. *Powder Technology*, Vol. 102, 2, (March 1999), pp. (151-165), ISSN 0032-5910

Andry, M. C., Edwards-Lévy, F. & Lévy, M. C. (1996). Free amino group content of serum albumin microcapsules. III. A study at low pH values. *International Journal of Pharmaceutics*, Vol. 128, 1-2, (February 1996), pp. (197-202), ISSN 0378-5173

Andry, M. C. & Lévy, M. C. (1997). In vitro degradation of serum albumin microcapsules: Effect of process variables. *International Journal of Pharmaceutics*, Vol. 152, 2, (June 1997), pp. (145-151), ISSN 0378-5173

Anton, N., Benoit, J. P. & Saulnier, P. (2008). Design and production of nanoparticles formulated from nano-emulsion templates - A review. *Journal of Controlled Release*, Vol. 128, 3, (June 2008), pp. (185-199), ISSN 0168-3659

Belgacem, M. N. & Gandini, A. (Ed(s).). (2008). *Monomers, polymers and composites from renewable resources*, Elsevier, ISBN 978-0-08-045316-3, Oxford

Carr, R. L. (1965). Classifying flow properties of solids. *Chemical Engineering*, Vol. 72, 3, 1965), pp. (69-72)

Chang, T. M. S. (1964). Semipermeable microcapsules. *Science*, Vol. 146, 364, (October 1964), pp. (524-&), ISSN 0036-8075

Couvreur, P., Barratt, G., Fattal, E., Legrand, P. & Vauthier, C. (2002). Nanocapsule technology: A review. *Critical Reviews in Therapeutic Drug Carrier Systems*, Vol. 19, 2, (March 2002), pp. (99-134), ISSN 0743-4863

Daggupati, V. N., Naterer, G. F., Gabriel, K. S., Gravelsins, R. J. & Wang, Z. L. (2011). Effects of atomization conditions and flow rates on spray drying for cupric chloride particle formation. *International Journal of Hydrogen Energy*, Vol. In Press, Corrected Proof, pp. 0360-3199, ISSN 0360-3199

Daus, S. & Heinze, T. (2010). Xylan-based nanoparticles: Prodrugs for ibuprofen release. *Macromolecular Bioscience*, Vol. 10, 2, (November 2010), pp. (211-220), ISSN 1616-5195

Den Haan, R. & Van Zyl, W. H. (2003). Enhanced xylan degradation and utilisation by Pichia stipitis overproducing fungal xylanolytic enzymes. *Enzyme and Microbial Technology*, Vol. 33, 5, (October 2003), pp. (620-628), ISSN 0141-0229

Drotleff, S., Lungwitz, U., Breunig, M., Dennis, A., Blunk, T., Tessmar, J. & Gopferich, A. (2004). Biomimetic polymers in pharmaceutical and biomedical sciences. *European Journal of Pharmaceutics and Biopharmaceutics*, Vol. 58, 2, (September 2004), pp. (385-407), ISSN 0939-6411

Ebringerová, A. (2005). Structural diversity and application potential of hemicelluloses. *Macromolecular Symposia*, Vol. 232, 1, (February 2005), pp. (1-12), ISSN 1521-3900

Ebringerova, A. & Heinze, T. (2000). Xylan and xylan derivatives - Biopolymers with valuable properties, 1 - Naturally occurring xylans structures, procedures and properties. . *Macromolecular Rapid Communications*, Vol. 21, 9, (June 2000), pp. (542-556), ISSN 1022-1336

Ebringerova, A. & Hromadkova, Z. (1999). Xylans of industrial and biomedical importance, In: *Biotechnology and Genetic Engineering Reviews*, pp. (325-346), Intercept Ltd Scientific, Technical & Medical Publishers, ISBN 0264-8725, Andover

Ebringerova, A., Hromadkova, Z., Alfodi, J. & Hribalova, V. (1998). The immunologically active xylan from ultrasound-treated corn cobs: extractability, structure and properties. *Carbohydrate Polymers*, Vol. 37, 3, (November 1998), pp. (231-239), ISSN 0144-8617

Ebringerova, A., Hromadkova, Z. & Hribalova, V. (1995). Structure and mitogenic activities of corn cob heteroxylans. *International Journal of Biological Macromolecules*, Vol. 17, 6, (December 1995), pp. (327-331), ISSN 0141-8130

Ebringerova, A., Hromadkova, Z., Kacurakova, M. & Antal, M. (1994). Quaternized xylans: Synthesis and structural characterization. *Carbohydrate Polymers*, Vol. 24, 4, (May 1994), pp. (301-308), ISSN 0144-8617

Edwards-Lévy, F., Andry, M. C. & Levy, M. C. (1994). Determination of free amino group content of serum-albumin microcapsules. II. Effect of variations in reaction-time and terephthaloyl chloride concentration. *International Journal of Pharmaceutics*, Vol. 103, 3, (March 1994), pp. (253-257), ISSN 0378-5173

Friend, D. R. (2005). New oral delivery systems for treatment of inflammatory bowel disease. *Advanced Drug Delivery Reviews*, Vol. 57, 2, (January 2005), pp. (247-265), ISSN 0169-409X

Garcia, R. B., Ganter, J. & Carvalho, R. R. (2000). Solution properties of D-xylans from corn cobs. *European Polymer Journal*, Vol. 36, 4, (April 2000), pp. (783-787), ISSN 0014-3057

Garcia, R. B., Nagashima Jr, T., Praxedes, A. K. C., Raffin, F. N., Moura, T. F. A. L. & Egito, E. S. T. (2001). Preparation of micro and nanoparticles from corn cobs xylan. *Polymer Bulletin*, Vol. 46, 5, (May 2001), pp. (371-379), ISSN 1436-2449

Geldart, D., Abdullah, E. C., Hassanpour, A., Nwoke, L. C. & Wouters, I. (2006). Characterization of powder flowability using measurement of angle of repose. *China Particuology*, Vol. 4, 3-4, (July 2006), pp. (104-107), ISSN 1672-2515

Gouin, S. (2004). Microencapsulation: Industrial appraisal of existing technologies and trends. *Trends in Food Science & Technology*, Vol. 15, 7-8, (July-August 2004), pp. (330-347), ISSN 0924-2244

Grootaert, C., Delcour, J. A., Courtin, C. M., Broekaert, W. F., Verstraete, W. & Van de Wiele, T. (2007). Microbial metabolism and prebiotic potency of arabinoxylan oligosaccharides in the human intestine. *Trends in Food Science & Technology*, Vol. 18, 2, (February 2007), pp. (64-71), ISSN 0924-2244

Habibi, Y. & Vignon, M. R. (2005). Isolation and characterization of xylans from seed pericarp of Argania spinosa fruit. *Carbohydrate Research*, Vol. 340, 7, (May 2005), pp. (1431-1436), ISSN 0008-6215

Hausner, H. H. (1967). Friction conditions in a mass of metal powders. *International Journal of Powder Metallurgy*, Vol. 3, (February 1967), pp. (7-13), ISSN 0888-7462

Heinze, T., Petzold, K. & Hornig, S. (2007). Novel nanoparticles based on xylan. *Cellulose Chemistry and Technology*, Vol. 41, 1, January 2007), pp. (13-18), ISSN 0576-9787

Hromadkova, Z., Kovacikova, J. & Ebringerova, A. (1999). Study of the classical and ultrasound-assisted extraction of the corn cob xylan. *Industrial Crops and Products*, Vol. 9, 2, (Januar 1999), pp. (101-109), ISSN 0926-6690

Jiang, B. B., Hu, L., Gao, C. Y. & Shen, J. C. (2006). Cross-linked polysaccharide nanocapsules: Preparation and drug release properties. *Acta Biomaterialia*, Vol. 2, 1, (January 2006), pp. (9-18), ISSN 1742-7061

Jyothi, N. V. N., Prasanna, P. M., Sakarkar, S. N., Prabha, K. S., Ramaiah, P. S. & Srawan, G. Y. (2010). Microencapsulation techniques, factors influencing encapsulation efficiency. *Journal of Microencapsulation*, Vol. 27, 3, (June 2010), pp. (187-197), ISSN 0265-2048

Kacurakova, M., Capek, P., Sasinkova, V., Wellner, N. & Ebringerova, A. (2000). FT-IR study of plant cell wall model compounds: pectic polysaccharides and hemicelluloses. *Carbohydrate Polymers*, Vol. 43, 2, (October 2000), pp. (195-203), ISSN 0144-8617

Kayserilioglu, B. S., Bakir, U., Yilmaz, L. & Akkas, N. (2003). Use of xylan, an agricultural by-product, in wheat gluten based biodegradable films: mechanical, solubility and water vapor transfer rate properties. *Bioresource Technology*, Vol. 87, 3, (May 2003), pp. (239-246), ISSN 0960-8524

Kulkarni, N., Shendye, A. & Rao, M. (1999). Molecular and biotechnological aspects of xylanases. *FEMS Microbiology Reviews*, Vol. 23, 4, (July 1999), pp. (411-456), ISSN 0168-6445

Kumar, M. (2000). Nano and microparticles as controlled drug delivery devices. *Journal of Pharmacy and Pharmaceutical Sciences*, Vol. 3, 2, (May-August 2000), pp. (234-258), ISSN 1482-1826

Kumar, V., de la Luz Reus-Medina, M. & Yang, D. (2002). Preparation, characterization, and tabletting properties of a new cellulose-based pharmaceutical aid. *International Journal of Pharmaceutics*, Vol. 235, 1-2, (March 2002), pp. (129-140), ISSN 0378-5173

Levy, M. C. & Andry, M. C. (1990). Microcapsules prepared through interfacial cross-linking of starch derivatives. *International Journal of Pharmaceutics*, Vol. 62, 1, (July 1990), pp. (27-35), ISSN 0378-5173

Li, B.-z., Wang, L.-j., Li, D., Chiu, Y. L., Zhang, Z.-j., Shi, J., Chen, X. D. & Mao, Z.-h. (2009). Physical properties and loading capacity of starch-based microparticles crosslinked with trisodium trimetaphosphate. *Journal of Food Engineering*, Vol. 92, 3, (June 2009), pp. (255-260), ISSN 0260-8774

Li, X., Shi, X., Wang, M. & Du, Y. (2011). Xylan chitosan conjugate - A potential food preservative. *Food Chemistry*, Vol. 126, 2, (May 2011), pp. (520-525), ISSN 0308-8146

Liu, Z., Jiao, Y., Wang, Y., Zhou, C. & Zhang, Z. (2008). Polysaccharides-based nanoparticles as drug delivery systems. *Advanced Drug Delivery Reviews*, Vol. 60, 15, (December 2008), pp. (1650-1662), ISSN 0169-409X

Luo, J. Y., Zhong, Y., Cao, J. C. & Cui, H. F. (2011). Efficacy of oral colon-specific delivery capsule of low-molecular-weight heparin on ulcerative colitis. *Biomedicine & Pharmacotherapy*, Vol. 65, 2, (March 2011), pp. (111-117), ISSN 0753-3322

Marques, A. P. C., H. R.; Coutinho, O. P.; Reis, R. L. (2005). Effect of starch-based biomaterials on the in vitro proliferation and viability of osteoblast-like cells. *Journal of Materials Science : Materials in Medicine*, Vol. 16, 1, (September 2005), pp. (833-842), ISSN 0957-4530

Morris, K. R., Nail, S. L., Peck, G. E., Byrn, S. R., Griesser, U. J., Stowell, J. G., Hwang, S.-J. & Park, K. (1998). Advances in pharmaceutical materials and processing. *Pharmaceutical Science & Technology Today*, Vol. 1, 6, (September 1998), pp. (235-245), ISSN 1461-5347

Muller, R. H., Gohla, S. & Keck, C. M. (2011). State of the art of nanocrystals - Special features, production, nanotoxicology aspects and intracellular delivery. *European Journal of Pharmaceutics and Biopharmaceutics*, Vol. 78, 1, (May 2011), pp. (1-9), ISSN 0939-6411

Nagashima, T., Oliveira, E. E., Silva, A. E., Marcelino, H. R., Gomes, M. C. S., Aguiar, L. M., Araujo, I. B., Soares, L. A. L., Oliveira, A. G. & Egito, E. S. T. (2008). Influence of the lipophilic external phase composition on the preparation and characterization of xylan microcapsules - A technical note. *AAPS PharmSciTech*, Vol. 9, 3, (September 2008), pp. (814-817), ISSN 1530-9932

Nunthanid, J., Laungtana-Anan, M., Sriamornsak, P., Limmatvapirat, S., Puttipipatkhachorn, S., Lim, L. Y. & Khor, E. (2004). Characterization of chitosan acetate as a binder for sustained release tablets. *Journal of Controlled Release*, Vol. 99, 1, (September 2004), pp. (15-26), ISSN 0168-3659

Oliveira, E. E., Silva, A. E., Nagashima Jr, T., Gomes, M. C. S., Aguiar, L. M., Marcelino, H. R., Araujo, I. B., Bayer, M. P., Ricardo, N. M. P. S., Oliveira, A. G. & Egito, E. S. T. (2010). Xylan from corn cobs, a promising polymer for drug delivery: Production and characterization. *Bioresource Technology*, Vol. 101, 14, (July 2010), pp. (5402-5406), ISSN 0960-8524

Qv, X. Y., Zeng, Z. P. & Jiang, J. G. (2011). Preparation of lutein microencapsulation by complex coacervation method and its physicochemical properties and stability. *Food Hydrocolloids*, Vol. 25, 6, (August 2011), pp. (1596-1603), ISSN 0268-005X

Raffin, R. P., Jornada, D. S., Ré, M. I., Pohlmann, A. R. & Guterres, S. S. (2006). Sodium pantoprazole-loaded enteric microparticles prepared by spray drying: Effect of the scale of production and process validation. *International Journal of Pharmaceutics*, Vol. 324, 1, (October 2006), pp. (10-18), ISSN 0378-5173

Rao, J. P. & Geckeler, K. E. (2011). Polymer nanoparticles: Preparation techniques and size-control parameters. *Progress in Polymer Science*, Vol. 36, 7, (July 2011), pp. (887-913), ISSN 0079-6700

Ribeiro, A. J., Neufeld, R. J., Arnaud, P. & Chaumeil, J. C. (1999). Microencapsulation of lipophilic drugs in chitosan-coated alginate microspheres. *International Journal of Pharmaceutics*, Vol. 187, 1, (September 1999), pp. (115-123), ISSN 0378-5173

Rubinstein, A. (1995). Approaches and opportunities in colon-specific drug-delivery. *Critical Reviews in Therapeutic Drug Carrier Systems*, Vol. 12, 2-3, 1995), pp. (101-149), ISSN 0743-4863

Saha, B. C. (2000). Alpha-L-arabinofuranosidases: Biochemistry, molecular biology and application in biotechnology. *Biotechnology Advances*, Vol. 18, 5, (August 2000), pp. (403-423), ISSN 0734-9750

Sedlmeyer, F. B. (2011). Xylan as by-product of biorefineries: Characteristics and potential use for food applications. *Food Hydrocolloids*, Vol. In Press, Corrected Proof, pp. ISSN 0268-005X

Shallom, D. & Shoham, Y. (2003). Microbial hemicellulases. *Current Opinion in Microbiology*, Vol. 6, 3, (June 2003), pp. (219-228), ISSN 1369-5274

Silva, A. K. A., Silva, E. L., Oliveira, E. E., Nagashima, J. T., Soares, L. A. L., Medeiros, A. C., Araujo, J. H., Araujo, I. B., Carriço, A. S. & Egito, E. S. T. (2007). Synthesis and characterization of xylan-coated magnetite microparticles. *International Journal of Pharmaceutics*, Vol. 334, 1-2, (April 2007), pp. (42-47), ISSN 0378-5173

Simó, C., Cifuentes, A. & Gallardo, A. (2003). Drug delivery systems: Polymers and drugs monitored by capillary electromigration methods. *Journal of Chromatography B*, Vol. 797, 1-2, (November 2003), pp. (37-49), ISSN 1570-0232

Stuart, M. A. C. (2008). Supramolecular perspectives in colloid science. *Colloid and Polymer Science*, Vol. 286, 8-9, (August 2008), pp. (855-864), ISSN 0303-402X

Sun, R., M. Fang, J., Goodwin, A., M. Lawther, J. & J. Bolton, A. (1998). Fractionation and characterization of polysaccharides from abaca fibre. *Carbohydrate Polymers*, Vol. 37, 4, (December 1998), pp. (351-359), ISSN 0144-8617

Sun, X. F., Xu, F., Sun, R. C., Geng, Z. C., Fowler, P. & Baird, M. S. (2005). Characteristics of degraded hemicellulosic polymers obtained from steam exploded wheat straw. *Carbohydrate Polymers*, Vol. 60, 1, (April 2005), pp. (15-26), ISSN 0144-8617

Tewa-Tagne, P., Briançon, S. & Fessi, H. (2007). Preparation of redispersible dry nanocapsules by means of spray-drying: Development and characterisation. *European Journal of Pharmaceutical Sciences*, Vol. 30, 2, (April 2007), pp. (124-135), ISSN 0928-0987

Tewa-Tagne, P., Briançon, S. & Fessi, H. (2006). Spray-dried microparticles containing polymeric nanocapsules: Formulation aspects, liquid phase interactions and particles characteristics. *International Journal of Pharmaceutics*, Vol. 325, 1-2, (November 2006), pp. (63-74), ISSN 0378-5173

Torchilin, V. P. (2006). Multifunctional nanocarriers. *Advanced Drug Delivery Reviews*, Vol. 58, 14, (December 2006), pp. (1532-1555), ISSN 0169-409X

Tran, V. T., Benoît, J. P. & Venier-Julienne, M. C. (2011). Why and how to prepare biodegradable, monodispersed, polymeric microparticles in the field of pharmacy? *International Journal of Pharmaceutics*, Vol. 407, 1-2, (December 2011), pp. (1-11), ISSN 0378-5173

Ünlu, C. H., Günister, E. & Atici, O. (2009). Synthesis and characterization of NaMt biocomposites with corn cob xylan in aqueous media. *Carbohydrate Polymers*, Vol. 76, 4, (May 2009), pp. (585-592), ISSN 0144-8617

Vehring, R. (2008). Pharmaceutical particle engineering via spray-drying. *Pharmaceutical Research*, Vol. 25, 5, (May 2008), pp. (999-1022), ISSN 0724-8741

Villanova, J. C. O., Orefice, R. L. & Cunha, A. S. (2010). Pharmaceutical applications of polymers. *Polimeros - Ciência e Tecnologia*, Vol. 20, 1, (January-March 2010), pp. (51-64), ISSN 0104-1428

Wang, Y. & Zhang, J. (2006). A novel hybrid process, enhanced by ultrasonication, for xylan extraction from corncobs and hydrolysis of xylan to xylose by xylanase. *Journal of Food Engineering*, Vol. 77, 1, (November 2006), pp. (140-145), ISSN 0260-8774

Yang, R., Xu, S., Wang, Z. & Yang, W. (2005). Aqueous extraction of corn cob xylan and production of xylooligosaccharides. *LWT - Food Science and Technology*, Vol. 38, 6, (September 2005), pp. (677-682), ISSN 0023-6438

5

Hydrodynamic Properties of Gelatin – Studies from Intrinsic Viscosity Measurements

Martin Alberto Masuelli[1] and Maria Gabriela Sansone[2]
[1]*Instituto de Física Aplicada, CONICET, FONCyT*
Cátedra de Química Física II, Área de Química Física
[2]*Área de Tecnología Química y Biotecnología, Departamento de Química*
Facultad de Química, Bioquímica y Farmacia, Universidad Nacional de San Luis
Argentina

1. Introduction

Gelatin is a natural polymer widely used in pharmaceutical, cosmetic, photographic, and food industries. It is obtained by denaturation and partial hydrolysis of fibrous collagen. Collagen is the most abundant structural protein of animals and by far the main organic component of skin and bone of vertebrates. However, collagen is standing for a family of proteins with 21 different types of aminoacids described to date. Skin and bones, the raw materials for gelatin manufacture, mainly consist of type I collagen and a small fraction of type III collagen. Each type I collagen molecule consists of polypeptide chains are twisted into left-handed helices. The rod like triple-helical collagen molecules is arranged in a parallel but staggered orientation to form fibrils (Meyer & Morgenstern 2003).

To convert insoluble native collagen into gelatin requires pre-treatment to breakdown the non-covalent bonds and disorganize the protein structure, allowing swelling and cleavage of intra and intermolecular bonds to solubilize the collagen. Subsequent heat treatment cleaves the hydrogen and covalent bonds to destabilize the triple helix, resulting in helix-to-coil transition and conversion into soluble gelatin (Harrington & von Hippel 1962). The degree of conversion of the collagen into gelatin is related to the severity of both the pre-treatment and the extraction process, depending on the pH, temperature, and extraction time. Two types of gelatin are obtainable, depending on the pre-treatment. These are known commercially as type-A gelatin (isoelectric point from 7.5-9.2), obtained in acid pre-treatment conditions, and type-B gelatin (IP from 4.5-5.1), obtained in alkaline pre-treatment conditions. Industrial applications call for one or the other gelatin type, depending on the degree of collagen cross-linking in the raw material, in turn depending on a number of factors, such as collagen type, tissue type, species, animal age, etc. (Karim & Bhat 2009). The most important property of gelatin is its ability to form stiff gels at about 30°C, when cooling a hot gelatin solution with sufficient high concentration (>1%). In contrast to soluble collagen, gelatins show a broad molar mass distribution on solution. The gelatins fractions can be prepared by coacervation, precipitation, or fractionated dissolution

Gelatin has long been used as a food ingredient (e.g., gelling and foaming agent), in the preparation of pharmaceutical products (e.g., soft and hard capsule, microspheres), in the biomedical field (wound dressing and three-dimensional tissue regeneration) and in numerous non-food applications (e.g., photography) (Chatterjee & Bohidar 2008).

Gels are commonly classified into two groups: chemical gels and physical gels, which are different essentially through the nature of the bonds linking the polymer molecules together. If a gel is formed by chemical reaction, the bonds created are covalent bonds and the gel formed is irreversible. It is possible to treat the viscoelastic properties of chemical gels through percolation theory because of the very nature of the bonding in chemical gelation. However, the physical gels achieve solution stability through an array of possible secondary forces, like hydrogen bonds, van der Waals' forces, dipolar interactions and hydrophobic interactions, etc. Since these are weaker bonds the physical gels are reversible when thermodynamic parameters such as pH, ionic strength, and/or temperature are modified. Unlike chemical gels, physical gels are fragile and the viscoelastic properties cannot be described through models like percolation theory. Regardless, all the gels have a characteristic temperature, called gelation temperature T_{gel} below which gelation occurs and, a threshold polymer concentration. Gelatin gels are made of an ensemble of physically interconnected triple helices, which are held together by intermolecular hydrogen bonding. The gelation transition in gelatin has been modeled through a variety of theoretical models to name a few; percolation model, kinetic aggregation model and Smoluchowski aggregation model, etc. In these gels, the continuous phase, which is usually water, stays in equilibrium with the dispersed gelatin network in a state of dynamic equilibrium and exhibits two distinct physical states, namely; water bound to the gelatin chains and interstitial water trapped inside the gel network. The hydrodynamic environment of the network characterizes the possible relaxation modes that the gel can execute at temperature, $T < T_{gel}$. The structure of bound and interstitial water undergoes significant change as the network structure is altered affecting the hydrodynamic environment of the network and thus modulating its relaxation features, which can be easily captured in experiments (Karim, & Bhat 2008).

Biodegradable films made from edible biopolymers from renewable sources could become an important factor in reducing the environmental impact of plastic waste. Proteins, lipids, and polysaccharides are the main biopolymers employed to make edible films and coatings. Which of these components are present in different proportions and determine the properties of the material, as a barrier to water vapor, oxygen, carbon dioxide, and lipid transfer in food systems (Gomez-Guillen et al. 2002 and 2009).

Influence of molecular weight heterogeneity and drug solubility, drug loading and hydrodynamic conditions on drug release kinetics from gelatin nanoparticles as a potential intravascular probe for diagnostic purposes and in improving the biodelivery of drug (Saxena et al. 2005).

The thermodynamic properties of the gelatin-water system have been studied using methods of static light scattering (LS), osmosis, mixing calorimetry, differential scanning calorimetry, ultracentrifugation, swelling and swelling pressure. At temperatures below 35°C the system undergoes a coil-helix transition, which is clearly demonstrated by temperature-dependent measurements of LS and optical rotation. If the concentration of gelatin is above its critical value for network formation, the solutions (sols) undergo a phase transition to the gel state (Kaur et al. 2002).

One of the techniques to determine the molecular weight of macromolecules is size exclusion chromatography (SEC) with size exclusion chromatography-multi-angle laser light scattering (SEC-MALLS) and sedimentation equilibrium (Borchard 2002). The determination of the molar mass distribution of this acid soluble collagen (ASC or gelatin) using the technique SEC-MALS by direct measurement of M_w (from 19,000-61,000g/mol). According to the next scaling law $<s^2>^{1/2} = KM^{\alpha}$, $\alpha = 0.78$ was determined for the gelatin. This α could reflect a structure in solution, which is more similar to an ellipsoid than to a random coil (Meyer & Morgenstern 2003).

Although intrinsic viscosity is a molecular parameter that can be interpreted in terms of molecular conformation, it does not offer as high resolution on molecular structure as other methods, but intrinsic viscosity measurement is a very economical alternative and easy determination with a few experiences. Viscosity of protein water solution depends on intrinsic biopolymer characteristics (such as molecular mass, volume, size, shape, surface charge, deformation facility, estherification degree in polysaccharides, and aminoacids content) and on ambient factors (such as pH, temperature, ionic strength, solvent, etc.). The method of choice has been capillary viscometry because it is a simple and useful method that requires low cost equipment and yields useful information on soluble macromolecules. Although in literature there is much information on hydrodynamic measurements from determinations of viscosity, very few of them evaluate the situation at different temperatures. The importance of this type of study lies in analyzing the behavior of the protein in industrial processes so as to reduce energy requirements and avoid flow problems and product quality control. (Masuelli 2011)

Pouradier & Venet (1952), showed that an equation of the type $[\eta] = 0.166M^{0.885}$ exists between the intrinsic viscosity and molecular weight of gelatin. Whereas k and a were each the same for two alkali-processed gelatins, they were different for an acid-processed gelatin (Zhao 1999).

Boedker and Doty (1954), worked with gelatin B and $[\eta]$ in water solution at 40.2°C was 38.1cm³/g for an M_w of 90,000g/mol and R_g of 17nm.

Nishihara and Doty (1958), prepared soluble calf-skin collagen by extraction in citrate buffer at pH 3.7, and was fractionated by different time of sonication from 10 at 440 minutes. They obtained $[\eta]$ (measured at 24.8°C) varied from 1075 to 220 cm³/g with M_w calculated from 336,000 to 137,000g/mol. They proposed Mark-Houwink equation as following $[\eta] = 1.23 \times 10^{-7} M^{1.80}$. This value of the a exponent is that of the upper limit for ellipsoids having a constant minor axis or for cylinders of constant diameter, and confirms assumption concerning the homologous, rod-like nature of the collagen sonicates.

Olivares et al. (2006), study dilute gelatin (M_n 133,000 g/mol) prepared to pH 9.8, ionic strength of 0.110M (NaCl) and mature 20hs, where intrinsic viscosity with an range temperature of 5-35°C was of 116.2 to 56.8 cm³/g.

Haug et al. (2004), worked with fish gelatin at 30°C and 0.1M NaCl, M_w 199,000 to 140,000 g/mol and $[\eta]$ from 42 to 47 cm³/g.

Domenek et al. (2008), released measurement of the intrinsic viscosity of gelatin type B in water solution at 42°C, where given values were of 27 cm³/g and 70 cm³/g.

Works where study the hydrodynamic properties of a biopolymers in aqueous solution at different temperatures are made by Guner (1999), and Guner & Kibarer (2001) for dextran; Chen & Tsaih (1998) and Kasaii (2008) for chitosan, Bohidar for gelatin (1998), and Monkos for serum proteins (1996, 1997, 1999, 2000, 2004 and 2005).

Bohidar (1998), study the aggregation properties of Gelatin chains in neutral aqueous solutions, were reported in the temperature range of 35-60°C. Data measured to 35°C were: intrinsic viscosity $[\eta]$ = 384cm^3/g, diffusion coefficient D = 1.12x10^{-7}cm^2/s, molecular weight M_w = 410,000 g/mol, and hydrodynamic radius of R_H = 28nm. The Mark-Houwink equation of $[\eta] = 0.328\ n^{0.69}$, with monomer molecular weight of 110 g/mol to the repeating unit, and n values of 3,727. Bohidar concluded that the intermolecular interaction was found to be repulsive which showed significant decrease as the temperature was reduced; and random coil shape was confirmed with date obtained.

In this work, an experimental study was conducted on gelatin in semi-dilute region in water solution; and research the effect of temperature, pH, zeta potential, and ionic strength on hydrodynamic properties by viscometry, in order to determine the conformational characteristic, and phase transition (T_{gel}).

2. Materials and methods

2.1 Sample preparation

Gelatin B from cow bone was supplied by Britannia Lab Argentine with isoelectric point of 5.10. Gelatin was dried and sealed in zip plastic bags and then kept in desiccators. Finally, gelatin was dissolved in distilled water preparing a solution of 0.25, 0.5, 0.75 and 1% (wt.).

2.2 Density measurement

Density of solution and solvent were measurement with Anton Paar densimeter DMA5N. For determining the hydration value is used the following concentrations of gelatin in aqueous solution 0.2, 0.4, 0.5 and 0.6%.

2.3 Electrokinetic measurement

Electrokinetic measurements consisted of measuring the viscosity with and without NaCl (Carlo Erba, Argentine) (Figures 3-a and 3-b), while the isoelectric point (figure 2) and zeta potential (figure 3-c) were measured at different pH (HCl Ciccarelli and NaOH Tetrahedrom, Argentine).

3. Results and discussion

3.1 Intrinsic viscosity

If a solution tends to be independent of shear, then the measurement of viscosity (η) is based on Poiseuille's law can be made easy by grouping all those terms related to a specific viscometer as a calibration constant A.

$$\eta = A\,\rho\,t \tag{1}$$

where ρ is the density of solution. If we divide both sides by ρ, we have

$$\upsilon = \frac{\eta}{\rho} = A\,t \tag{2}$$

where υ is the cinematic viscosity of solution.

In macromolecular chemistry, the relative viscosity η_r is often measured. The relative viscosity is the ratio of the viscosity of the solution to that of the solvent:

$$\eta_r = \frac{\rho\,t}{\rho_0\,t_0} \tag{3}$$

The specific viscosity η_{sp} is obtained from the relative viscosity by

$$\eta_{sp} = \eta_r - 1 \tag{4}$$

3.1.1 Methods for determining the intrinsic viscosity

The intrinsic viscosity, denoted by [η], is defined as

$$[\eta] = \lim_{c\to 0} \frac{\eta_{sp}}{c} = \lim_{c\to 0} \frac{1}{c} \ln \eta_r \tag{5}$$

or as

$$[\eta] = \lim_{\eta_{sp}\to 0} \frac{\eta_{sp}}{c} \tag{6}$$

where c is the concentration of the polymer in grams per 100 cm^3 or grams per milliliter of the solution. The quantity η_{sp}/c is called the reduced viscosity. The unit of intrinsic viscosity is deciliters per gram (dL/g), milliliters per gram (mL/g) or (cm^3/g) depending on the concentration unit of the solution. The intrinsic viscosity is also called the limiting viscosity number. The plot of η_{sp}/c versus c or 1/c ln η_r versus c often gives a straight line, the intercept of which is [η].

3.1.2 Huggins, Kraemer and Schulz-Blaschke methods

Huggins (1942) showed that the slope is

$$\frac{d\eta_{sp}/c}{dc} = k_H [\eta]^2 \tag{7}$$

Rearranging and integrating the resulting equation is

$$\frac{\eta_{sp}}{c} = [\eta] + k_H [\eta]^2 c \tag{8}$$

where k_H is a dimensionless constant, called the Huggins constant. The value of k_H is related to the structures of polymers or biopolymers. Some other equations for the determination of $[\eta]$ are:

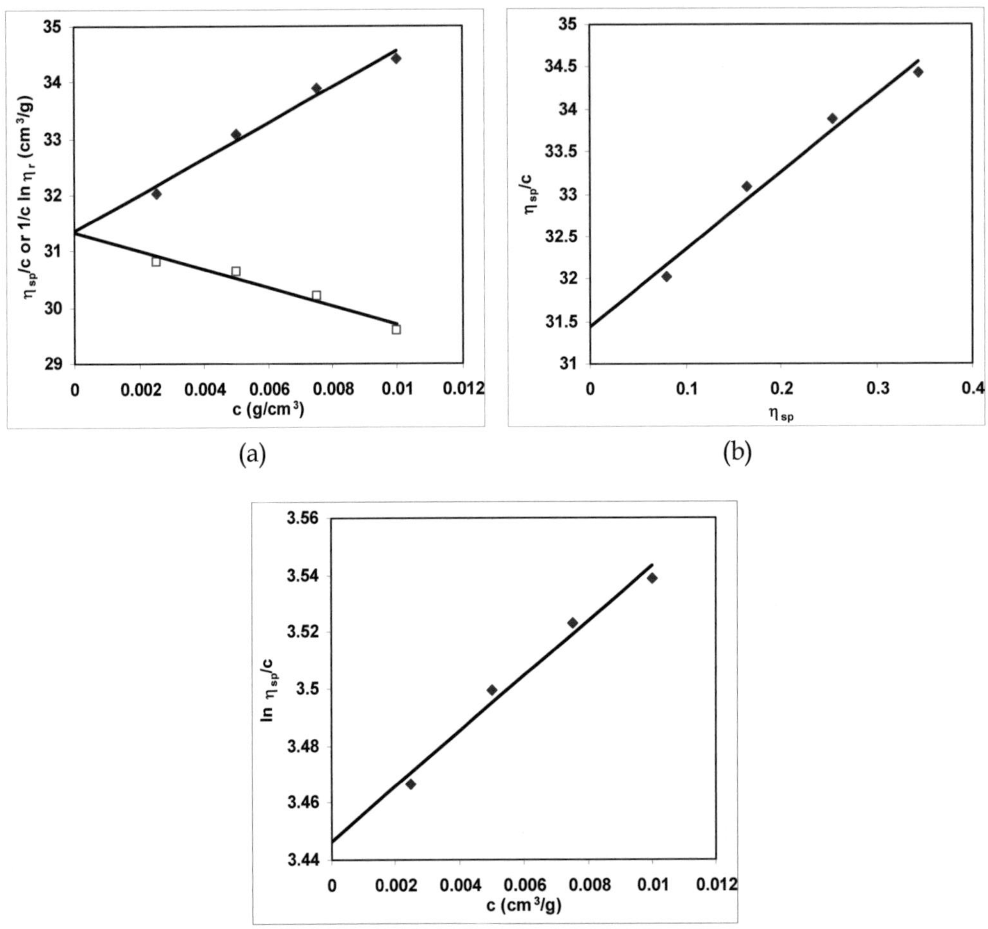

Fig. 1. Methods of Huggins (positive slope), Kraemer (negative slope) (a), Schulz-Blaschke (b), and Martin (c). Data obtained from experimental measures for gelatin B at 37.4°C.

Kraemer (1938)
$$\frac{1}{c}\ln\eta_r = [\eta] + k_K[\eta]^2 c\,, \tag{9}$$

Schulz-Blaschke (1941)
$$\frac{\eta_{sp}}{c} = [\eta] + k_{S-B}[\eta]\,\eta_{sp} \tag{10}$$

and Martin (1942)
$$\frac{\eta_{sp}}{c} = [\eta]\,e^{k_M[\eta]c} \tag{11}$$

For molecules of high intrinsic viscosity a correction must be made for the effect of the rate of shear strain. For relatively low intrinsic viscosity, the rate of shear strain does not have any appreciable effect.

Temperature	37.4°C			
Method	Huggins	Kramer	Schulz-Blaschke	Martin
C (g/cm³)	η_{sp}/c	$1/c \ln \eta r$	η_{sp}	$\ln \eta_{sp}/c$
0.0025	32.02	30.80	0.0800	3.46
0.0050	33.09	30.62	0.1654	3.50
0.0075	33.89	30.20	0.2542	3.52
0.0100	34.43	29.59	0.3443	3.54
[η] (cm³/g)	31.35	31.32	31.44	31.39
Constant	k_H	k_K	k_{S-B}	k_M
	0.34853	0.1660	0.3139	0.3078
σ^2	0.9787	0.9712	0.9751	0.9757
$\%E_R$	-	0.0957	0.2552	0.1286

Table 1. Data obtained applying the classic methods from experimental measurement.

The intrinsic viscosity data calculated from the methods of Huggins, Kramer, Schultz-Blaschke and Martin are shown in Table 1, and how to perform these steps are shown in Figure 1.

All values were calculated for intrinsic viscosity of gelatin B solutions to 37.4°C and compared against the value of Huggins, normally used as standard. It is noteworthy that each method has a relative error percentage and low for methods of more than four pair's values ($E_R\% > 0.30$).

3.1.3 Single point methods

Frequently occurs that extrapolations do not have a common value at their origin ordinates. These deviations may be caused by inadequate lineal extrapolations. The above mentioned is the routine method used for [η] determination. The procedure is laborious and consumes a considerable amount of time and reactive; because of this, several equations were developed which estimate intrinsic viscosity at one single concentration and do not require a graphic. They are known as "single-point" methods.

Single-point equations suppose that k_H, k_K and k_{SB} are constants and that $k_H + k_K = 0.5$, as is indicated by the combination of equations Huggins and Kraemer. They all include the values for relative viscosity, increment of viscosity and concentration. For example, Solomon-Ciuta (1962) proposes:

$$[\eta] \cong \frac{1}{c}\sqrt{2\eta_{sp} - 2\ln\eta_r} \tag{12}$$

In 1968, Deb and Chatterjee (1968 and 1969) suggested that:

$$[\eta] \cong \frac{1}{c}\sqrt[3]{3\ln\eta_r + 1.5\eta_{sp}^2 - 3\eta_{sp}} \tag{13}$$

More recently, Rao and Yanseen (1986) gave a simplified expression:

$$[\eta] \cong \frac{1}{2c}\{\eta_{sp} + \ln\eta_r\} \tag{14}$$

Kuwahara (1963) uses the expression:

$$[\eta] \cong \frac{1}{4c}\{\eta_{sp} + 3\ln\eta_r\} \tag{15}$$

While Palit and Kar (1967) suggest:

$$[\eta] \cong \frac{1}{c}\sqrt[4]{4\eta_{sp} - 4\ln\eta_r + 1.33\eta_{sp}^3 - 2\eta_{sp}^2} \tag{16}$$

and the following equation due to Maron (1961) makes use of the previously calculated parameters:

$$[\eta] \cong \frac{\eta_{sp} + \gamma\ln\eta_r}{c\{1+\gamma\}} \tag{17}$$

$$\gamma = \frac{k_H}{k_k} \qquad 18)$$

Chee (1987) and Rao & Yaseen (1986) have examined the applicability of the single-point method and have found that some equations are inadequate or applicable only to some specific macromolecule-solvent systems.

Curvale and Cesco suggest a double point equation

$$[\eta] = \frac{0.5c_2}{c_2 - c_1}\left\{\frac{\eta_{sp,1}}{c_1} + \frac{\ln\eta_{r,1}}{c_1}\right\} - 0.5\frac{c_1}{c_2 - c_1}\left\{\frac{\eta_{sp,2}}{c_2} + \frac{\ln\eta_{r,2}}{c_2}\right\} \tag{19}$$

where subscript 1 and 2 referrer to concentrations measured. The application of different methods of single point are shown and compared against the method of Huggins (Table 2).

For single point methods fit with minor errors of 3.54%, but the method of Solomon-Ciuta is only valid for concentrations below 0.25% wt of gelatin. Double point error is low of 2.00%.

While all methods of single and double point are used to characterize a polymer solution is always advisable to increase the statistical weight (to reduce errors) with at least four different concentrations of polymer in a given solvent.

Methods						
c (g/cm³)	Solomon-Ciuta	Deb-Chaterjee	Ram, Mohan, Rao, Yanseen	Kuwahara	Palit & Kar	Maron
0.0025	31.97	31.41	31.41	31.10	31.17	31.19
0.0050	16.23	31.83	31.85	31.24	31.90	31.42
0.0075	10.94	31.99	32.04	31.12	32.24	31.39
0.0100	8.26	31.93	32.01	30.79	32.31	31.15
$\%E_R$						
0,0025	2.44	0.65	0.64	0.33	0.12	0.05
0,0050	47.99	2.01	2.07	0.09	2.20	0.67
0,0075	64.94	2.52	2.67	0.28	3.30	0.57
0,0100	73.52	2.30	2.56	1.32	3.54	0.19

Table 2. Data obtained applying the methods of single point from experimental measurement.

3.2 Hydrodynamic properties

In physics, fluid dynamics is a sub-discipline of fluid mechanics that deals with fluid flow—the natural science of fluids (liquids and gases) in motion. It has several subdisciplines itself, including aerodynamics (the study of air and other gases in motion) and hydrodynamics (the study of liquids in motion). Fluid dynamics offers a systematic structure that underlies these practical disciplines, that embraces empirical and semi-empirical laws derived from flow measurement and used to solve practical problems. The solution to a fluid dynamics problem typically involves calculating various properties of the fluid, such as velocity, pressure, density, viscosity and temperature, as functions of space and time.

3.2.1 Mark-Houwink parameters

Staudinger (1932) suggested that the molecular weight M of polymers is proportional to the reduced viscosity:

$$[\eta] = k_0 M \tag{20}$$

where k_0 is proportionality constant. Mark (1938) and Houwink (1940) independently correlated the intrinsic viscosity with molecular weight:

$$[\eta] = k\, M^a \tag{21}$$

where k and a both are constants. The Mark-Houwink equation is applicable to many polymers and is extensively used to determine molecular weight. The constants k and a both vary with polymers and solvents.

Equation (21) describes the relationship between viscosity and molecular weight. Since molecular weight is related to the size of the polymer chain. Generally, for proteins using the following equation:

$$[\eta]=k\left(\frac{M}{M_0}\right)^a \tag{22}$$

Where M_0 is the molecular weight of the repeating unit, gelatin is 540g/mol, according Pouradier-Venet (1954), and Bohidar is 110g/mol. The calculation of Mark-Houwink (M-H) parameters is carried out by the graphic representation of the following equation:

$$\ln[\eta]=\ln k+a\ln M_w \tag{23}$$

Where k and a are M-H constants, these constant depend of the type of polymer, solvent, and temperature of viscometric determinations. The exponent a is a function of polymer geometry, and varies from 0.5 to 2. These constants can be determined experimentally by measuring the intrinsic viscosity of several polymer samples for which the molecular weight has been determined by an independent method (i.e. osmotic pressure or light scattering). Using the polymer standards, a plot of the ln $[\eta]$ vs ln M_w usually gives a straight line. The slope is a value and intercept is equal to ln k value. The M-H exponent bears the signature of a three-dimensional configuration of a polymer chain in the solvent environment. For a values are from 0-0.5 rigid sphere in ideal solvent, from 0.5-0.8 random coil in good solvent, and from 0.8-2 rigid or rod like (stiff chain). The fact that the intrinsic viscosity of a given polymer sample is different according to the solvent used gives and insight into the general shape of polymer molecules in solution. A long-chain polymer molecule in solution takes on a somewhat kinked or curled shape, intermediate between a tightly curled mass (coil) and a rigid linear configuration. All possible degrees of curling may be displayed by any molecule, but there will be an average configuration which will depend on the solvent. In a good solvent which shows a zero or negative heat of mixing with the polymer, the molecule is fairly loosely extended, and the intrinsic viscosity is high. The Mark-Houwink "a" constant is close to 0.75 or higher for these "good" solvents. In a "poor" solvent which shows a positive heat of mixing, segments of a polymer molecule attract each other in solution more strongly than attract the surrounding solvent molecules. The polymer molecule assumes a tighter configuration, and the solution has a lower intrinsic viscosity. The M-H "a" constant is close to 0.5 in "poor" solvents. For a rigid or rod like polymer molecule that is greatly extended in solution, the M-H "a" constant approaches a value of 2.0 (Masuelli 2011).

Table 3 shows the classical values calculating for Mark-Houwink parameters, a and k for temperature. These studies on M-H parameters are usually carried out at a given temperature, obtaining a consistent result but in a very limited range of temperature (for gelatin: Pouradier & Venet 1954 and Bohidar 1998; Monkos for serum proteins 1996, 1997, 1999, 2000, 2004 and 2005). This value shows a clear functionality between these parameters and temperature.

T(°C)	[η] (cm³/g)	k (cm³/g)	*a*
20.00	62.12	0.1681	0.9119
25.00	48.65	0.1660	0.8850
26.60	44.46	0.1631	0.8737
28.30	41.15	0.1626	0.8621
31.00	39.28	0.1621	0.8554
34.00	35.53	0.1618	0.8382
37.40	31.35	0.1614	0.8198

Table 3. Data obtained of intrinsic viscosity and Marck-Houwink parameters of gelatin B at different temperatures.

The molecular weight calculated for gelatin is 333,000g/mol. The value of *a* given at different temperatures shows that this biopolymer in aqueous solution behaves in a conformation predominantly confined to the rod-like, different as observed by Bohidar 1998.

3.2.2 Hydration values

It is well known that biopolymers adsorbed water during dry storage and its quality depends on water content. For example the length of keratin depends on water content and therefore it is used as a hygrometer. The amount of adsorbed water depends on temperature and pressure of water vapor.

On the other hand, biopolymers in solution exhibit the phenomenon of hydration due to the polar properties of water molecules. The electronic formula of water shows that the center of charge of the negatively charged electrons is nearer to the oxygen atom to the positively charged hydrogen nuclei. The center of the positive charges is nearer the two hydrogen atoms. Assuming a molecule, in which the centers of the positive and negative charges do not coincide, a polar molecule or dipole. Always, dipoles are attracted by ions. In the fits phase, the ion attracts the opposite pole and repels the pole of same sign; in a second phase, attraction is stronger than repulsion because the attracted pole is nearer the ion than the repelled pole. For similar reasons attraction takes place between two dipoles.

Hydration consists of the binding of water dipoles to ions or ionic groups, to dipoles, or polar groups. Hydration takes place in solid substances as well as in solution.

Since the components of a compound are linked to each other in such a way that they have lost same of their free translational mobility, the volume of hydrated molecule is always smaller than the sum of the volumes of its components, the hydration is accompanied by a decrease of the total volume.

The amount of water bound to the proteins and polysaccharides depends primarily on the ratio of water to the biopolymer in the investigated system. The two extreme cases are the dry biopolymer (water content tend to zero) and highly diluted aqueous solutions of the biopolymers. The dry biopolymer undergoes hydration if is exposed to the water vapor of increased vapor pressure. The extent of hydration can be determined y measuring the increment in weight. It is much more difficult to determine the extent of hydration in aqueous solutions of biopolymers. Although hydration is accompanied by a volume contraction of the solute and the solvent, this change in volume is very small and difficult to measure directly. It is customary to measure the density of biopolymer solution.

The amount of hydrated biopolymer and of free water in the biopolymer-water system, the thermodynamic notion of partial specific volume has been introduced and is frequently determined. The relation to v_{sp}, the specific volume, is shown by the equation:

$$v_{sp} = g_p \bar{v} + g_0 \bar{v}_0 \tag{24}$$

where g_p and g_0 are the amount of biopolymer and water, respectively, in 1g of the mixture.

The magnitude of $\bar{v}$ can be determined by varying the biopolymer-water ratio (g_p/g_0) and plotting v_{sp} agains g_p, the slope is $\bar{v}$.

However is not to measure solution density at each concentration since the correction of Tanford (1955) can be applied:

$$[\eta] = [\eta]_0 + \left\{ \frac{1 - \rho_0 \bar{v}}{\rho_0} \right\} \tag{25}$$

$$\text{or} \quad \left(\frac{\eta_{sp}}{c} \right) = \left(\frac{\eta_{sp}}{c} \right)_0 + \left\{ \frac{1 - \rho_0 \bar{v}}{\rho_0} \right\} \tag{26}$$

Of course if this latter is not known for the solvent conditions being used, or cannot be calculated from the chemical composition of the macromolecule then solution density measurements are required:

$$\bar{v} = \left\{ \frac{1 - \partial \rho / \partial c}{\rho_0} \right\} \tag{27}$$

ρ_0 and ρ are density of solvent and solution, respectively, and can be measured using densimeter or picnometer.

The swollen specific volume v_{sp} (cm^3/g) is defined when an anhydrous biomacromolecules essentially expand in suspended or dissolved in solution because of solvent association, and

$$v_{sp} = \frac{v_H M}{N_A} \tag{28}$$

where v_H is swollen or hydrodynamic volume (cm^3), M the molecular weight (Da or g/mol), and N_A is Avogadro's number. This associated solvent which we consider in more detail below can be regarded as which is either chemically attached or physically entrained by the biomacromolecules. v_{sp} can be related to a popular term called the hydration value δ, by the relation

$$v_{sp} = \bar{v} + \frac{\delta}{\rho_0} \tag{29}$$

The corresponding value of the 'hydration' δ of the molecule (see table 4), defined by

$$\delta = (v_{sp} - \bar{v})\rho_0 \tag{30}$$

where v_{sp} is specific volume (cm³/g). Although, because of the approximations we have made, the actual numerical value must be treated with very great caution, this treatment does however suggest that polysaccharide is highly expanded, but perhaps not to the same extent as found for coil-like polysaccharide structures to is ~50g/g solvent bound per g of solute. For globular proteins a δ value of 0.3-0.4 has been inferred by RMN, IR, and simulation computer. Hydration value from sedimentation or diffusion data varied from 0.1 to 1.

3.2.3 Perrin number

Most biological polymers, such as proteins and nucleic acids and some synthetic polymers, have relatively inflexible chains. For rigid particles, the size is no longer of predominant importance, because the polymer chain is no longer in the form of a flexible random coil; instead, shape becomes an important parameter. Following are some theoretical proposals for the estimation of the shape factor p from the viscosity measurement (table 4). The term f/f_0 is sometimes denoted as p, Perrin constant.

Combination of the Perrin function, p often referred as the 'frictional ratio due to shape' with the frictional ratio (f/f_0) enables the degree of expansion of the molecule ($v_H/\bar{v}$) to be estimated, where v_H, (cm³/g) is the volume of the swollen molecule (Polysaccharide or protein + associated solvent) per unit mass of polysaccharide and $\bar{v}$ is the partial specific volume (essentially the anhydrous molecule):

$$\frac{f}{f_0} = P\left(\frac{v_H}{\bar{v}}\right)^{1/3} \tag{31}$$

When the biopolymer is contracted, term of expansion is negligible.

3.2.4 Einstein viscosity increment (Simha number)

There are two molecular contributions to the intrinsic viscosity: one from shape, the other from size or volume, as summarized by the relation

$$[\eta] = \upsilon_{a-b} v_{sp} \tag{32}$$

Where $\upsilon_{a\text{-}b}$ is a molecular shape (Simha 1940) parameter known as viscosity increment and v_{sp} defined in equation 28.

The viscosity increment $\upsilon_{a\text{-}b}$ is referred to as a universal shape function or Simha number (table 4); it can be directly related to the shape of a particle independent of volume. For its experimental measurement it does however require measurement of v_{sp}, $\bar{v}$, δ, ρ_0, as well as of course [η].

In a study of the viscosity of a solution of suspension of spherical particles (colloids), suggested that the specific viscosity η_{sp} is related to a shape factor $\upsilon_{a\text{-}b}$ in the following way:

$$\eta_{sp} = \upsilon_{a-b}\phi \tag{33}$$

where ϕ is the volume fraction;

$$\phi = \frac{n\, v\, \upsilon_{a-b}}{V} \tag{34}$$

where n is the number of no interacting identical particles, v is the volume of each particle, and V is the volume of the solution or suspension. Assume that the molecules are of a spherical shape, rigid and large relative to the size of the solvent molecules, and that the particles are small enough to exhibit Brownian motion but large enough to obey the laws of macroscopic hydrodynamics (Teraoka 2002). Then $\upsilon_{a\text{-}b} = 2.5$ is for spherical particle.

The Einstein equation is now used as a reference to estimate the shape of macromolecules. Any deviation can be interpreted as the fact that the molecules are not a sphere.

3.2.5 Scheraga-Mandelkern parameter

The first attempt to the problem of the hydration for ellipsoids of revolution, suggests a combination graphic $\upsilon_{a\text{-}b}$ with the contribution of the form of the Perrin function "p" (ratio of friction). This was followed in for Flory and Scheraga-Mandelkern describing and analytical combination of $\upsilon_{a\text{-}b}$ with p to yield a function β, which, with [η] in cm^3/g is given by:

$$\beta \equiv \frac{[\eta]^{1/3}\eta_0}{M^{2/3}\left(1-\rho_0\bar{v}\right)100^{1/3}} = \frac{N_A^{1/3}}{\left(16200\pi^2\right)^{1/3}}\frac{\upsilon_{a-b}^{1/3}}{p} \tag{35}$$

Unfortunately the β-function proved very insensitive to shape change. Fortunately further combination of $\upsilon_{a\text{-}b}$ with other universal shape parameters have proved more successful.

Scheraga-Mandelkern equations (1953), for effective hydrodynamic ellipsoid factor β (Sun 2004), suggested that [η] is the function of two independent variables: p, the axial ratio, which is a measure of shape, and V_e, the effective volume. To relate [η] to p and V_e, introduced f, the frictional coefficient, which is known to be a direct function of p and V_e. Thus, for a sphere we have

$$\frac{\eta_{sp}}{c} \equiv [\eta] = \upsilon_{a-b}\frac{N_A}{100}\frac{V_e}{M} \tag{36}$$

$$f_0 = 6\pi\eta_0\left(\frac{3V_e}{4\pi}\right)^{1/3} \tag{37}$$

Using the Stokes-Einstein equation of diffusion coefficient

$$D = \frac{k_B T}{f} \tag{38}$$

where k_B is Boltzmann constant and T is the temperature.

And Svedberg equations of sedimentation coefficient

$$S = \frac{M\left(1-\rho\bar{v}\right)}{N_A f} \tag{39}$$

obtain

$$\beta \equiv \frac{D[\eta]^{1/3} M^{1/3} \eta_0}{k_B T} \tag{40}$$

or

$$\beta \equiv \frac{N_A S[\eta]^{1/3} \eta_0}{M^{2/3}\left(1-\rho\bar{v}\right)} \tag{41}$$

The value of β is a measure of the effective hydrodynamic ellipsoid (table 4).

3.2.6 Flory parameters

The classical size-independent combinations are the Flory parameters that combine the intrinsic viscosity, $[\eta]$, and the radius of gyration, R_g:

$$\phi_0 = \frac{[\eta] M_w}{6^{3/2} R_g^3} \tag{42}$$

and another combining the friction coefficient with the radius of gyration:

$$P_0 = \frac{f}{6\eta R_g} \tag{43}$$

These quantities have been proposed along the years, at different times by Flory. As a consequence of the diversity in their origin, the set of classical universal size independent quantities suffers some inconveniences. Two of them, unimportant but somehow cumbersome, are related to the diversity not only in the symbols employed to represent them, but mainly in the disparity of their numerical values and the order of magnitude for typical cases; thus, while the values for these two structures in the case of the ϕ_0 are $9.23\text{x}10^{23}\text{mol}^{-1}$ and $2.60\text{x}10^{23}\text{mol}^{-1}$. Thus, it is accepted that, for every flexible-chain polymer in a Θ (ideal) solvent, there is a universal value of $\phi_0 = 2.50\text{x}10^{23}\text{mol}^{-1}$ (Ortega & Garcia de la Torre 2007). The values obtained for gelatin B are show in table 4.

3.3 pH effect and Isoelectric point

The IEP for aminoacid can be calculated from the ionization constant according to the equation:

$$IEP = \frac{pK_1 + pK_2}{2} \tag{44}$$

If the acid and basic groups of aminoacid were ionized to the same extent, its salt-free water solution would have the same pH value as pure water. Since the ionization of carboxyl groups is higher than of amino groups the IEP are neighborhood of pH 6. An aqueous solution of aminoacid contains small amounts of hydrogen ions and of anions $H_2N{\cdot}R{\cdot}COO^-$ in addition to large amounts of dipolar ion $H_3N^+{\cdot}R{\cdot}COO^-$. Since the aminoacid are weak acids and weak bases at the same time, their mixtures with strong acids and bases are used as buffer solutions.

Since the binding of extraneous ions considerably alters the value of IEP of an aminoacid or protein, this point is not a constant. The term isoionic point (IP) is used to designate the pH value of a pure protein in salt-free water. The direct determination of this constant is difficult and because many proteins are insoluble in the absence of salts. The isoionic point is usually determined indirectly, that is, by measuring the IEP at different concentrations of the neutral salts and extrapolating to zero concentration. The value of the isoionic point may differ from IEP by more than a pH unit (Haurowitz 1963).

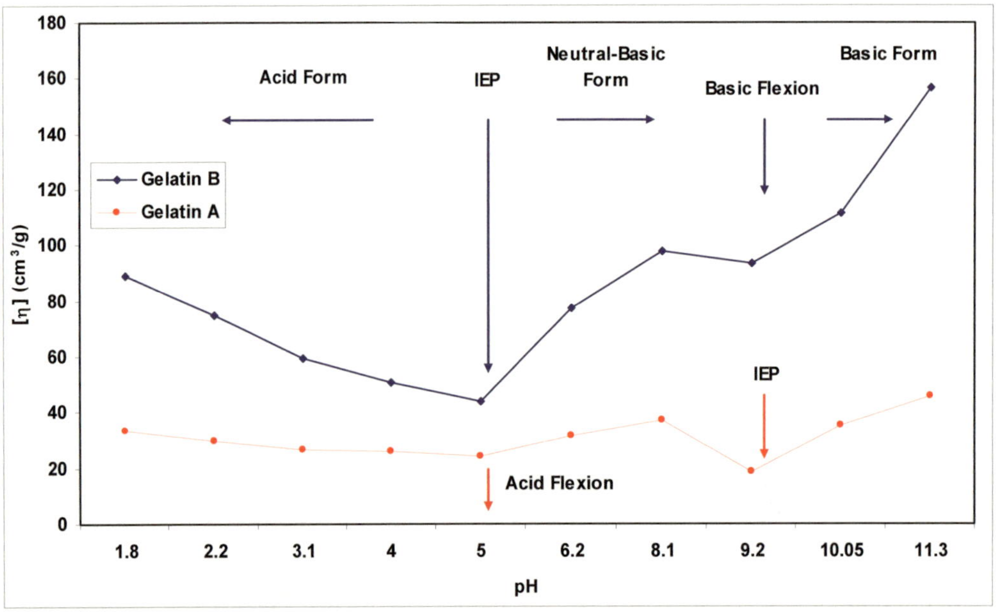

Fig. 2. Influence of pH on the intrinsic viscosity of gelatin A and B.

On point of interest at the pH at which a protein or order polyampholite has zero charge (isoelectric point) the mobility should be zero. At lower pH values a protein will be positively charged and move toward the negative electrode, and at higher pH transport will be in the opposite direction. Thus the measure of isoelectric point (IEP) can be made simply and unambiguously by means electrophoresis (van Holde 1971).

Gelatins are classified according to whether an acid or an alkali is used in the final preextraction step. If an acid solution is used as the final solvent, type-A gelatin (acid process) is obtained. In case of alkali as the final solvent, type-B gelatin (alkali process) is obtained. Type-A gelatin's isoelectric point is higher compared to that of type-B gelatin, as a milder acid process does not remove the amide nitrogen of glutamine and aspargine, therefore, the resulting gelatin's isoelectric point might be as high as 9.4. If a more severe acid treatment is required, then some of the amide groups are hydrolyzed and the isoelectric point would be similar to that of the original collagen molecule, which generally lies between 6 and 8. Type-B gelatin's isoelectric point might be as low as 5.2 or 4.8, as the alkali process results in the loss of the amide groups. Benson 1963, prepared protein solutions at various pH and determining by plott the viscosity relative or intrinsic viscosity as a function of pH is a minimum corresponding to the isoelectric point.

Kenchington and Ward 1954, conducted studies of gelatin obtained by titration of acid and basic process. They claim the IEP gelatin A to the amide groups in addition to carboxyl groups. In the case of gelatin B, awarded the IEP conversion of arginine to ornithine, where the conversion of the guanidine groups in the amino acid occurs during treatment extent slight alkaline of the collagen.

The figure clearly shows two minimum for a gelatin B corresponds to the isoelectric point, pH 5.1, and the other corresponds to a flexion in the basic medium at pH 9.1; in the case of gelatin A the isoelectric point corresponds to pH 9.2 and at pH 5 an flexion. The isoelectric point is presented as the more compact form of gelatin, i.e. less drainage time, evidenced with smaller hydrodynamic radius. Both the basic or acidic flexion is due to compact forms of gelatin but not so small as to be isoelectric point. No less are the forms they take in acidic or basic, identified as linear gain widespread forms of gelatin A or B. In the neutral pH from the isoelectric points bending is the neutral form. A basic pH after the isoelectric point 9.2 shall register a new or extended basic shape.

3.4 Solvent effect

Solubility of protein in water varies within wide limits. While some proteins dissolve easily in salt-free water, others dissolve only in the presence of certain concentrations of salts, a third group is soluble in mixtures of water and ethanol, and insoluble in any solvent.

The solubility protein in water is determined by chemical structure, number of aminoacids that forming the molecule, and folding of the peptide chains in the molecule. Peptide structure and charged ionic groups increase the affinity of the protein for water and its solubility. The solubility of proteins depends to a great extent on pH and on the concentration of salts present on solution. The minimum solubility is found at the isoelectric point. Neutral salts have a two-fold effect on protein solutions. Low concentration of the salts increase the solubility of proteins phenomena denominated as salting-in, while high concentrations of neutral salts reduce the solubility and give rise to the formation of precipitates (salting-out). Salting-out effect is evidently due to a competition between the salt and protein for molecules of the solvent.

Bohidar et al. 1998, realized studies of Sol and Gel state properties of aqueous gelatin solutions of concentrations 4%, 6%, 8% and 10% (w/v) were investigated through dielectric relaxation studies done at various temperatures in the range from 20 to 60°C carried out over a frequency range 20Hz-10MHz and no relaxation of any nature was observed.

Quite generally, high dielectric constants are found in polar molecules, low constant in non-polar molecules. The proteins are highly polar substances; one would expect zwitterions to have high dielectric constants. The dielectric constant of protein aqueous solutions is increase and proportional to the dissolved substance.

3.5 Ionic strength and electroviscous effect

The ionic strength dependence of intrinsic viscosity is function of molecular structure and protein folding. It is well known that the conformational and rheological properties of charged biopolymer solutions are dependent not only upon electrostatic interactions between macromolecules but also upon interactions between biopolymer chains and mobile ions. Due electrostatic interactions the specific viscosity of extremely dilute solutions seems to increase infinitely with decreasing ionic concentration. Variations of the intrinsic viscosity of a charged polyampholite with ionic strength have problems of characterization.

It was found earlier by experiment and theory that the viscosity intrinsic of polyelectrolyte solutions is nearly linear with the reciprocal square root of the ionic strength over a certain range, such as

$$[\eta]=[\eta]_{\infty}+\varepsilon\, I^{-1/2} \tag{45}$$

Where the slope ε determined by the plot is the extension coefficient. The $I^{-1/2}$ dependence viewed in terms of an increased Debye length can be explained as the electrostatic excluded volume contribution.

To study the electroviscous effect should address the theory of the electrical double layer developed by Smolowchoski, Helmholtz, Guy and Stern (Lyklema 1995). According to the Stern double layer consists of two parts: one, which is about the thickness of anion remains almost fixed to the solid surface. In this layer there is a definite fall of potential. The second part extends some distance within the liquid phase and is diffuse in this region, thermal motion allows the free movement of particles, but the distribution of positive and negative ions is not uniform, because the electrostatic field in the surface would cause preferential attraction of the opposite charge. The result is a fall of potential within the liquid where the charge distribution is uniform. The approximate boundary between the bulk liquid and bulk solution is the electrokinetic potential or zeta potential. The magnitude of the zeta potential could be alter, or even sign, the presence of ions in solution. The higher the valence of the ion of opposite sign is greater effect to reduce the potential electrokinetic. Increasing the concentration of electrolyte on a surface of negative charge, cations tend to accumulate on the fixed layer, reducing the thickness of the double layer and consequently charge density and zeta potential. To study the stability of colloids are essential trace amounts of electrolytes, at least in water, but larger quantities cause aggregation of particles and precipitate formation. The phenomenon of precipitation, flocculation, coagulation, and conservation depends on what kind of effect on the type of colloid electrolyte concentration exercises. Given the concentration of salts and the concentration of colloid is possible to determine whether precipitation occurs or not. As a result of these studies it is clear that the ion is effective for the coagulation is the opposite to that of the colloidal particle, and that the clot can grow considerably increases the valence of the ion (Schulze-Hardy rule).

The stability of a colloid such as gelatin in water is determined by the electric charge and hydration. The addition of large amounts of electrolytes to colloids (biopolymers) causes precipitation dispersed substance, a phenomenon known as salting. That is, high concentrations of salts dehydrate the solvated biopolymer and reduce the zeta potential. The salificantion effect depends on the nature of the ion, and salts of a given cation. Can be arranged according to decreasing ability to remove the lyophilic substances on colloidal solution, these series have been called Hofmeister or lyotropic series.

Electroviscous effect occurs when a small addition of electrolyte a colloid produces a notable decrease in viscosity. Experiments with different salts have shown that the effective ion is opposite to that of the colloid particles and the influence is much greater with increasing oxidation state of the ion. That is, the decrease in viscosity is associated with decreased potential electrokinetic double layer. The small amount of added electrolyte can not appreciably affect on the solvation of the particles, and thus it is possible that one of the determinants of viscosity than the actual volume of the dispersed phase is the zeta potential.

The electroviscous effect present with solid particles suspended in ionic liquids, to increase the viscosity over that of the bulk liquid. The primary effect caused by the shear field distorting the electrical double layer surrounding the solid particles in suspension. The secondary effect results from the overlap of the electrical double layers of neighboring particles. The tertiary effect arises from changes in size and shape of the particles caused by the shear field. The primary electroviscous effect has been the subject of much study and has been shown to depend on (a) the size of the Debye length of the electrical double layer compared to the size of the suspended particle; (b) the potential at the slipping plane between the particle and the bulk fluid; (c) the Peclet number, i.e., diffusive to hydrodynamic forces; (d) the Hartmann number, i.e. electrical to hydrodynamic forces and (e) variations in the Stern layer around the particle (Garcia-Salinas et al. 2000).

The primary electroviscous effect occurs, for a dilute system, when the complex fluid is sheared and the electrical double layers around the particles are distorted by the shear field. The viscosity increases as a result of an extra dissipation of energy, which is taken into account as a correction factor ``p_i'' to the Einstein equation:

$$\eta_r = 1 + \nu_{a-b}(1 + p_i)\phi \tag{46}$$

The effective viscosity of a suspension of particles in a fluid medium is greater than that of the pure fluid, owing to the energy dissipation within the electrical double layers.

Finally, p_i is the primary electroviscous coefficient which is a function of the charge on the particle or, more conventionally, the electrostatic potential, ζ , on the "slip-ping plane" which defines the hydrodynamic radius of the particle, and properties (charge, bulk density number, and limiting conductance) of the electrolyte ions (Rubio–Hernandez et al. 2000).

Equation 46 suggests that, maintaining p_i constant, η_r must depend linearly on ϕ if only a first-order electroviscous effect exists, and an increase in the electrolyte concentration implies a decrease in the thickness, $1/\kappa$, of the electrical double layer,

$$\frac{1}{\kappa} = \sqrt{\frac{\varepsilon_r \varepsilon_0 k_B T}{4\pi e^2 \sum_i c_i z_i^2}} \tag{47}$$

where ε_r is the dielectric constant of the liquid medium, ε_0 is the vacuum permittivity, e is the electron charge, k_BT is the Boltzmann energy, and c_i and z_i are the concentrations and the valencies, respectively, of the various ionic species in the solution, far away from any particle (Oshima 2008 and Overveek 1976).

The slopes of the different curves correspond to the full electrohydrodynamic effect, $\phi + \phi\, p_i$, where the first term expresses the hydrodynamic effect, and the second is the consequence of the distortion of the electrical double layer that surrounds the particles. To determine this second term and, more exactly, the primary electroviscous coefficient, p_i.

The calculation of zeta potential from electoviscous effect measures (Rubio-Hernandez et al. 1998 and 2004), is given by the equation

$$\eta_{sp} = \nu_{a-b}\phi\left[1+\frac{1}{\eta_o \kappa R_H^2}\left(\frac{\varepsilon_r \varepsilon_0 \zeta}{2\pi}\right)^2\right] \tag{48}$$

Proteins and polysaccharides are biopolymers carrying charges due to dissociation of their ionizables groups in an aqueous solution and have been widely used in paints, cosmetics, and film industries. Different from neutral polymers, polyelectrolytes have a comparatively extended conformation, owing to the repulsive intrachain electrostatic interaction screened by the surrounding small ions in the solution. As a result, their rheological properties in a dilute solution due to an applied flow will change accordingly. This is called a tertiary electroviscous effect. In addition, the imposed flow distorts the ionic cloud around each polyelectrolyte from its equilibrium state, leading to two effects. First, it results in additional energy dissipation associated with the electrical interaction between each polyelectrolyte and small ions. This direct effect, depending on the chain configuration, is analogous to the primary electroviscous effect of charged rigid spheres, rods, or polyions of arbitrary shape but has sometimes been ignored in the field of flexible polyelectrolyte rheology. Second, an indirect effect changes the polyelectrolyte conformation through the modification of the intrachain electrostatic interaction, compared to that with equilibrium double layer. Therefore, the primary and tertiary effects are indeed coupled for polyelectrolytes (Hunter 1981).

The increasing dilution of flexible polyelectrolytes at low ionic strength, the reduced viscosity may increase first, reach a maximum, and then decrease. Since a similar behavior can also be observed even for solutions of polyelectrolyte lattices at low salt concentration, the primary electroviscous effect was thought as a possible explanation for the maximum, as opposed to conformation change.

It is very difficult and scarce to find literature to study the electrokinetic phenomena of proteins or macromolecules in solution therefore limit us to the basic concepts of electrokinetic changes observed, they are conformational change because of the presence of salts and the zeta potential change in pH.

For polyelectrolyte solutions with added salt, prior experimental studies found that the intrinsic viscosity decreases with increasing salt concentration. This can be explained by the tertiary electroviscous effect. As more salts are added, the intrachain electrostatic repulsion is weakened by the stronger screening effect of small ions. As a result, the polyelectrolytes are more compact and flexible, leading to a smaller resistance to fluid flow and thus a lower viscosity. For a wormlike-chain model by incorporating the tertiary effect on the chain conformation to predict the intrinsic viscosity in zero-shear limit. The effect of the intrachain electrostatic repulsion on the chain conformation is obtained based on the equilibrium interaction, which depends on the Debye screening length (Li Jiang et al. 2001).

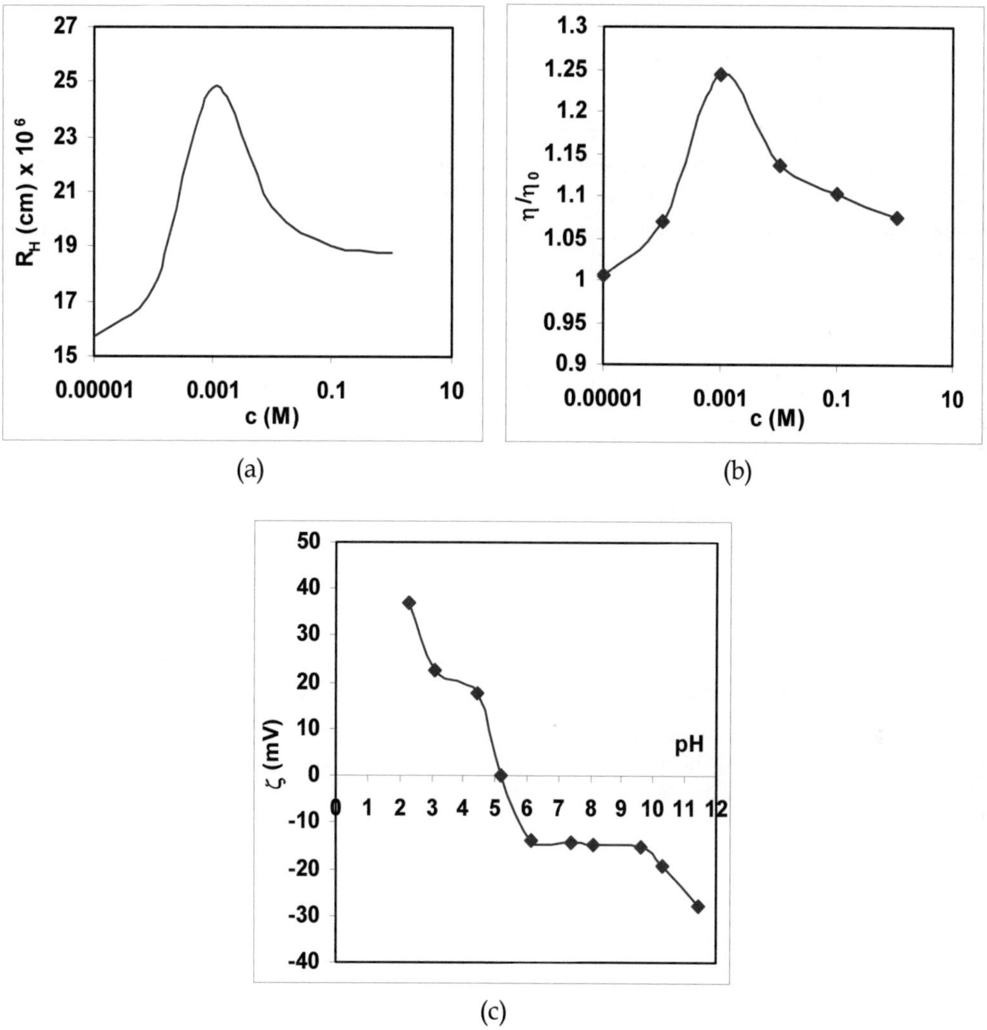

Fig. 3. Study of the electroviscous effect of NaCl on gelatin B. a-Hydrodynamic radius, b-η/η_0. c- Zeta potential at different pH (0.001M NaCl).

Note that when the concentration of added salt is very low, Debye length needs to be modified by including the charge contribution of the dissociating counterions from the polyelectrolytes. Because the equilibrium interaction is used, their theory predicts that the intrinsic viscosity is independent of ion species at constant ionic strength. At very high ionic strength, the intrachain electrostatic interaction is nearly screened out, and the chains behave as neutral polymers. Aside from the tertiary effect, the intrinsic viscosity will indeed be affected by the ionic cloud distortion and thus cannot be accurately predicted by their theory.

The effects of ion valence and polyelectrolyte charge density showed that at very low ionic strength found that when the counterion valence of added salt changes from monovalent (NaCl) to divalent ($MgSO_4$), the reduced viscosity decreases by a factor of about 4.5. If $La(NO_3)_3$ is used, the reduced viscosity will be further decreased although not drastically. As for polyelectrolyte charge density, the intrinsic viscosity was found to increase with it because of an enhanced intrachain electrostatic repulsion (Antonietti et al. 1997).

The data on the electroviscous effect of NaCl in gelatin B can be seen in Figure 3. Figure 3 b shows that in 0.001M is a maximum, this maximum defines the most expanded form of gelatin B, at higher concentrations this value is constant from 0.1M, a phenomenon known as salting-out. At lower concentrations found in the salting-in. Figure 3 c, we find the zeta potential and pH where it is zero corresponds to the isoelectric point to a concentration of 0.001M NaCl.

3.6 Temperature effect

Theta temperature (Flory temperature or ideal temperature) is the temperature at which, for a given polymer-solvent pair, the polymer exists in its unperturbed dimensions. The theta temperature, Θ, can be determined by colligative property measurements, by determining the second virial coefficient. At theta temperature the second virial coefficient becomes zero. More rapid methods use turbidity and cloud point temperature measurements. In this method, the linearity of the reciprocal cloud point temperature ($1/T_{cp}$) against the logarithm of the polymer volume fraction (ϕ) is observed. Extrapolation to $\log \phi = 0$ gives the reciprocal theta temperature (Guner and Kara 1998).

Theta temperature is one of the most important thermodynamic parameters of polymer solutions. At theta temperature, the long-range interactions vanish, segmental interactions become more effective and the polymer chains assume their unperturbed dimensions. It can be determined by light scattering and osmotic pressure measurements. These techniques are based on the fact that the second virial coefficient, A_2, becomes zero at the theta conditions.

Another method is known as the cloud point/turbidity measurements yielding more rapid and accurate determination of the theta temperature. This technique is mainly based on the observed linearity of the reciprocal of the cloud point temperature when plotted versus the logarithm of the polymer volume fraction and the extrapolation to ϕ. The information on unperturbed dimensions has been accessed via extrapolation methods by viscometry yielding intrinsic viscosity values (Gouinlock et al. 1955). In these methods, it is also possible to determine the long-range interactions, and evaluate the theta temperature by the temperature dependence of long-range interaction parameter. Simply, the intrinsic viscosity may be given in the form:

$$[\eta]_T = [\eta]_\Theta \alpha_\eta^3 \tag{49}$$

where $[\eta]_T$ and $[\eta]_\Theta$ are the intrinsic viscosities determined at the studied temperature interval (20-60°C) and the theta temperature. α_η is the expansion of the polymer for the employed solvent system. The relation between hydrodynamic linear expansion factor, α, and α_η is given by $\alpha_\eta^3 = \alpha^{5/2}$.

The thermodynamic linear expansion factor has been related to Flory or thermodynamic interaction parameter, χ, and the entropy of dilution parameter, χ_S, through the Flory-Fox [10] equations,

$$\alpha^5 - \alpha^3 = 2C_M\left(0.5 - \chi_s\right)\left(1 - \Theta / T\right)M^{1/2} \tag{50}$$

where Θ is the theta temperature. The coefficient C_M is given by

$$C_M = \frac{27v_{sp}^2 M^{3/2}}{2N_A v_1\left(2\pi < r^2 >_0\right)^{3/2}} \tag{51}$$

where v_{sp} is the specific volume of the polymer, N_A is the Avogadro's number, v_1 is the molar volume of the solvent (for water, 18 cm^3/mol) and $<r^2>_0^{1/2}$ denotes the unperturbed root-mean-square end-to-end distance.

α depends on the factor $(0.5-\chi_s)(1-\Theta/T)$, measuring the intensity of the thermodynamic interactions and also representing the power of the solvent, i.e. the higher is this factor, the better is the solvent. The coefficient C_M seems to be less dependent on temperature, however, the parameter C_M involves the unperturbed dimension end-to-end distance, $<r^2>_0^{1/2}$, which depends on temperature through the effective bond character of the chain. Consequently, C_M is indirectly governed by temperature through $<r^2>_0^{1/2}$ term present in Eq. (51). So, the temperature dependence of α is much more governed by this factor, whereas temperature dependence of C_M is not negligible. The decrease of alpha with increasing temperature is for different molecular weights of sample. Considering the structure of gelatin, it is strongly expected that hydrogen bonding will form between polymer segments, and obviously these bonds will break at increased temperature and hydrophobic interactions between polymer segments will be more dominant and can be related to hydrodynamic systems. Hydration, hydrogen bonding/molecular association between polymer segments and water molecules are destroyed with an increment of temperature for water-soluble polymers. It is observed that the strong interaction between polymer segments and solvent molecules through hydrogen bonding will form for gelatin/water system. Therefore, χ_H must be negative. Experimental χ values seem to be higher than 0.5 and χ_S values are the only dominant driving force in setting χ numerically to 0.5 which is believed to be the ideal condition of polymer solutions. The relation between χ_H, χ_S and the interaction parameter χ is defined as $\chi-0.5 = \chi_H-(0.5-\chi_s)$.

This stipulation of the interaction parameter to be equal to 0.5 at the theta temperature is found to hold with values of χ_H and χ_S equal to $0.5 - \chi < 2.7 \times 10^{-5}$, and this value tends to decrease with increasing temperature. The values of Θ = 308.6 K were found from the temperature dependence of the interaction parameter for gelatin B. Naturally, determination of the correct theta temperature of a chosen polymer/solvent system has a great physic-chemical importance for polymer solutions thermodynamically. It is quite well known that the second virial coefficient can also be evaluated from osmometry and light scattering measurements which consequently exhibits temperature dependence, finally yielding the theta temperature for the system under study. However, the evaluation of second virial coefficient from Zimm plot is really a time consuming and difficult task although quite reliable (Guner and Kibarer 2001).

Gelatin is easily dissolved in water by heating the solutions at about 40 to 50ºC. Then the gelatin chains are believed to be in the coil conformation. When solutions are cooled below 30ºC a reverse coil→helix transition takes place which can be detected by important modifications of the optical rotation, mainly due to the left-handed helix conformation. The helices have to be stabilized by hydrogen bonds which are perpendicular to their axes. Thus, at very low concentration 10^{-4} to 10^{-3} w/w intramolecular hydrogen bonds are formed preferentially by a back refolding of single chains. A higher concentrations > 1% w/w, can be proved the helix growth indices chain association, and three-dimensional network formation. Models of a chain have been proposed: 1- a conformation coil→helix transition by local association of three different chains (intermolecular bonds) along short helical sequences; 2- a crystallization mechanism leading to fiber growth, similar to fringe micelle model of synthetic polymer crystallization. The diameter of fiber would depend on temperature and concentration.

Eysturskard et al. (2009), found that the gelation temperature and helix content are very affected (increase) by increasing molecular weight above 250,000 g/mol. This phenomenon is found in the present work, which accounts for a sol-gel transition different from those found by the team of Djabourov (1988a, 1988b, 1991, 1995, 2007). In this work this phenomenon is awarded to the high molecular weight gelatin B studies; also this biopolymer has high numbers Simha and Perrin, which realizes that this macromolecule is related to the rod-like shape than the random-coil shape.

Olivares et al. (2006), studies performed viscometers very dilute gelatin solutions with concentrations between 10^{-5} and 10^{-3} g/cm^3, where either intermolecular aggregation or intramolecular folding are possible, respectively, and the sol–gel transition is not observed.

Djabourov (1988a, 1988b, 1991, 1995, 2007) proposes a T_g of 26-30°C or below 30ºC in this study we found a T_g of about 30°C.

In this work, the linear relation between viscosity and temperature, where the values obtained in the temperature range of 20-29ºC is E_{avf} 8,556.62 cal/mol, and A_{vf} 8.59x10^{-9} g/cm s, with σ^2 0.9983, and range of 31-37.4ºC is E_{avf} 5,649.81 cal/mol, and A_{vf} 7.55x10^{-7} g/cm s, with σ^2 0.9971. This occurs due to the higher resistance to flow of biopolymers requiring, therefore, more energy in gel state. The increment of activation energy of viscous flow ($\Delta E_{avf} = E_{avf} - E_{avf0}$) occurs due to the higher resistance to flow of biopolymer respect to solvent; where ΔE_{avf} is 4,454.2 and 1,547.4cal/mol, respectively.

Noting the influence of temperature on the intrinsic viscosity is given by the parameter of chain flexibility ($d\ln[\eta]/dT$), which gives information about the conformation of the macromolecule chain in solution (Kasaii 2007, Chen and Tsaih 1998). The chain flexibility parameter in the temperature range of 20-29ºC is $d\ln[\eta]/dT$ = 4,404.11K^{-1}, σ^2 0.9993; and in the range of 31-37.4ºC is 2,987.89, σ^2 0.9845. This phenomenon indicates that the chain flexibility, for this gelatin molecular weight (333,000 g/mol) is rigid for temperature range of 20-29ºC and flexible for range of 31-37.4ºC (figure 4).

Figure 4 shows that the intrinsic viscosity is influenced by temperature for gelatin. Where the influence of temperature is manifested in a phase transition at 30°C, presented as a change of slopes given between 20-29°C and another between 31-37.4°C, with gelation temperature, T_{gel} of 30°C. This phenomenon is repeated by looking at other hydrodynamic properties as seen in the change in hydrodynamic radius as in Figure 4. The hydrodynamic radius and intrinsic viscosity for proteins decrease with increasing temperature (Bohidar for gelatin 1998, and Monkos for serum proteins 1996, 1997, 1999, 2000, 2004 and 2005).

According to Stokes-Einstein equation, the diffusion coefficient is inversely proportional to the solution viscosity which increases with temperature. Hence, a lower diffusion coefficient corresponds to a lower size molecule.

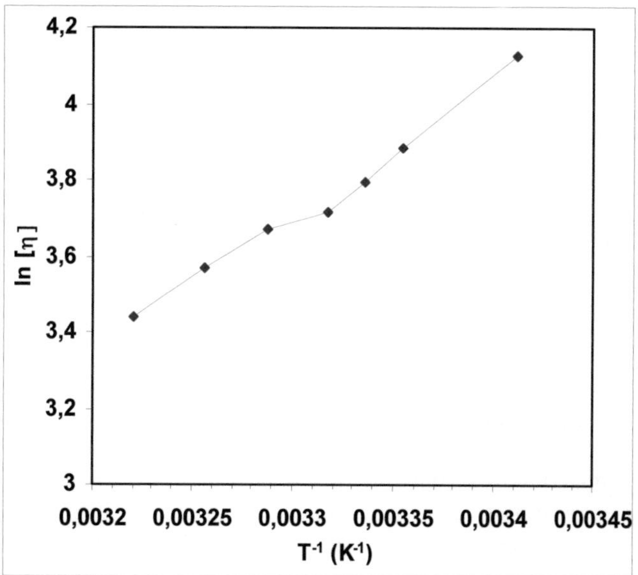

Fig. 4. Ln [η] i function of 1/T.

Analyzing the values of the hydrodynamic properties of gelatin in aqueous solution shows that all values vary with the temperature (table 4). β values increases from 2.24 to 2.83x10^{6} with temperature increase, indicating a coil structure. The values of ϕ_0 and P_0 decreases from 12.70 to 6.37x10^{21}mol^{-1} and 8.88 to 5.56 demonstrating a low flexibility of the colloid. The value of p decreases from 6.70 to 5.28, and $\nu_{(a/b)}$ with 14.6 which confirms that gelatin in aqueous solution is a biopolymer with a rod-like conformation with tendency to compaction with increasing temperature (R_H decreases). The value of δ as expected decreases from 4.07 to 1.50g/g with increases of temperature, this phenomenon is due to loss of water due to compression of gelatin by the effect of increasing temperature (Durand 2007 and Morris et al. 2009).

The parameters of Mark-Houwink for biopolymers may be varied with solvent and temperature (Chen et al. 2009, Chen & Tsai 1998). This is because the macromolecule changes hydrodynamic radius with type solution and temperature via change in their chain flexibility. In a good solvent and sol phase, a temperature increase results in an intrinsic viscosity decrease and in a less-extended conformation (*D*> and R_H<), because the entropy value increases with an increase in temperature and it is unfavorable for an extended conformation (ΔE_{avf}>).

T (°C)	[η] (cm³/g)	R_H (cm)	δ (g/g)	β	ϕ_0 (mol^{-1})	$p = f/f_{hyd}$	P_0
20.00	62.12	3.25E-06	4.07	3.40E+06	1.27E+22	6.70	8.88
25.00	48.65	3.00E-06	3.04	3.56E+06	9.95E+21	6.39	7.81
26.60	44.46	2.91E-06	2.72	3.82E+06	9.09E+21	5.95	7.05
28.30	41.15	2.83E-06	2.45	3.94E+06	8.42E+21	5.78	6.68
31.00	39.28	2.79E-06	2.25	4.00E+06	8.03E+21	5.68	6.47
34.00	35.53	2.70E-06	1.91	4.21E+06	7.27E+21	5.41	5.95
37.40	31.12	2.58E-06	1.50	4.31E+06	6.37E+21	5.28	5.56

Table 4. Gelatin B hydrodynamic parameters, measurement to different temperatures.

Mark-Houwink values confirm that for these conditions gelatin is behaves rod-like conformation. Such of empirical equations can be relating the parameters of Mark-Houwink with *T*, which ultimately describe this type of thermodynamic parameters are relations between properties the solute with the solvent and temperature dependence.

4. Conclusions

The temperature influence is manifested in a phase transition about at 30°C (gelation temperature) presented as a change of slopes given between 20-29°C and another between 31-37.4°C. The ΔE_{avf} and chain flexibility parameter values obtained in the gel range temperature of 20-29°C is slightly less than twice the sol range of 31-37.4°C. This phenomenon indicates that the chain flexibility is rigid and low flows for temperature range of 20-29°C, and is flexible and more fluid for range of 31-37.4°C.

The Mark-Houwink parameters are influenced by temperature. The numerical value of *a* indicates that gelatin acquire a shape of a rod-like in aqueous solution with temperature increases; and *k* demonstrates that under water their value increases with temperature.

Due to the lack of data on the uniformity of intrinsic viscosity measurements in the gelatin/water system, clearly shows a decrease in "*a*" with temperature, and this M_w is 333,000g/mol.

Gelatin behavior in this system indicates that it behaves rod-like that tends to contract with increasing temperature. This conclusion is supported by the observed data from the hydrodynamic properties analyzed.

An increase in temperature causes the gelatin/water system to show that the biopolymer tends to compaction (decreasing in R_H and $[\eta]$), which requires an increase of energy consumption due to a difficulty in flowing (increase in D and high ΔE_{avf}). This phenomenon is observed in the case of ideal solvents, evidencing a decrease of a with temperature.

The effect of pH on the intrinsic viscosity testing gives a minimum at the isoelectric point at pH 5.1 for gelatin B to pH 9.1 for gelatin A. from electroviscous effect analysis shows that 0.001 M ionic strength the hydrodynamic radius is at its maximum.

5. Acknowledgment

The author thanks Universidad Nacional de San Luis, FONCyT, and CONICET for their contribution. And thanks Dr. R. Curvale, Dr. J. Marchese (CONICET) and Dr. Ariel Ochoa (CONICET) for their help.

6. References

Antonietti M., Briel A., Fosrster S. Quantitative Description of the Intrinsic Viscosity of Branched. Polyelectrolytes. Macromolecules 1997, 30, 2700-2704.

Benson J.E., Viscometric detrmination of isoelectric point of a protein. J. Chem. Educ. 40 (1969) 468-469).

Boedtker H., Doty P. 1954. A study of gelatin molecules aggregates and gels. Journal of American Chemical Society 58, 668-983.

Bohidar H.B. 1998. Hydrodynamic properties of gelatin in dilute solutions. International Journal of Biological Macromolecules 23, 1-6.

Bohidar H. B., Maity S., Saxena A., Jena S. Dielectric behaviour of Gelatin solutions and Gels. Journal Colloid & Polymer Science 276, 1 (1998) 81-86.

Borchard W., Colfen H., Kisters D., Straatmann A. 2002. Evidence for phase transitions of aqueous gelatin gels in a centrifugal field. Progress in Colloid Polymers Science 119, 101-112.

Chatterjee S., Bohidar H.B. 2005. Effect of cationic size on gelation temperature and properties of gelatin hydrogels. International Journal of Biological Macromolecules 35, 81-88.

Chee K.K. A critical evaluation of the single-point determination of intrinsic viscosity. Journal of Applied Polymer Science 34, 3 (1987) 891-899.

Chen R.H., Tsaih M.L. 1998. Effect of temperature on the intrinsic viscosity and conformation of chitosans in dilute HCl solution. International Journal of Biological Macromolecules 23, 135-141.

Chen R.H., Chen W.Y., Wang S.T., Hsu C.H., Tsai M.L. 2009. Changes in the Mark-Houwink hydrodynamic volume of chitosan molecules in solutions of different organic acids, at different temperatures and ionic strengths. Carbohydrate Polymers 78, 902-907.

Curvale, R.A., Cesco, J.C. 2009. Intrinsic viscosity determination by "single-point" and "double-point" equations. Applied Rheology 19, 5, 53347.

Curvale, R.; Masuelli, M.; Perez Padilla, A. 2008. Intrinsic viscosity of bovine serum albumin conformers. International Journal of Biological Macromolecules 42, 133-137.

Deb P.C., Chatterjee S.R. On polynomial expansion of log relative viscosity. Die Makromolekulare Chemie 125, 1 (1969) 283-285.

Deb P.C., Chatterjee S.R. Unperturbed dimension of polymer molecules from viscosity measurements. Die Makromolekulare Chemie 120, 1 (1968) 49-57.

Djabourov M., Leblond J., Papon P. Gelation of aqueos gelatin solutions. I. Structural investigation. J. Phys. France 49 (1988a) 319-332.

Djabourov M., Leblond J., Papon P. Gelation of aqueos gelatin solutions. II. Rheology of the sol-gel transition. J. Phys. France 49 (1988b) 333-343.

Djabourov M. Gelation- A review. Polymer International 25 (1991) 135-143.

Djabourov M., Grillon Y., Leblond J. The sol-gel Transition in gelatin viewed by Diffusing colloidal probes. Polymer Gels and Networks 3 (1995) 407-428.

Domenek S., Petit E., Ducept F., Mezdoura S., Brambati N., Ridoux C., Guedj S., Michon C. 2008. Influence of concentration and ionic strength on the adsorption kinetics of gelatin at the air/water interface. Colloids and Surfaces A: Physicochem. Eng. Aspects 331, 48-55.

Durand, A. 2007. Aqueous solutions of amphiphilic polysaccharides: Concentration and temperature effect on viscosity. European Polymer Journal 43, 1744-1753.

Elharfaoui N., Djabourov M., Babel W.. Molecular weight influence on gelatin gels : structure, enthalpy and rheology. Macromolecular Symposia 256 (2007) 149-157.

Eysturskiard J., Haug I., Elharfaoui N., Djabourov M., Draget K. Structural and mechanical properties of fish gelatin as function of extraction conditions. Food Hydrocolloids 23 (2009) 1702-1711.

García de la Torre J., Carrasco B. 1999. Universal size-independent quantities for the conformational characterization of rigid and flexible macromolecules. Progress in Colloid Polymers Science 113, 81-86.

García-Salinas M. J., Romero-Cano M. S., de las Nieves F. J.. Zeta potential study of a polystyrene latex with variable surface charge: influence on the electroviscous coefficient. Progr Colloid Polym Sci (2000) 115 : 112-116.

Gomez-Guillen M.C., Perez-Mateos M., Gomez-Estaca J., Lopez-Caballero E., Gimenez B., Montero P. 2009. Fish gelatin: a renewable material for developing active biodegradable films. Trends in Food Science & Technology 20, 3-16.

Gomez-Guillen M.C., Turnay J., Fernandez-Diaz M.D., Ulmo N., Lizarde M.A., Montero P. 2002. Structural and physical properties of gelatin extracted from different marine species: a comparative study. Food Hydrocolloids 16, 25-34.

Gouinlock E. V., Flory P.J., Scheraga H.A.. Molecular configuration of gelatin. Journal of Polymer Science 16 (1955) 383-395.

Guner A. 1999. Unperturbed dimensions and theta temperature of dextran in aqueous solutions. Journal of Applied Polymer Science 72, 871-876.

Guner, A., Kibarer, G. 2001. The important role of thermodynamic interaction parameter in the determination of theta temperature, dextran/water system. European Polymer Journal, 37, 619-622.

Guner A. and Kara M. Cloud points and Θ temperatures of aqueous poly(N-vinyl-2-pyrrolidone) solutions in the presence of denaturing agents. Polymer 39, 8 9, 1569-1572, 1998.

Hunter R. Zeta potential in colloid science. Academic Press 1981.

Harding, S.E.; Day, K.; Dham, R.; Lowe, P.M. 1997a. Further observations on the size, shape and hydration of kappa-carrageenan in dilute solution. Carbohydrate Polymers 32, 81-87.

Harding, Stephen E. 1997b. The Viscosity Intrinsic of Biological Macromolecules. Progress in Measurement, Interpretation and Application to Structure in Dilute Solution. Progress in Biophysical Molecules Biological 68, 207-262.

Harrington, W.F. & von Hippel P.H. The stricture of collagen and gelatin. Advances in Protein Chemistry, 16 (1962) 1-138.

Haug I.J., Draget K.I., Smidsrød O. 2004. Physical and rheological properties of fish gelatin compared to mammalian gelatin. Food Hydrocolloids 18, 203-213.

Haurowitz F., The Chemistry and Functions of Proteins, Academic Press, 1963.

Houwink, R., Zusammenhang zwischen viscosimetrisch und osmotisch bestimm- ten polymerisationsgraden bei hochpolymeren. J. Prakt. Chem., 157 (1940) 15.

Huggins M.L. The Viscosity of Dilute Solutions of Long-Chain Molecules. IV. Dependence on Concentration. J. Am. Chem. Soc., 64, 11 (1942) 2716–2718.

Kasaai M.R. 2007. Calculation of Mark-Houwink-Sakurada (MHS) equation viscometric constants for chitosan in any solvent-temperature system using experimental reported viscometric constants data. Carbohydrate Polymers 68, 477-488.

Karim A.A., Bhat R. 2009. Fish gelatin: properties, challenges, and prospects as an alternative to mammalian gelatins. Food Hydrocolloids 23, 563-576.

Karim A.A., Bhat R. 2008. Gelatin alternatives for the food industry: recent developments, challenges and prospects. Trends in Food Science & Technology 19, 644-656.

Kaur M., Jumel K., Hardie K.R., Hardman A., Meadows J., Melia C.D. 2002. Determining the molar mass of a plasma substitute succinylated gelatin by size exclusion chromatography-multi-angle laser light scattering, sedimentation equilibrium and conventional size exclusion chromatography. Journal of Chromatography A 957, 139-148.

Kenchington A. W. & Ward A.G. 1954, The titration curve of gelatin. J. Biochemistry 58 (1954) 202-207.

Kraemer, E. O., Molecular Weights of Celluloses and Cellulose Derivates. Ind. Eng. Chem. 30, (1938) 1200-1204.

Kuwahara N. On the polymer–solvent interaction in polymer solutions. Journal of Polymer Science Part A.1, 7, (1963) 2395-2406.

Li Jiang, Dahong Yang, and Shing Bor Chen. Electroviscous Effects of Dilute Sodium Poly(styrenesulfonate) Solutions in Simple Shear Flow. Macromolecules 2001, 34, 3730-3735.

López Martínez, M.C.; Díaz Baños, F.G.; Ortega Retuerta, A.; García de la Torre, J. 2003. Multiple Linear Least-Squares Fits with a Common Intercept: Determination of the Intrinsic Viscosity of Macromolecules in Solution. Journal of Chemical Education 80(9), 1036-1038.

Lyklema J. Fuedamentals of inteface and colloid science II, Academic Press 1995.

Mark, H. in Der feste Körper (ed. Sänger, R.), 65–104 (Hirzel, Leipzig, 1938).

Maron, Samuel H.. Determination of intrinsic viscosity from one-point measurements. Journal of Applied Polymer Science 5, 15 (1961) 282-284.

Martin, A. F. Abstr. 103rd Am. Chem. Soc. Meeting, p. 1-c ACS (1942).

Masuelli, Martin Alberto. Viscometric study of pectin. Effect of temperature on the hydrodynamic properties., International Journal of Biological Macromolecules 48 (2011) 286-291.

Matsuoka S., Cowman M.K. 2002. Equation of state for polymer solution. Polymer 43, 3447-3453.

Meyer M., Morgenstern B. 2003. Characterization of Gelatine and Acid Soluble Collagen by Size Exclusion Chromatography Coupled with Multi Angle Light Scattering (SEC-MALS). Biomacromolecules 4, 1727-1732.

Monkos K. 1996. Viscosity of bovine serum albumin aqueous solutions as a function of temperature and concentration. International Journal of Biological Macromolecules 18, 61-68.

Monkos, Karol 1997. Concentration and temperature dependence of viscosity in lysozyme aqueous solutions. Biochimica et Biophysica Acta 1339, 304-310.

Monkos K., Turczynski B. 1999. A comparative study on viscosity of human, bovine and pig IgG immunoglobulins in aqueous solutions. International Journal of Biological Macromolecules 26, 155-159.

Monkos, Karol 2000. Viscosity analysis of the temperature dependence of the solution conformation of ovalbumin. Biophysical Chemistry 85, 7-16.

Monkos, Karol 2004. On the hydrodynamics and temperature dependence of the solution conformation of human serum albumin from viscometry approach. Biochimica et Biophysica Acta 1700, 27-34.

Monkos, Karol 2005. A comparison of solution conformation and hydrodynamic properties of equine, porcine and rabbit serum albumin using viscometric measurements. Biochimica et Biophysica Acta 1748, 100-109.

Morris G.A., Patel T.R., Picout D.R., Ross-Murphy S.B., Ortega A., Garcia de la Torre J., Harding S.E. 2008. Global hydrodynamic analysis of the molecular flexibility of galactomannans. Carbohydrate Polymers 72, 356-360.

Morris, G. A.; Castile, J.; Smith, A.; Adams, G.G.; Harding, S.E. 2009. Macromolecular conformation of chitosan in dilute solution: A new global hydrodynamic approach. Carbohydrate Polymers 76, 616-621.

Nishihara T., Doty P. 1958. The Sonic Fragmentation of Collagen Macromolecules. Proceeding National Academic Sciences 44, 411-417.

Olivares M.L., Peirotti M.B., Deiber J.A. 2006. Analysis of gelatin chain aggregation in dilute aqueous solutions through viscosity data. Food Hydrocolloids 20, 1039-1049.

Ortega, A.; García de la Torre, J. 2007. Equivalent Radii and Ratios of Radii from Solution Properties as Indicators of Macromolecular Conformation, Shape, and Flexibility. Biomacromolecules 8, 2464-2475.

Ohshima H. Primary Electroviscous Effect in a Dilute Suspension of Soft Particles. Langmuir 2008, 24, 6453-6461.

Overbeek J. Th. G. Polyelectrolytes, Past, Present and Future. Pure & Appi Chem., 46 (1976) 91-101.

Palit S.R., Kar I. Polynomial expansion of log relative viscosity and its application to polymer solutions. Journal of Polymer Science Part A-1, 5, 10 (1967) 2629-2636.

Pouradier, J. & Venet, M. (1952). Contribution a l'etude de la structure des gélatines V. - Dégradation de la gélatine en solution isoélectrique, Journal de Chimie Physique et de Physico-Chimie Biologique 49, 85-91.

Ram Mohan Rao M. V., Yaseen M. Determination of intrinsic viscosity by single specific viscosity measurement. Journal of Applied Polymer Science 31, 8 (1986) 2501-2508.

Rubio-Hernandez F.J., Carrique F., Ruiz-Reina E. The primary electroviscous effect in colloidal suspensions. Advances in Colloid and Interface Science 107 (2004) 51–60.

Rubio-Hernandez F. J., Ruiz-Reina E., and Gomez-Merino A. I. Primary Electroviscous Effect with a Dynamic Stern Layer: Low κa Results. Journal of Colloid and Interface Science 226, 180–184 (2000).

Rubio-Hernandez F.J., Gomez-Merino A.I., Ruiz-Reina E., Carnero-Ruiz C. The primary electroviscous effect of polystyrene latexes. Colloids and Surfaces A: Physicochemical and Engineering Aspects 140 (1998) 295–298.

Saxena A., Sachin K., Bohidar H.B., Verma A.K. 2005. Effect of molecular weight heterogeneity on drug encapsulation efficiency of gelatin nano-particles. Colloids and Surfaces B: Biointerfaces 45, 42-48.

Scheraga, H. A., and Mandelkern L. Consideration of the Hydrodynamic Properties of Proteins. J. Am. Chem. Soc. 75 (1953) 179-184.

Schulz, G. V. and Blaschke, F. (1941) Eine Gleichung zur Berechnung der Viskositatszahl fur sehr kleine Konzentrationen. J. Prakt. Chem., 158 (1941) 130-135.

Simha, R. The Influence of Brownian Movement on the Viscosity of Solutions. J. Phys. Chem. 44, (1940) 25-34.

Solomon, O. F.; Ciută, I.Z. 1962. Détermination de la viscosité intrinsèque de solutions de polymères par une simple détermination de la viscosité. Journal of Applied Polymer Science 6, 683-686.

Staudinger, H., Die hochmolekularen organischen Verbindungen. Berlin: Julius Springer, 1932.

Sun S.F. Physical chemistry of Macromolecules. John Wiley & Sons, 2004.

Tanford, Charles. The electrostatic free energy of globular protein ions in aqueous salt solution. Journal of Physical Chemistry 59 (1955) 788-793.

Teraoka, Iwao. Polymer Solutions: An Introduction to Physical Properties. John Wiley & Sons, 2002.

van Holde K.E. Physical Biochemistry, Foundations of Modern Biochemitry Series, Prentice-Hall, 1971.

Zhao W. B. Gelatin. Polymer Data Handbook. Oxford University Press, 1999.

Part 2

Applications of Biopolymers

6

Role of Biopolymers in Green Nanotechnology

Sonal I. Thakore
Department of Chemistry, Faculty of Science, The Maharaja Sayajirao University of Baroda, Vadodara, Gujarat India

1. Introduction

1.1 Green nanotechnology

There is currently considerable interest in processing polymeric composite materials filled with nanosized rigid particles. This class of material called "nanocomposites" describes two-phase materials where one of the phases has at least one dimension lower than 100 nm [13]. Because the building blocks of nanocomposites are of nanoscale, they have an enormous interface area. Due to this there are a lot of interfaces between two intermixed phases compared to usual microcomposites. In addition to this, the mean distance between the particles is also smaller due to their small size which favors filler-filler interactions [14]. Nanomaterials not only include metallic, bimetallic and metal oxide but also polymeric nanoparticles as well as advanced materials like carbon nanotubes and dendrimers, However considering environmetal hazards, research has been focused on various means which form the basis of green nanotechnology.

1.2 Dual role of biopolymers

Various biopolymers such as starch and cellulose have been of increased interest due to more environmentally aware consumers, increased price of crude oil and global warming. Due to various advantages like renewability, non-toxicity and biocompatibility, their **biocomposites** are used in variety of applications, like therapeutic aids, medicines, coatings, food products and packing materials. **Biocomposites** are composite materials comprising one or more phase(s) derived from a biological origin. In terms of the reinforcement, this could include plant fibers such as cotton, flax, hemp or fibers from recycled wood or waste paper, or even by-products from food crops. The manufacturing of true biocomposites demands that the matrix be made predominantly from renewable resources, although the current state of biopolymer technology dictates that synthetic thermoplastics and thermosets dominate commercial biocomposite production.

This chapter describes two important and diverse roles played by polysaccharides in development of biocomposites viz. **as reinforcing agents in polymers matrix and second as matrix for synthesis of green metal nanocomposites**

1.3 Polysaccharides as reinforcing agents in bionanocomposites

The variety of nanofillers for development of nanocomposites is restricted due to various reasons such as

- limited availability
- cost and
- tendency to aggregate which may prevents high level of dispersion

Hence polysaccharides have been viewed as a potential renewable source of nanosized reinforcement. Being naturally found in a semicrystalline state, aqueous acids can be employed to hydrolyze the amorphous sections of the polymer. As a result the crystalline sections of these polysaccharides are released, resulting in individual monocrystalline nanoparticles [13]. The concept of reinforced polymer materials with polysaccharide nanofillers has known rapid advances leading to development of a new class of materials called **Bionanocomposites**, which successfully integrates the two concepts of biocomposites and nanometer sized materials. The first part of the chapter deals with the synthesis of polysaccharide nanoparticles and their performance as reinforcing agents in bionanocomposites.

1.4 Metal-polysaccharide nanocomposites

In the past few decades, many efforts have been made in the synthesis of metal nanoparticles because of their unusual properties and potential applications in optical, electronic, catalytic, and magnetic materials. Conventional methods of their synthesis involves chemical agents like sodium borohydride and hydrazine hydrate. All these chemicals are highly reactive and pose potential environmental and biological risks. Over the past decade, increasing awareness about the environment has led researchers to focus on green synthetic approaches. Utilization of nontoxic chemicals, environmentally benign solvents and renewable materials are some of the key issues that merit important consideration in a green synthesis strategy. The later part of the chapter describes the vital role of biopolymers as a matrix for the fabrication of metal nanoparticles, eventually leading to synthesis of green metal nanocomposites.

2. Bionanocomposites of natural rubber with polysaccharides

2.1 Composites and fillers

Composites consist of two (or more) distinct constituents or phases, which when combined result in a material with entirely different properties from those of the individual components. Typically, a manmade composite would consist of a reinforcement phase of stiff, strong material, embedded in a continuous matrix phase. This reinforcing phase is generally termed as filler. The matrix holds the fillers together, transfers applied loads to those fillers and protects them from mechanical damage and other environmental factors. The matrix in most common traditional composites comprises either of a thermoplastic or thermoset polymer [1].

The properties of composites are dictated by the intrinsic properties of the constituents. The two most important factors that affect the performance of composites are

i. **Filler architecture :** Filler geometry to some extent is influenced by the way in which the fillers are extracted and processed. The aspect ratio (the ratio of filler length to diameter) is an important characteristic for any material to be used as filler. Thus fillers with high aspect ratio are long and thin, while those with low aspect ratio are shorter in length and broader in the transverse direction [1].
ii. **The filler-matrix interface :** The interface between filler and matrix is also crucial in terms of composite performance. The interface serves to transfer externally applied loads to the reinforcement via shear stresses over the interface. Controlling the 'strength' of the interface is very important. Clearly, good bonding is essential if stresses are to be adequately transferred to the reinforcement and hence provide a true reinforcing function [1].

2.2 Polysaccharides as fillers

The concept of reinforced polymer materials with polysaccharides has known rapid advances in the last decade due to following advantages

- renewable nature
- availability
- diversity of the sources
- low density
- low energy consumption
- low cost
- high mechanical properties
- comparatively easy processability due to their nonabrasive nature, which allows high filling levels, resulting in significant cost savings
- relatively reactive surface, which can be used for grafting specific groups.

Owing to the above reasons, some biopolymers have been used directly or after modification, to replace the conventional fillers leading to partial biodegradation. A number of studies have been carried out with an aim to maximize the proportion of renewable resources used while retaining acceptable material properties.

For instance, natural rubber (NR) which is freely and naturally available at low expenses is widely used in industries. The unique mechanical properties of NR result from its highly stereoregular microstructure, the rotational freedom of the α-methylene C-C bonds and from the entanglements resulting from the high molecular weight which contributes to its high elasticity. The properties of NR can be tailored by the addition of fillers of varying surface chemistry and aggregate size/aspect ratio to suit the application concerned.

Carbon black (CB) manufactured by burning oil or natural gas in controlled conditions is the most important reinforcing agent [2]. But because of its origin from petroleum, it causes pollution and gives black color to the rubber. Hence research was focused on the development of other reinforcing agents to replace carbon black in rubber compounds. Silica and other types of fillers have a weaker polymer-filler interaction and are extensively used where a high degree of reinforcement is not essential [3,4]. The use of clay minerals such as montmorillonite (MMT) [8] and organoclays [6] have also been used as fillers in natural rubber. However, these inorganic fillers have a much reduced affinity toward the elastomer components and thus tend to form large aggregates, leading to drawbacks in processing and poor reinforcement. Hence, besides inorganic sources biomass is increasingly being looked upon as another potential source [7]. A variety of fibers like sisal [8], bamboo [9], short coir fibres [10] etc. have been used to prepare biocomposites of NR.

2.3 Biocomposites of natural rubber with polysaccharide fillers

Polysaccharides such as starch and cellulose have been used as reinforcing agents in natural rubber. Both solution blending and dry mixing methods have been employed for the development of biocomposites and the performance compared with the composites obtained using carbon black. Dry mixing method is more economically viable and environment friendly.

The results of mechanical properties (presented later in this section) showed that up to 20 phr, the biofillers showed superior strength and elongation behavior than CB, cellulose being the best. After 30 phr the mechanical properties of biocomposites deteriorated because of the poor compatibility of hydrophilic biopolymers with hydrophobic natural rubber(results not shown) . While increasing quantity of CB in composites leads to constant increase in the mechanical properties. Scanning electron micrographs revealed presence of polymer-filler adhesion in case of biocomposites at 20 phr.

Thermal stability is a crucial factor when polysaccharides are used as reinforcing agents because they suffer from inferior thermal properties compared to inorganic fillers. However, thermogravimetric analysis (TGA) of biocomposites suggested that the degradation temperatures of biocomposites are in close proximity with those of carbon black composites (Table-1).

Fillers	Degradation temperature (°C) for % wt. loss			Tg °C
	1	10	50	
Starch	128	303	351	-63.14
Cellulose	137	288	325	-62.96
Carbon black	114	304	364	-62.24

Table 1. Thermogravimetry data and Tg values of NR composites containing various fillers

Another major drawback of polysaccharides is their hydrophilic nature leading to low degrees of adhesion between fiber and matrix [11]. Moisture absorption takes place by three types of mechanisms namely diffusion, capillarity, and transport via micro cracks [2]. Among the three, diffusion is considered to be the major mechanism. Water absorption largely depends on the water-soluble or hygroscopic components embedded in the matrix, which acts as a semipermeable membrane. While, fiber/matrix adhesion and fiber architecture also affect the moisture absorption. The results of the water sorption experiment showed an interesting trend. The extent of water uptake was not very significant and also did not increase linearly with amount of filler (Table-2).

The unusual thermal stability and water uptake properties are due to the formation of a three-dimensional network in polysaccharides at high processing temperatures [12].

Owing to the general incompatibility between natural fillers and most polymers, methods of promoting adhesion are frequently needed. Several approaches have been explored, including chemical modification of the filler prior to composite manufacture and introducing compatibilizing agents to the polymer/filler during processing.

Fillers	mole % uptake		
	10 phr	20 phr	30 phr
Starch	1.2359	1.0382	1.0429
Cellulose	1.2065	1.055	1.0324
Carbon black	1.1938	1.0272	1.0108

Table 2. Water sorption of NR composites containing various fillers

The main hurdle for the use of starch as a reinforcing phase is its hydrophillicity leading to incompatibility with polymer matrix and poor dispersion causing phase separation. Two strategies have been adopted to improve the performance of polysaccharides.

1. Reduction in particle size of the biopolymers to obtain nanofillers which can result in more uniform distribution within the polymer matrix.
2. Organic modification of the nanofillers to obtain hydrophobic derivatives having improved compatibility with the polymer.

2.4 Polysaccharide nanoparticles and their organic modification

Native starch granules contain more or less concentric "growth rings" that are readily visible by optical or electronic microscopy as seen in Scheme-1 [15]. Acid treatment is needed to reveal the concentric lamellar structure of starch granules. The purpose of this treatment is to dissolve away regions of low lateral order so that the water-insoluble, highly crystalline residue may be converted into a stable suspension by subsequent vigorous mechanical shearing action. Dufresne et. al [16] obtained starch nanocrystals (SN) from waxy corn starch granules as per the conditions optimized as shown in Scheme-2. The insoluble hydrolyzed residue obtained from waxy corn was reported to be composed of crystalline nanoplatelets around 5-7 nm thick with a length of 20-40 nm and a width of 15-30 nm [16].

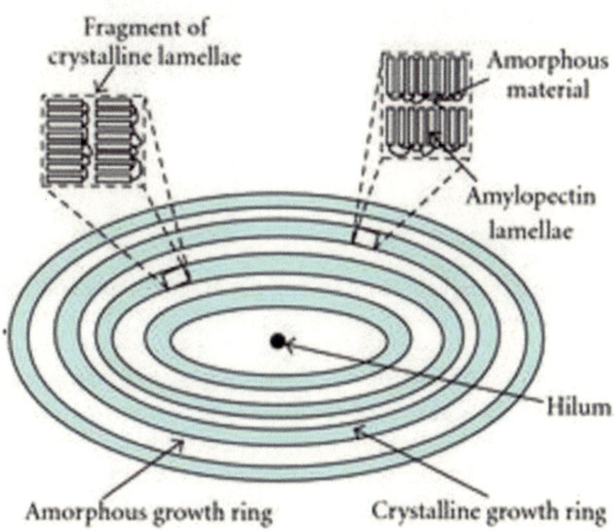

Scheme 1. Structure of starch granule

A similar procedure was adopted for synthesis of nanoparticles of cellulose (CelNPs). The polysaccharide nanoparticles were derivatised under ambient conditions to obtain nanosized hydrophobic derivatives. The challenge here is to maintain the nanosize even after derivatisation due to which less vigorous conditions are preferred. A schematic synthesis of acetyl and isocyanate modified derivatives of starch nanoparticles (SNPs) is shown in scheme 3. The organic modification was confirmed from X-ray diffraction (XRD) pattern which revealed that A- style crystallinity of starch nanoparticles (SNPs) was destroyed and new peaks emerged on derivatisation. FT-IR spectra of acetylated derivatives however showed the presence of peak at 3400 cm^{-1} due to -OH stretching indicating that the substitution is not complete.

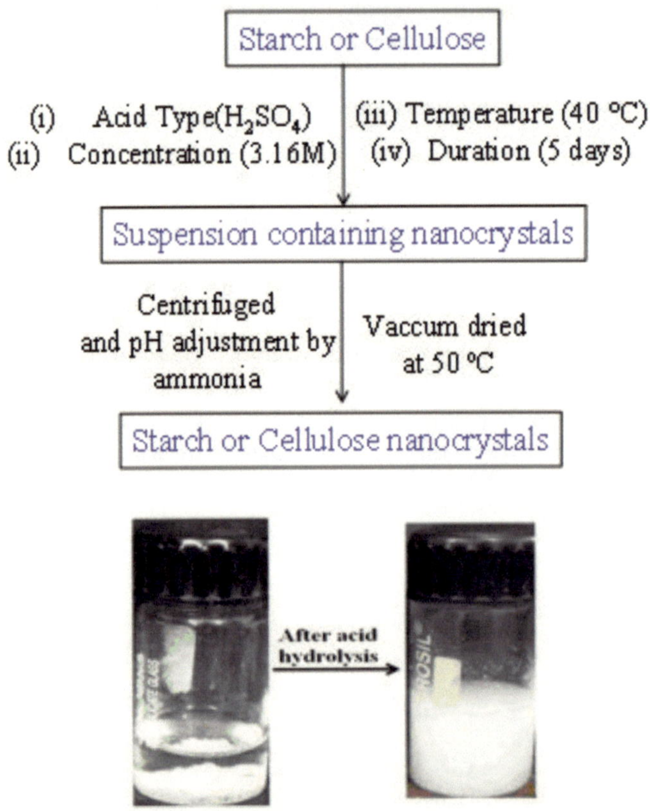

Scheme 2. Various conditions and steps involving general synthesis of polysaccharide nanoparticles

In the transmission electron microscopy (TEM) images, the starch nanoplatelets (SNPs) are believed to aggregate as a result of hydrogen bond interactions due to the surface hydroxyl groups [13] (Fig. 1A). Blocking these interactions by relatively large molecular weight molecules obviously improves the individualization of the nanoparticles. The acetylated starch and cellulose nanoparticles (SAcNPs and CelAcNPs) appeared more individualized and monodispersed than their unmodified counterparts with a size of about 50 nm (Fig. 1B & C).

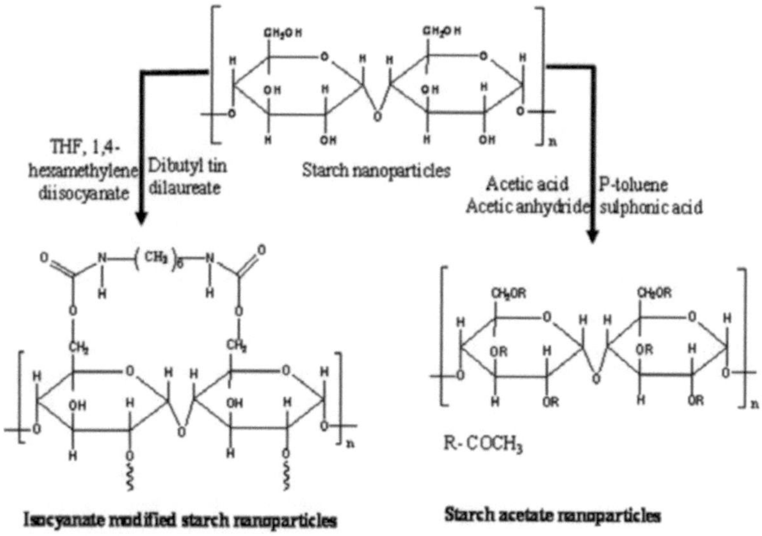

Scheme 3. Organic modification of starch nanoparticles

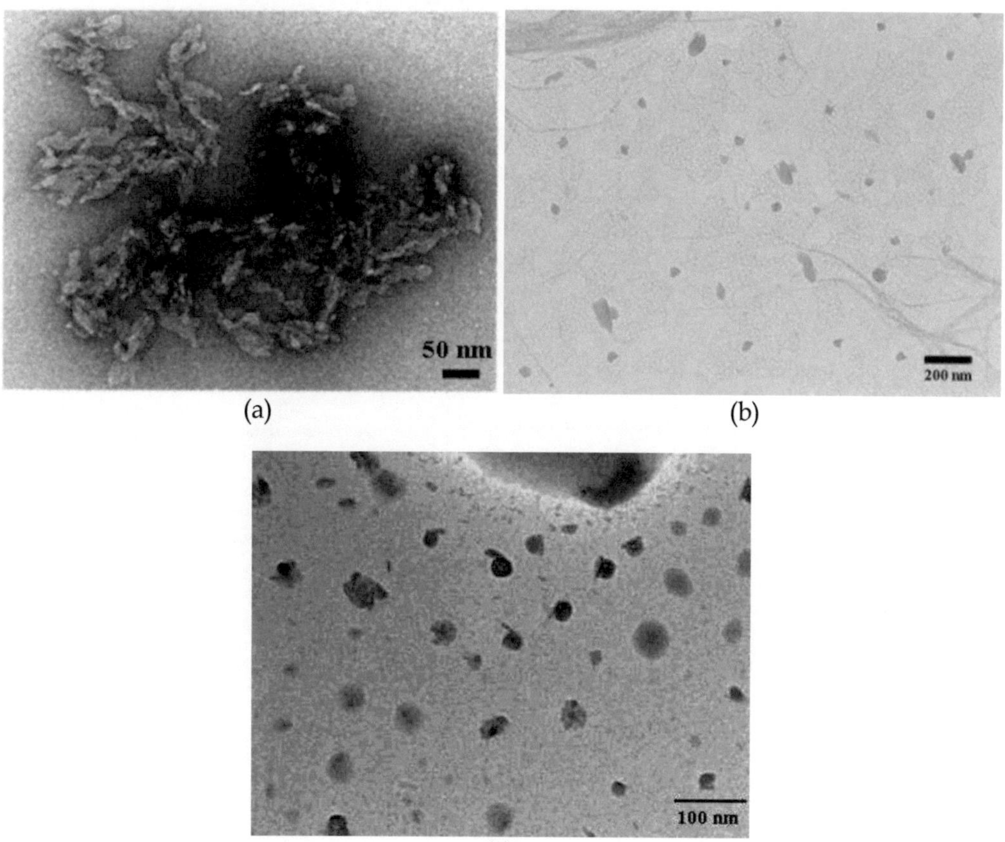

Fig. 1. TEM images of (A) SNPs taken from reference no. 16 (B) SAcNPs and (C) CelAcNPs

2.5 Polysaccharide nanoparticles as reinforcing agents

The unmodified and acetylated nanoparticles were used to develop nanobiocomposites of NR containing 40 phr polysaccharide content and their performance was evaluated. The results obtained from mechanical properties showed that for each measurement, the strain was macroscopically homogenous and uniform along the sample until it breaks. The lack of any necking phenomenon confirms the homogenous nature of these nanocomposites. The stress continuously increases with strain and the amount of fillers. Whereas Angellier et al. [17] observed a decrease in elongation as the amount of filler increases. The initial high stress is due to the reinforcement of the rubber with nanofillers [18]. As the strain increases, stress induced crystallization comes into role, which increases proportionally along with strain. The dispersion of nanofillers leads to an efficient reinforcement, which leads to improved stiffness. The unmodified nanoparticles imparted lower strength due to hydrophilic nature resulting in poor adhesion with NR. As a result the stress transfer from the matrix to the filler is poor and the mechanical properties of nanoparticles are not fully utilized. As the amount of nanofillers increases the tensile strength (T.S.) continued to increase as expected (figure 2). In case of CB composites the initial lower T. S. value rapidly increases from 10 to 30 phr loading but remains

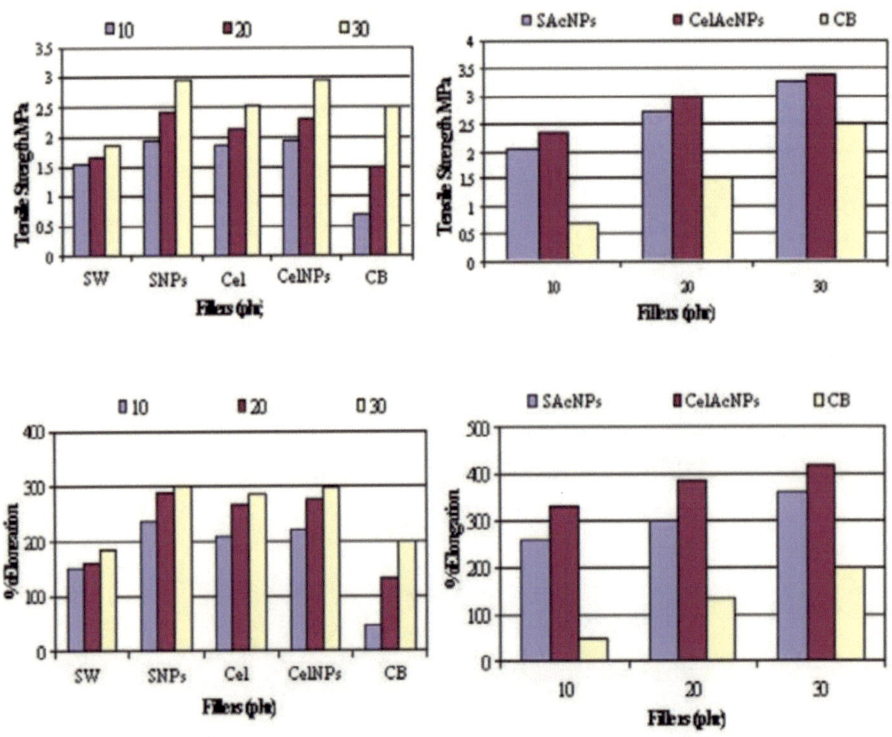

SW- waxy corn starch, SWNPs- starch nanoparticles, Cel- microcrystalline cellulose, CelNPs- cellulose nanoparticles, CB- carbon black, SWAcNPs- waxy starch acetate nanoparticles, CelAcNPs: cellulose acetate nanoparticles.

Fig. 2. Summary of mechanical properties of composites containing various fillers

lower than nanocomposites at all levels. Nanocomposites of acetylated polysaccharide nanoparticles showed not only higher strength but also greater elongation due to the improved dispersion of filler and better compatibility with NR owing to its hydrophobic nature and small particle size. The behavior is consistent with respect to all the nanofillers. This indicates that the nanofillers used retain the elastic property of natural rubber which is also concluded by dynamic mechanical analysis (discussed later).

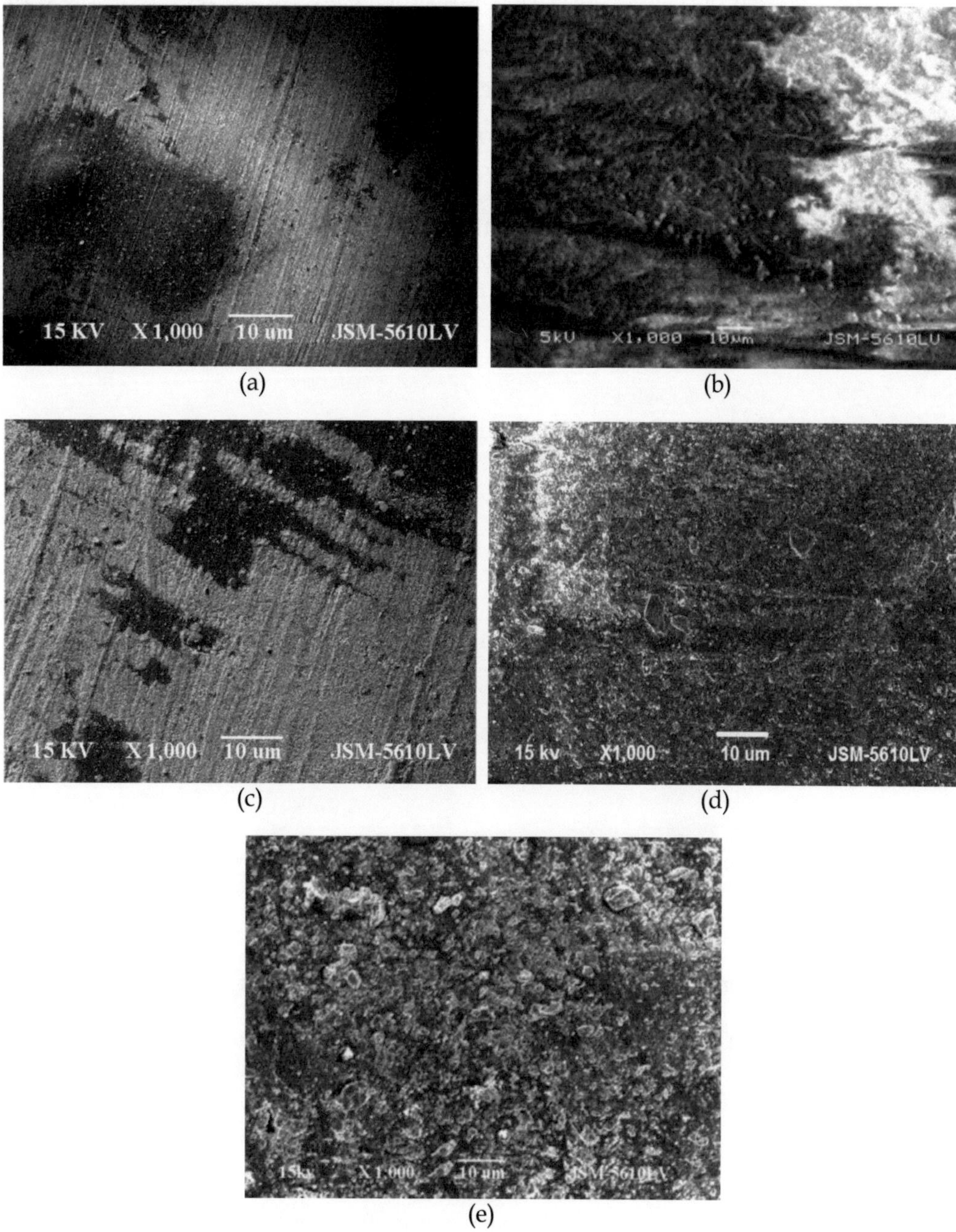

Fig. 3. SE Micrographs of NR composites at 40 phr loading of (A) SAcNPs, (B) CelAcNPs, (C) SNPs, (D) CelNPs and (E) CB

The results of the mechanical properties can be explained on the basis of morphology. The scanning electron micrographs (SEM) of fractured samples of biocomposites at 40 phr loading are shown in figure. 3. It can be seen that all the bionanofillers are well dispersed into polymer matrix without much agglomeration. This is due to the better compatibility between the modified polysaccharides nanoparticles and the NR matrix (Fig. 4A and B). While in case of unmodified polysaccharides nanoparticles the reduction in size compensates for the hydrophilic nature (Fig. 3C and D). In case of CB composites (Fig. 3E) relatively coarse, two-phase morphology is seen.

The tan δ curve (Fig. 4) at 1 Hz of nanocomposites showed a broad relaxation process from -80 °C to -10 °C. This may be due to the relaxation of rubber fraction confined inside the layers. The reduction in the tan δ maxima suggests a strong adhesion between NR and modified starch nanoparticles. Sliding along the exfoliated interlayer is suppressed [13]. In addition, chain slipping at the outer surfaces of the aggregates is also likely to be hampered. Therefore the loss maximum is smaller in case of the nanocomposite system with the strongest filler matrix coupling.

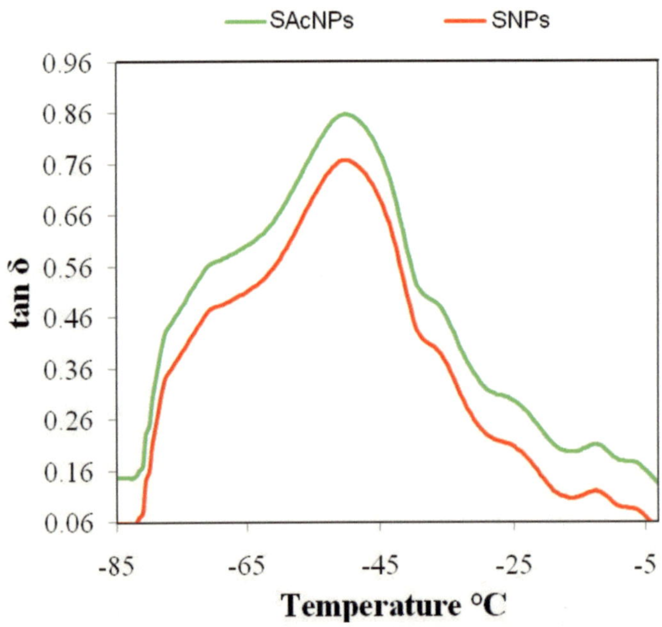

Fig. 4. Effect of nanocomposites on mechanical loss factor tan δ vs temperature

Due to the higher performance of cellulosic fillers, highly filled bionanocomposites were developed by incorporation of upto 60 phr (37.5 weight %) cellulosic nanofillers. Cellulose acetate nanoparticles (CelAcNPs) exhibit excellent reinforcing ability upto 50 phr (33.3 weight %) preserving the elastic behavior of nanocomposites (Fig. 5). At still higher loading the performance of carbon black was observed to be much superior to the cellulosic fillers. The results of mechanical properties also indicate that the combined effect of size reduction and organic modification drastically improves the performance of cellulose. Although, a drop in mechanical strength was observed above 50 phr, the cellulosic fillers proved to be potential reinforcing agents even at higher loadings.

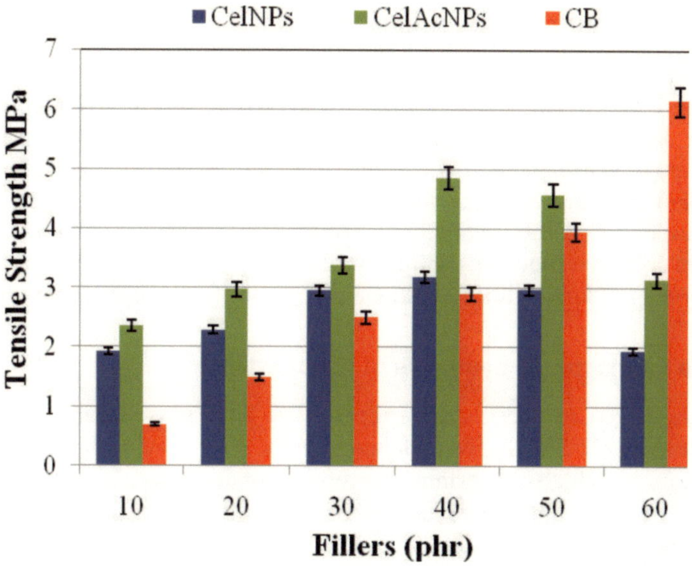

Fig. 5. Tensile strength of highly filled bionanocomposites containing cellulosic fillers

The results of thermal properties support the observation that increased hydrophobicity and reduced particle size of nanofillers imparts rigidity and strength to the network.

Fillers	% mole Uptake					
	Water			Toluene		
	10 phr	30 phr	50 phr	10 phr	30 phr	50 phr
SAcNPs	1.13	0.26	0.24	2.76	2.12	1.87
SNPs	3.10	2.44	0.61	1.22	0.34	0.31
CelNPs	1.39	0.79	0.38	2.79	2.36	1.82
CelAcNPs	1.12	0.51	0.21	3.12	2.56	2.02
CB	1.15	0.56	0.31	2.77	1.97	1.42

Table 3. Water and toluene sorption of NR composites containing various fillers

The interaction between polymer matrix and filler leads to the formation of a bound polymer in close proximity to the reinforcing filler, which restricts the solvent uptake [13]. The composites containing acetylated cellulose fillers exhibited higher uptake of toluene compared to water in accordance with their hydrophobic nature.

Overall the results led to the conclusion that acetylated nanoparticles of both starch and cellulose offer potential eco-friendly substitutes for the conventional filler carbon black upto 40 phr. They imparted high mechanical strength and elasticity with minimum compromise in themal stability and moisture absorption of the resulting bionanocomposites. Cellulose acetate nanoparticles afforded effective reinforcement even upto loadings as high as 50 phr.

3. Metal-polysaccharide nanocomposites

Polymers like polyvinyl pyrollidone (PVP) [19] and polyacrylamide [20] have been successfully used as the stabilizing agents for synthesis of various metal nanoparticles. Metal polymer nanocomposites exhibit interesting optical and conducting properties [21]. But in the present scenario, increasing awareness about the environment has led researchers to focus on 'green chemistry'. Biopolymers represent suitable matrices for the preparation of metal nanocomposites being generally low cost materials characterized by an easy processing. The inherent biocompatibility and biodegradability of these polymers enable their use in a variety of applications in biotechnology and in environmental protection.

3.1 Starch as a matrix for the synthesis of nanoparticles

Bioploymers like starch and chitosan are reported to have played the role of stabilizers for silver nanoparticles [22]. Starch, for instance, adopts right-handed helical conformation in aqueous solution, in which the extensive number of hydroxyl groups can facilitate the complexation of metal ions to the molecular matrix [23]. The concept of green nanoparticles preparation using starch as stabilizer was first reported by Raveendran et al. [23] where glucose was used as the reducing agent. The synthesis of stable Cu nanoparticles is of particular interest because of its tendency to get oxidised in an aqueous system [24]. Schematic representation of formation of copper nanoparticles in aqueous matrix of biopolymer such as starch can be seen in Fig.6. A typical TEM image of the Cu nanoparticles

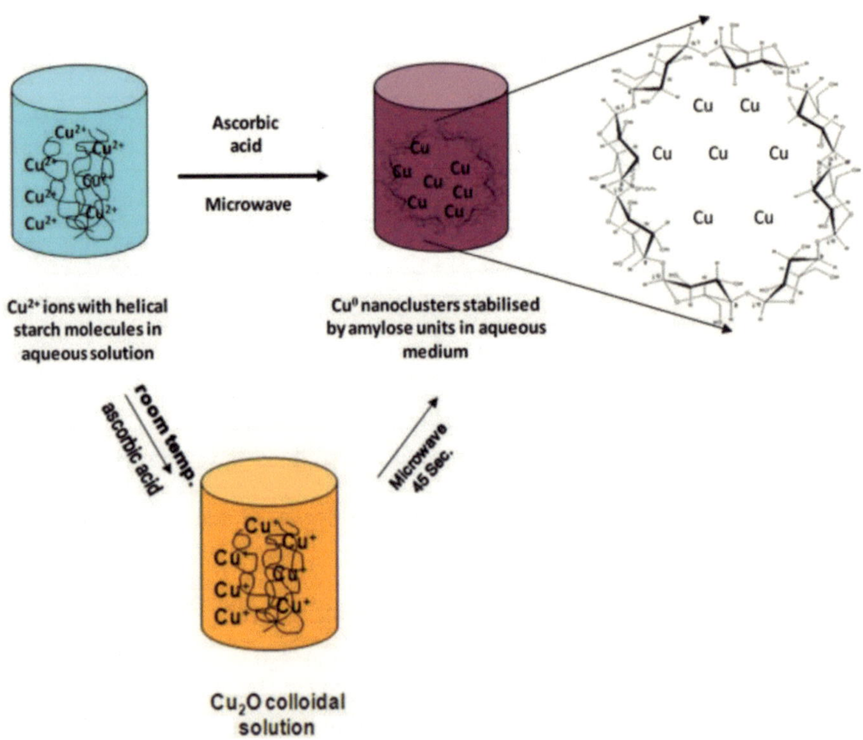

Fig. 6. Schematic representation of synthesis of starch capped copper nanoparticles under microwave conditions

(CuNPs) in Fig. 7 shows the monodisperse and uniformly distributed spherical particles of 10±5 nm diameter. The solution containing nanoparticles of silver was found to be transparent and stable for 6 months with no significant change in the surface plasmon and average particle size. However, in the absence of starch, the nanoparticles formed were observed to be immediately aggregated into black precipitate. The hydroxyl groups of the starch polymer act as passivation contacts for the stabilization of the metallic nanoparticles in the aqueous solution. The method can be extended for synthesis of various other metallic and bimetallic particles as well.

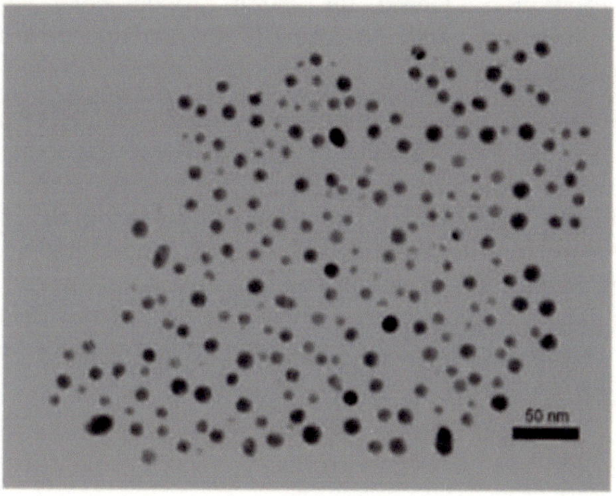

Fig. 7. TEM image of starch capped copper nanoparticles

3.2 Bimetallic nanoparticles

Bimetallic nanoparticles, either as alloys or as core-shell structures, exhibit unique electronic, optical and catalytic properties compared to pure metallic nanoparticles [24]. Cu–Ag alloy nanoparticles were obtained through the simultaneous reduction of copper and silver ions again in aqueous starch matrix. The optical properties of these alloy nanoparticles vary with their composition, which is seen from the digital photographs in Fig. 8. The formation of alloy was confirmed by single SP maxima which varied depending on the composition of the alloy.

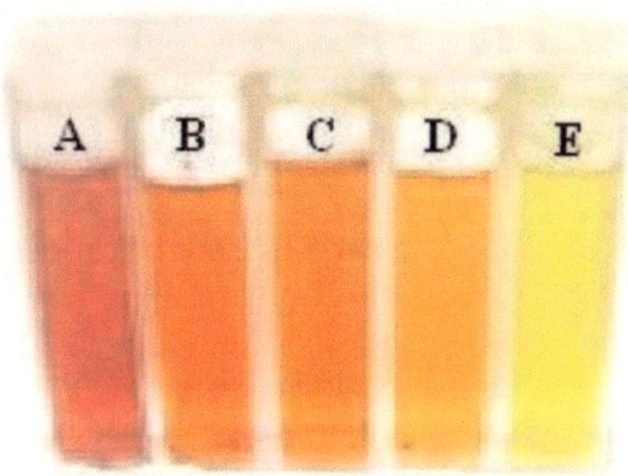

Fig. 8. Digital photograph of starch capped metallic and bimetallic nanoparticles in aqueous medium. A: Cu; B–D: Cu–Ag; E: Ag.

3.3 Biological applications of starch capped nanoparticles

Silver nanoparticles exhibits strong cytotoxicity towards a broad range of microorganisms, and its use as an antibacterial agent is well known [25]. for biological applications. The effective biocidal concentration of silver nanoparticles (AgNPs) is at a nanomolar level in contrast to a micromolar level of silver ions [26]. On the other hand, only a few studies have reported the antibacterial properties of copper nanoparticles. Neverthless, copper nanoparticles have a significant promise as bactericidal agent [27,28]. The biopolymer matrix offers additional advantages like water solubility and biocompatibility necessary for use in biological applications. Hence, the starch capped water soluble nanoparticles exhibited excellent antibacterial activity against both gram positive and gram negative bacteria at a very low concentration.

The mode of action of starch capped copper nanoparticles (SCuNPs) was compared with that of the well-known antibiotic amphicillin (Fig. 9). There was a drastic decrease in the optical density of compounds containing SCuNPs and ampicillin, ultimately reaching almost zero suggesting that there were no more bacteria present in the culture. Ampicillin at a concentration of 100 µg/ml has the ability to lyse *E.coli* almost immediately [29]. The same effect was produced by SCuNPs at 365 ng/ml concentration. The cell lysis occurs at the expense of the fact that at the point of cell division there occurs a deformation of the cell envelope. The decrease in optical density is possibly associated with the cell-envelope deformation occurring at the point of cell division [30].

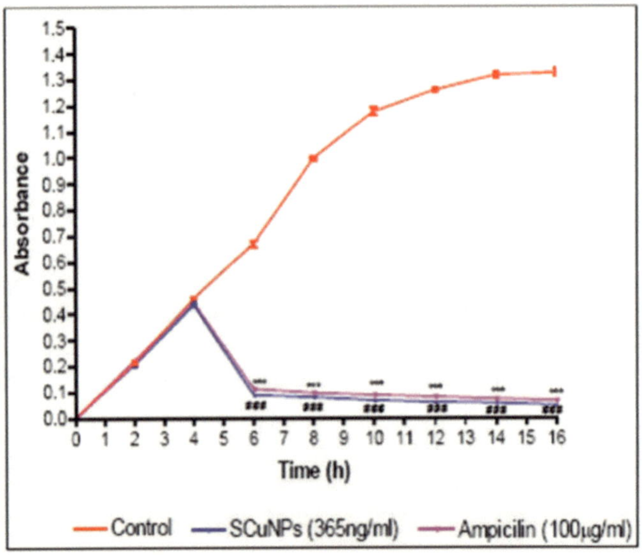

Fig. 9. Comparison of mode of action against *E.coli:* starch capped copper nanoparticles vs. amphicilin

The presence of nanoparticles suspended within the starch matrix would ensure continuous release of ions into the nutrient media. Copper ions released by the nanoparticles may attach to the negatively charged bacterial cell wall and rupture it, thereby leading to protein denaturation and cell death [31]. The attachment of both ions and nanoparticles to the cell wall caused accumulation of envelope protein precursors, which resulted in dissipation of the proton motive force. Thus it can be concluded that SCuNPs show lytic mode of action against *E.coli* similar to Ampicillin.

The biological impact of starch capped copper nanoparticles on mouse embryonic fibroblast (3T3L1) cells (*in vitro*) was also evaluated by various parameters. More than 85 % of the 3T3L1cells were found to be viable, even after 20 hours time exposure which implies minimum impact on cell viability and morphology. The study demonstrates dose dependent cytotoxic potential of SCuNPs, that is non cytotoxic in the nanogram dose and moderately cytotoxic in the microgram doses (Fig. 10). Comparison of SCuNPs with Cu^{2+} ions and uncapped copper nanoparticles (UCuNPs) revealed that, ions are more cytotoxic than SCuNPs. This observation supports the theory of slow release of ions from starch coated nanoparticles.

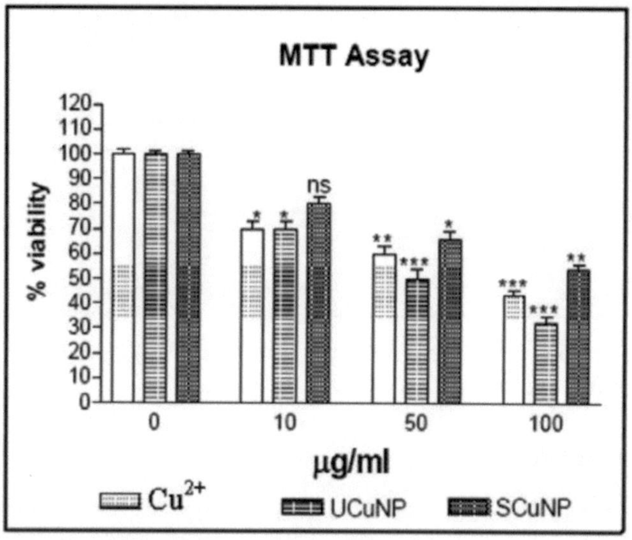

Fig. 10. Effect of starch capped copper nanoparticles on cell viability (MTT assay) in case of mouse embryonic fibroblast (3T3L1).

Visual imaging of cell population (*in vitro*) using low signal-to-noise ratio phase contrast microscopy can enable systematic monitoring measurements of cell quality, development and apoptosis. In the present study, microscopic evaluations as seen in Fig. 11 did not reveal any significant alteration in cellular morphology upto 1000 ng/ml.

Water soluble starch capped nanoparticles proved to be efficient non-cytotoxic bactericidal agents at nanomolar concentrations. The investigation also suggested that starch capped CuNPs have great potential for use in biomedical applications such as cellular imaging or photothermal therapy.

The cytotoxicity potential of the biopolymer capped nanoparticles was evaluated using various parameters like MTT cell viability assay and extracellular lactate dehydrogenase (LDH) release in human lung cancer cells. Other parameters that determine the oxidative stress viz., reactive oxygen species (ROS) generation, intracellular reduced glutathione (GSH), malondialdehyde (MDA), superoxide generation and acridine orange/ethidium bromide staining were also investigated. The study led to the conclusion that copper nanoparticles were toxic to human lung cancer cells in a dose dependent manner. Further work to use the biopolymer capped copper nanoparticles for targeted in vivo delivery to cancer cells is a challenging task for researchers.

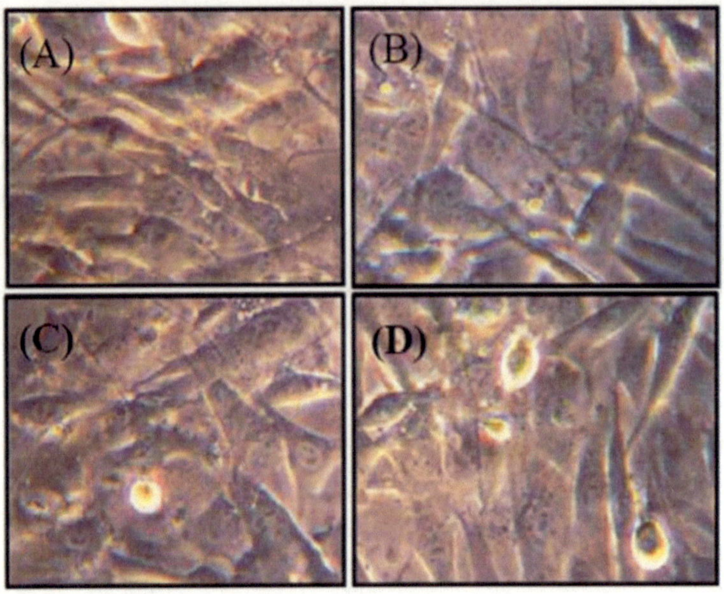

Fig. 11. Photomicrographs of mouse embryonic fibroblast (3T3L1) cells (A) untreated, treated with (B) control (C) 500ng CuNPs, (D) 1000ng CuNPs.

3.4 Starch as morphology directing agent

Silver nanoparticles have been prepared by using disaccharide sucrose and polysaccharides; amylose rich soluble starch and amylopectin-rich waxy corn starch so that the carbohydrates act as both reducing and stabilizing agent. The reduction was fastest in case of sucrose followed by waxy starch and soluble starch; while stability of nanoparticles followed the order soluble starch > waxy corn starch > sucrose [32]. Thus the size and molecular weight were important for stabilization of the nanoparticles. Also in case of starch the reduction occurred only after heating at 80 ^{o}C for 4 hrs. This is required because the more hydrolyzed the polysaccharide, the better is its ability to act as reducing agent. As such sucrose cannot act as a reducing agent but in presence of microwave sucrose gets hydrolyzed and the hydrolyzed products then reduce the metal. The time taken for reduction is also very short (30 sec).

The nanoparticles exhibited interesting morphology when synthesized under hydrothermal conditions depending upon the content of amylopectin in carbohydrates. The nanoparticles formed in waxy corn starch matrix were observed to have self-assembled into wire-like structures, (Fig. 12). Although the exact mechanism of the formation of the nanostructures is difficult to know, it was proposed that that the chain-shaped structure of starch could serve as a directing template for the growth of silver nanoparticles [32]. Starch is composed of a linear component, amylose and a branched component, amylopectin, The branching is due to 1, 6 acetal linkage in amylopectin which is absent in amylose. It is assumed that as a result of the bond angles in the alpha acetal linkage, amylose forms a spiral structure which helps in stabilization. While branched polymer might act as a morphology directing agent, facilitating the growth of silver nanowires.

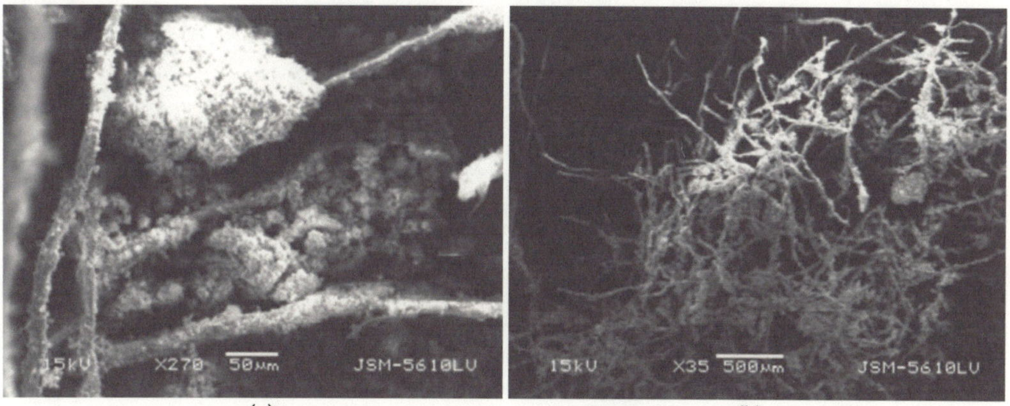

(a) (b)

Fig. 12. SEM images of Ag nanostructures in starch matrix (A) 270 x and (B) 35 x

Metal oxide nanostructures, as promising materials, have attracted much attention because of their extraordinary properties in different fields of optics, optoelectronics, catalysts, biosensors and so on. A simple and eco-friendly method was developed for the preparation of well-controlled, uniform Cu_2O dendrites using only starch as stabilizing, reducing as well as morphology directing agent in an aqueous solution by hydrothermal route. The digital photograph in Fig. 13 indicates the colour changes that occur as the reaction proceeds in a sealed pyrex tube. At 180 ^{0}C, as the time increased the self assembling of the nanoparticles into distinct continuous dendrites is clearly visible in figure 14A and B. The schematic diagram in figure 14C depicts the growth of three dimensional dendrites over the primary structure [33].

Fig. 13. Digital photograph showing various stages of the hydrothermal reduction of reaction of Cu_2O in a sealed tube.

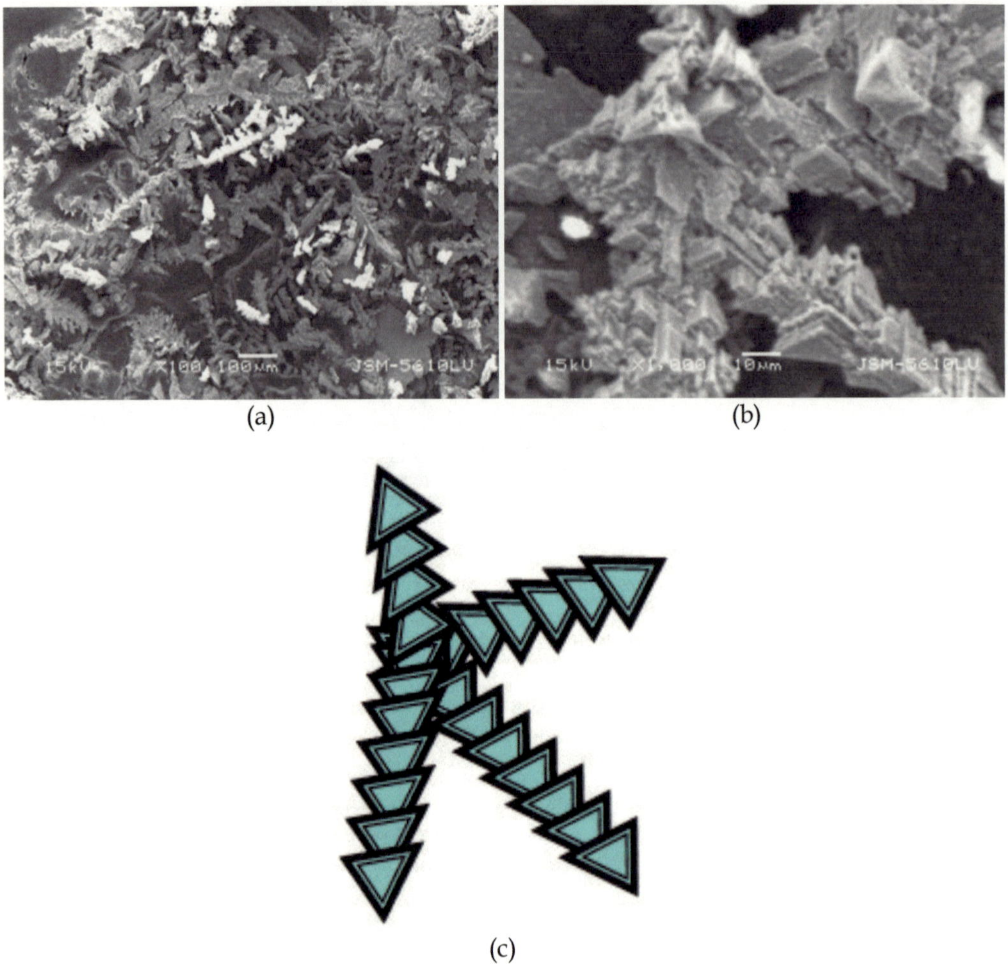

Fig. 14. SEM images of cuprous oxide nanostructures (A) 100 x, (B) 1,000 x, and (C) Schematic illustration of the dendrite structure formation process.

3.5 Metal-polysaccharide nanocomposites

Noble metal nanoparticles dispersed in insulating matrices have attracted the interest of many researchers from both applied and theoretical points of view [34]. The incorporation of metallic nanoparticles into easily processable polymer matrices offers a pathway for better exploitation of their characteristic optical, electronic and catalytic properties. On the other hand, the host polymers can influence the growth and spatial arrangement of the nanoparticles during the in situ synthesis, which makes them convenient templates for the preparation of nanoparticles of different morphologies. Furthermore, by selecting the polymer with certain favorable properties such as biocompatibility [35], conductivity [36] or photoluminescence [37], it is possible to obtain the nanocomposite materials for various technological purposes.

There are several reports of Ag nanocomposites with conducting polymers like polyaniline [38] and polypyrrole [39]. However, electrical conducting properties of green metal - starch nanocomposites would be of potential biomedical applications. Lyophilisation or vacuum drying of the aqueous solution of starch capped nanoparticles results in metal-starch nanocomposites with high thermal stability and interesting electrical conductivity. The TG Analysis in Fig. 15 shows that the thermal stability of starch can be improved by incorporation of AgNPs [40] making it suitable for conductivity measurements even at high temperature. As the concentration of silver goes on increasing the thermal stability goes on increasing.

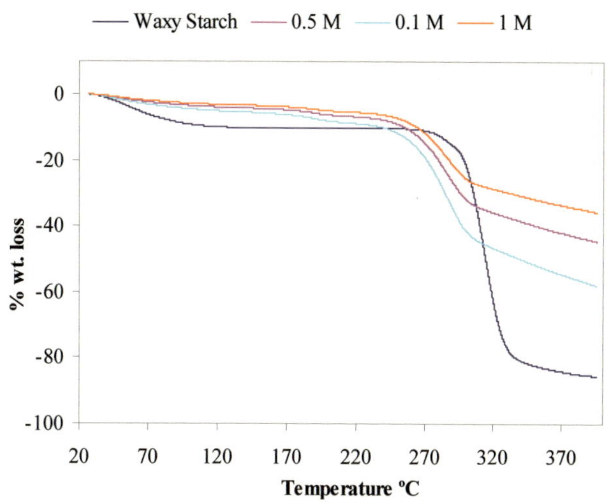

Fig. 15. TGA curve of pure waxy starch and Ag-waxy starch nanocomposites with different concentrations of silver ion precursor (moles of AgNO3)

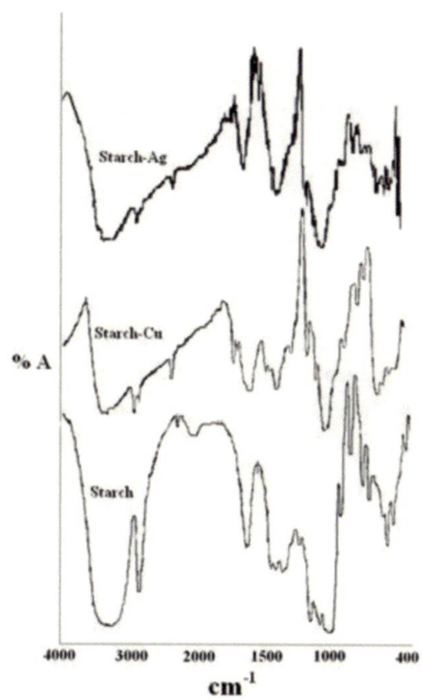

Fig. 16. FT-IR spectra of pure waxy starch and waxy starch metal nanocomposites

The FT-IR spectra of pure of nanocomposites exhibited shift in the absorption frequencies due to the interaction of metal with starch (Fig. 16).

Electrical conductivity measurements revealed that ionic conductivity of Ag-starch nanocomposites increased as a function of temperature (Fig.17) which is an indication of a thermally activated conduction mechanism [40]. This behavior is attributed to increase of charge carrier (Ag^+ ions) energy with rise in temperature. It is also found to increase with increasing concentration of Ag ion precursor (inset of Fig.17). This potentiality can lead to development of novel biosensors for biotechnological applications such as DNA detection.

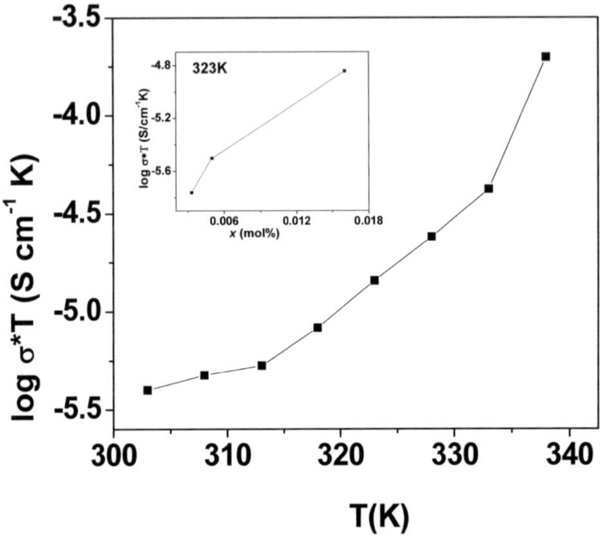

Fig. 17. Variation of conductivity of Ag-starch nanocomposites with temperature. Inset shows variation with concentration of silver nanoparticles.

4. Conclusions

Biopolymers have diverse roles to play in the advancement of green nanotechnology. Nanosized derivatives of polysaccharides like starch and cellulose can be synthesized in bulk and can be used for the development of bionanocomposites. They can be promising substitutes of environment pollutant carbon black for reinforcement of rubbers even at higher loadings (upto 50phr) via commercially viable process. The combined effect of size reduction and organic modification improves filler–matrix adhesion and in turn the performance of polysaccharides. The study opens up a new and green alternative for reinforcement of rubbers.

Starch can also be useful for completely green synthesis of various metallic nanoparticles serving as stabilizing agent as well as reducing agent in aqueous medium. Biopolymer capping offers additional advantages like water solubility, and biocompatibility for biological applications. Water soluble starch capped nanoparticles proved to be efficient non-cytotoxic bactericidal agents at nanomolar concentrations. The investigation also suggested that starch capped CuNPs have great potential for use in biomedical applications such as cellular imaging or photothermal therapy. Biopolymers are also promising biocompatible carrier of nanoparticles for destruction of tumor/cancer cells. Further attempts to use the biopolymer capped nanoparticles for targeted in vivo delivery to cancer cells are in progress.

Some frequently used Abbreviations

Sr. No	Material	Abbreviations
1.	Waxy corn starch	SW
2.	Cellulose	Cel
3.	Starch nanoparticles	SNPs
4.	Cellulose nanoparticles	CelNPs
5.	Starch acetate nanoparticles	SAcNPs
6.	Cellulose acetate nanoparticles	CelAcNPs
7.	Carbon black	CB
8.	Copper nanoparticles	CuNPs
9.	Silver nanoparticles	AgNPs
10.	Starch capped copper nanoparticles	SCuNPs
11.	Uncapped copper nanoparticles	UCuNPs
12.	Parts per hundred parts of rubber	Phr

5. References

[1] Fowler, P. A. Hughes, J. M. Elias, R. M. J. Sci. Food. Agric. 2006, 86, 1781.
[2] Valodkar, M. Thakore, S. J. Appl. Polym. Sci. 2010 (in press).
[3] Donnet, J.-B., Rubber Chem. Technol. 1998, 71, 323.
[4] Wang, M.-J., Rubber Chem. Technol. 1998, 71, 520–589.
[5] Valadares, L. F., Leite C.A.P. and Galembeck F., Polymer, 2006, 47, 672.
[6] Hrachova, J., Komadal P. and Chodak I., J., Mater. Sci., 2007, 43, 2012.
[7] Filson, P. B., Dawson-Andoh, B. E., Bioresour Technol 2009, 100, 2259.
[8] Ranby, B.G., 1952, 11, 158.
[9] Favier, V., Chanzy, H., Cavaille, J.Y., Macromolecules 1995, 28, 6365.
[10] Dong, X.M., Kimura, T., Revol, J.F., Gray, D.G., Langmuir 1996, 12, 2076–2082
[11] Valodkar, M. Thakore, S., Int. J. Polym. Anal. Charac. 2010, 15, 1.
[12] Bendahou, A., Kaddami, H., Dufresne, A., Eur. Polym. J. 2010, 46, 609-620.
[13] Valodkar, M., Thakore, S., Carbohydr. Polym. 2011, 86, 1244.
[14] Nair, K. G., Dufresne, A., Biomacromolecules. 2003, 4, 657.
[15] Corre, D. L., Bras, J., Dufresne, A., Biomacromolecules. 2003, 4, 657.
[16] Dufresne, A., Cavaille, J. Y., Helbert, W., Macromolecules 1996, 29, 7624.
[17] Angellier, H., Molina-Boisseau, S., Dufresne, A. Macromolecules, 2005, 38, 9161.
[18] Varghese, S., Karger-Kocsis, J. Polymer. 2003, 44, 4921–4927.
[19] Pal, A. Shah, S. Devi, S. Mater. Chem. Phys. 2009, 114, 530.
[20] Pal, A. Shah, S. Devi, S. Colloid. Surf. A: Physicochem. Eng. Aspect. 2007, 302, 51.

[21] Tyurin, G. De Filpo, D. Cupelli, F. P. Nicoletta, A. Mashin, G. Chidichimo Exp. Polym. Lett. 2010, 4, 71.
[22] Merga, G. Wilson, R. Lynn, G. Milosavljevic, B. H. Meisel, D. J. Phys. Chem. C 2007, 111, 12220–12226.
[23] Raveendran, P. Fu, J, Wallen, S. L. J. Am. Chem. Soc. 2003, 125, 13940–13941.
[24] Valodkar M., Modi S., Pal A.and Thakore S., Materials Research Bulletin 46, 384, 2011
[25] Maneerung, T. Tokura, S. Rujiravanit, R. Carbohydr. Polym. 2008, 72, 43–51.
[26] Kong, H. Jang, J. Langmuir 2008, 24, 2051–2056.
[27] Ruparelia, J. P. Chatterjee, A.K. Duttagupta, S. P. Mukherji, S. Acta Biomater., 2008, 4, 707-716.
[28] Esteban-Tejeda, L. Malpartida, F. Esteban-Cubillo, A. Pecharroman, C. Moya, J. S. Nanotechnol. 2009, 20, 6-11.
[29] Guliy, O. I. Ignatov, O. V. Markina, L. N. Bunin, V. D. Ignatov, V. V. Int. J. Envi. Anal. Chem. 2005, 85, 981-992.
[30] Greenwood, D. O'Grady, O. J. Med. Microbiol. 1969, 2, 435-441.
[31] Lin, Y. E. Vidic, R. D. Stout, J. E. Mccartney, C. A. Yu, V. L. Water. Res. 1998, 32, 1997–2000.
[32] Valodkar, M. Bhadoria, A. Pohnerkar, J. Mohan, M. Thakore, S. Carbohydr. Res. 2010, 345, 1767-1773.
[33] Valodkar, M., Pal, A. Thakore, S. J. Alloy. Compd. 2011, 509, 523-528.
[34] Bozanic, D. K. Trandafilovic, L. V. Luyt, A. S. Djokovic, V. React.Funct. Polym. 2010, 70, 869-873.
[35] Liu, B. S. Huang, T.-B. Macromol. Biosci. 2008, 8, 932-940.
[36] Balamurugan, A. Ho, K. C. Chen, S. M. Synth. Met. 2009, 159, 2544-2548.
[37] Yang, Y. Zhang, S. Gautam, L. Liu, J. Dey, W. Chen, R.P. Mason, C.A. Serrano, K.A. Schug, L. Tang, Proc. Natl. Acad. Sci. USA 2009, 106, 11818-11823.
[38] Shengyu, J., Shangxi, X., Lianxiang, Y., Yan, W.; Zhao, C. Mater. Lett. 2007, 61, 2794–2797.
[39] Jing, S. Xing, S. Yu, L. Zhao, C. Mater. Lett. 2007, 61, 4528– 4530.
[40] Valodkar, M. Sharma, P. Kanchan, D. K. Thakore, S. Int. J. Green Nanotechnol. 2, 2010, 10-19.

7

Stable Isotope Applications in Bone Collagen with Emphasis on Deuterium/Hydrogen Ratios

Katarina Topalov, Arndt Schimmelmann, P. David Polly and Peter E. Sauer
Department of Geological Sciences, Indiana University, Bloomington
USA

1. Introduction

The broad scope of isotopic applications of bone collagen ranges from modern ecological and physiological investigations to learning about living conditions of animals in the past (Ambrose & DeNiro, 1989; Chisholm et al., 1982; DeNiro & Epstein 1978, 1981; DeNiro & Weiner, 1988; Hare et al., 1991; Hedges & Law, 1989; Leyden et al., 2006; Lis et al., 2008; Reynard & Hedges, 2008). Bone collagen represents one of the best preserved proteins in fossilized animal remains that in some cases has been chemically preserved for up to 120,000 years (Bocherens et al., 1999). Collagen is protected from degradation by being encapsulated into the bone mineral bioapatite. In comparison to other biopolymers found in the archaeological record, bone collagen is relatively abundant, easily extracted, and offers a long-term record of the life of humans and animals (Collins et al., 2002). Stable isotope ratios in the living biomass directly relate to the stable isotope ratios of life-supporting substrates in the environment. Animals integrate stable isotopes into their biomass through the air we breathe, water we drink and food we eat. Despite the apparent permanence of mineral constituents of bone, the organic components undergo constant turnover through the life of a vertebrate and its collagen content represents a 'running average' of months, years, or a lifetime, depending on the specific bone and the animal. Bone collagen thus offers a valuable insight into the environmental conditions during part of or the entire life of an animal (Lee-Thorp, 2008; Tuross et al., 2008). Carbon and nitrogen stable isotopes of bone collagen have been in broad use in determining diets, paleodiets, trophic levels, and paleoenvironments associated with modern and fossilized bone samples (Ambrose & DeNiro 1989; Chisholm et al., 1982; Hare et al., 1991; Schwarcz, 2000; Walker & DeNiro, 1986; Walter & Leslie, 2009), while other isotopes are only recently finding useful applications. Hydrogen is unique among all elements in terms of the doubling of the atomic mass from the light hydrogen nuclide protium ^{1}H to the less abundant, but heavier nuclide deuterium ^{2}H (or traditionally also abbreviated as D). This large mass difference between the two stable hydrogen isotopes creates strong fractionation effects (i.e. the preferential use or participation of one of the two isotopes in chemical and physical processes) and generates an extremely wide natural isotopic range of hydrogen stable isotope ratios (Sessions et al., 1999). Available hydrogen stable isotope data from bone collagen suggest greater variability than the most of the earlier studies had presumed, a situation that would compromise the diagnostic value of individual forensic collagen δD values for many species.

2. Structure, chemistry and preservation potential of bone collagen

Bone represents a complex system composed of 70% mineral and 30% organic material by weight when dried, and about 90% of the organics are collagen, a fibrous, high-tensile-strength protein functioning as a supportive framework (Hedges et al., 2006) and effectively playing the role of strands of steel in reinforced concrete. Remaining organic material in live bone reflects cells, lipids and other proteins (Price et al. 1985). Although the metabolic rates of bone tissues are low relative to those of the musculature, digestive, or dermal tissues, exchange of metabolites does occur and bone cells are constantly created and reabsorbed. The mineral phase of bone (bioapatite) protects collagen and other bone proteins from rapid *post-mortem* decomposition by physically shielding them from the external environment, especially from microbial access. The encapsulation of bone collagen by mineral crystallites is termed 'mineral stabilization" (Collins et al., 2002). In fact, the entire bone may be considered to be a mineralized collagen (Collins et al., 2002).

Collagens are the most common proteins in the matrix of connective tissues such as bone, cartilage, and dentine (Fig. 1) and form a group of over 20 fibrillar and microfibrillar proteinaceous macromolecules that share a triple helix structure that gives them great tensile strength (Gelse et al., 2003). Three collagen helices are coiled into a super helix in the collagen molecule (Price et al., 1985). Collagen found in bone is so-called type I collagen (Gelse et al., 2003) and can also be found in skin, tendons and other tissues, which makes it the most abundant of approximately 20 collagen species. The three helices in bone are α1 and α2 chains. Two identical α1 chains and one α2 chain form a highly organized superstructure, in which the fibrils are tightly packed in parallel orientation to yield a tissue with excellent tensile strength, pressure resistance and torsional stiffness (Gelse et al., 2003). A typical sequential order of amino acids is $(Gly\text{-}X\text{-}Y)_n$ where Gly stands for glycine, the most frequent amino acid in collagen, and X and Y are other amino acids, most commonly

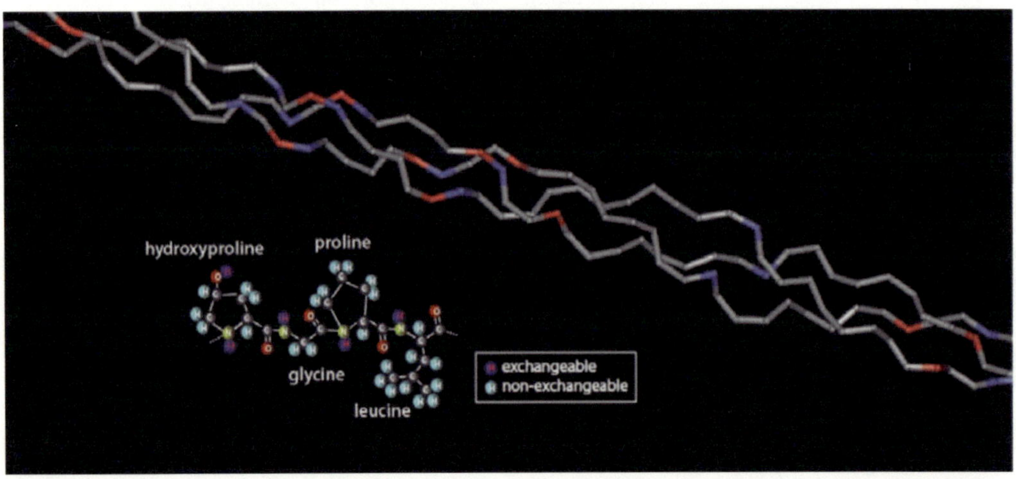

Fig. 1. Triple helix structure of Type I collagen from *Rattus norvegicus*. Three helical collagen proteins are intertwined. Color coding in the triple helix shows the charge of individual residues, positive in blue, negative in red, and neutral in grey (produced from data in GenBank, MMDB ID: 38933). The positions of exchangeable and non-exchangeable hydrogen atoms in 4 linked amino acids are simplistically depicted in the lower insert.

proline and hydroxyproline (Eyre, 1980). The Gly-X-Y triplets in this type of collagen repeat to about 300 times forming chains of approximately 1000 amino acid monomers (Gelse et al., 2003). Glycine is the smallest amino acid having one carbon-linked hydrogen atom in place where other amino acids carry a side chain. Proline, the second most common amino acid in collagen, has a cyclical structure that enables coiling of the helices (Balzer et al., 1997). Collagen also contains two amino acids uncommonly found elsewhere - hydroxyproline and hydroxylysine. The two do not occur freely outside collagen, but are derived from hydroxylation of proline and lysine already integrated into the collagen chain (Epstein, 1970). Hydroxyproline is responsible for linking the three α-helices with hydrogen bonds, and hydrolysine forms covalent bonds in the cross-linking of monomers (Balzer et al., 1997), increasing the strength and thermal stability of the superhelix (van Klinken, 1991). Smaller glycine is usually positioned towards the center of the helix axis, while the larger amino acids are placed towards the outside of the structure. This arrangement enables tighter coiling of the helix and a more compact and stronger macromolecule (Gelse et al., 2003). The collagen biopolymer is a water-insoluble, resilient, folded structure that is resistant to chemical decay (Balzer et al., 1997).

There are approximately 200 other proteins present in bone, though most of them are present only in trace amounts (Delmas et al., 1984; Linde et al., 1980, as cited in van Klinken, 1991). The second most common bone protein, osteocalcin, comprises 1-2 weight % of total fresh bone. Osteocalcin bonds with both the bone mineral fraction and bone collagen, but it seems to be unstable in solutions. Due to its small molecular size and strong mineral stabilization, osteocalcin can survive up to 50.000 years (C.I. Smith et al., 2005), and it may offer an alternative to the use of collagen in paleoenvironmental stable isotope research. However, osteocalcin's role and importance in this field of study has yet to be defined (Collins et al., 2002).

3. Preservation of bone collagen

The value of bone collagen as an indicator of past ecological and environmental conditions depends on its durability, a quality that directly results from being embedded in resistant inorganic biominerals that protect it from physical, chemical, and microbiological attack (Ambrose & DeNiro, 1989; Collins et al., 2002; Tuross et al., 1988). While preservation varies dramatically with sedimentary conditions and is favored by cold (<0°C), dry, and anoxic conditions (i.e., those that preclude biodegrading microorganisms), bone collagen is the longest lasting organic component of animal remains even in microbially active depositional environments such as soils (Child, 1995). Moreover, collagen is water-insoluble, abundant (in contrast to DNA material), homogenous (highly repetitive amino acid chains), and easy to extract from bone (Collins et al., 1995; Hedges & Law, 1989). The enhanced preservation potential of bone collagen stands in contrast to the relatively unprotected proteinaceous keratin (the dominant protein in feathers and skin) and other biopolymers like chitin and cellulose, which made collagen widely utilized in archaeological research (Collins et al., 2002; Lee-Thorp, 2008). However, the preservation of bone collagen is variable, and several methods have been developed to test bone integrity (Balzer et al., 1997). The preservation of collagen in bone depends on environmental and burial conditions, degree of decomposition, contact with soil or ground water, and the extent of leaching and other diagenetic factors (Balzer et al., 1997).

The physical and chemical *post mortem* alteration of bone is called diagenesis. Collins et al. (2002) recognize three pathways in bone degradation, namely organic-chemical deterioration of collagen, bone mineral breakdown, and microbial (biochemical) deterioration of collagen. Chemical deterioration of collagen represents a gradual loss of collagen matter and an increase in bone porosity due to depolymerization of collagen and chemical loss of low-molecular organic compounds. This is a dominant process in bone fossilization whereby secondary pores are created in place of lost collagen and a bone remnant in this phase is called a "mineral ghost". Secondary minerals crystallize inside the pores replacing the lost collagen fibrils. In the authors' opinion, this is the slowest and therefore the least likely degradation pathway albeit the most influential in fossilization processes. Mineral breakdown and microbial biodegradation are faster and thus more likely to be ultimately responsible for the long-term loss of bone collagen. Tuross (2002) analyzed a set of 30,000-year old bones in an effort to estimate the level of collagen degradation. She observed a preferential loss in some of the collagen amino acids and suggested that the chemical degradation may play a greater role in collagen loss than was previously believed. Bone mineral breakdown proceeds via dissolution in water depending on the hydrology of the environment. Bone mineral dissolution will rapidly expose bone collagen and subsequently accelerate the chemical and biochemical degradation of bone collagen. While chemical collagen breakdown is accelerated by extreme pH and elevated temperatures, the biochemical (microbial) decay of collagen most effectively occurs at circum-neutral pH. The latter is a dominant diagenetic pathway because it occurs rapidly after death of an animal (Bell et al., 1996) due to the abundance of microorganisms in the environment, and lack of physiological fluids that inhibit tissue deterioration normally found within a body during life. Also, unlike organic-chemical collagen decay that occurs evenly throughout the bone, microbial attack focuses on isolated areas of the bone where microbes have access to collagen (Collins et al., 2002).

The rate of bone collagen turnover in living organisms includes terms for synthesis and catalysis, but the effective turnover rate is poorly constrained due to the difficulty of conducting the necessary *in vivo* experiments (Babraj et al., 2005). The reported estimates range from less than a year (Hobson & Clark, 1992) to 10 years (van Klinken, 1991) or over (Hedges et al., 2007). According to Waterlow (2006), the collagen turnover rate varies within the bone. The immature bone collagen (procollagen) is degraded within hours, while the mature collagen shows almost no turnover. The average rate for all bone collagen then depends on the ratio of the mature and immature collagen fractions, which are in turn dependent on life stage and level of stress (Waterlow, 2006).

4. Stable isotopes

The ratios of heavy to light stable isotopes of hydrogen $^{2}H/^{1}H$ (or D/H), carbon $^{13}C/^{12}C$, nitrogen $^{15}N/^{14}N$, oxygen $^{18}O/^{16}O$, and sulfur $^{34}S/^{32}S$ show distinctive patterns of distribution in the living world that can be used to interpret respective data from fossil samples. Stable isotope ratios in modern organisms are determined by (i) water, air and food intake, (ii) metabolic isotopic fractionations occurring inside the body, and (iii) evapotranspirative and excretionary loss of material. Stable isotope ratios are typically expressed as δ (delta) values:

$$\delta(‰) = \left(\frac{R_{unknown}}{R_{standard}} - 1 \right)$$

where R is the isotope abundance ratio for sample ($R_{unknown}$) or standard ($R_{standard}$) (Coplen, 2011). Stable isotope scales use δ units and are defined by internationally recognized anchor points: Vienna-Pee Dee Belemnite (VPDB) for ^{13}C, air nitrogen for ^{15}N, Vienna Standard Mean Ocean Water (VSMOW) for ^{18}O and ^{2}H (e.g., Coplen, 1996). Delta values are expressed in ‰ (permil). A sample with a δD value of 20‰ is enriched in D (deuterium) by 20‰ relative to the VSMOW water standard. δD = -10‰ indicates a D-depletion of 10‰ relative to VSMOW.

4.1 Isotopic fractionation

Mass differences between the nuclei of pairs of stable isotopes of an element result in isotopic fractionations during physical transitions, transport processes, and chemical reactions that are typically expressed in terms of slight preferences for either the lighter or the heavier isotope of an element. As a result, natural systems offer a rich spectrum of systematic natural abundances of stable isotopes (Faure & Mensing, 2004). Isotopic fractionation along physical and (bio)chemical processes can be controlled kinetically or be based on equilibrium conditions. Biochemical syntheses of lipids, carbohydrates, proteins, etc. rely on different low-molecular organic source compounds ('building blocks'), enzymes, and reaction pathways. Different biochemical heritage is reflected in contrasting isotopic compositions. For examples, organic hydrogen in lipids tends to be systematically depleted in deuterium relative to organic hydrogen in other compound classes (Sessions et al., 1999). Moreover, various types of biological tissues are influenced or even dominated by certain biochemical compound classes that result in distinct isotopic characteristics of entire tissue types. Some tissue types or compound classes within tissues can exhibit enhanced turnover rates, for example lipid reserves during times of reduced food availability. Half of the organic carbon in quail muscle and bone collagen was turned over within 12.4 and 173.3 days, respectively, indicating that bone collagen holds a relatively long-term biochemical signal (Hobson & Clark, 1992).

4.2 Carbon isotopes

The main source of carbon for terrestrial autotrophic life is atmospheric carbon dioxide CO_2, whereas marine autotrophs typically rely on dissolved CO_2 and bicarbonate ions (HCO_3^-). On average these two global environments with contrasting inorganic carbon pools express $\delta^{13}C$ values of -7.5 and +1.5‰, respectively. The isotopic difference permeates from primary producers to higher levels of food chains, although the effect is often obfuscated by additional carbon isotope fractionation mechanisms superimposing stronger isotopic variations (Hayes, 2001; Lee-Thorp et al., 1989; van Klinken, 1991). The uptake of inorganic carbon by autotrophs discriminates more or less strongly against ^{13}C depending on the biochemical mode of photosynthesis. Most plants utilize the Calvin (C3) cycle and consequently express $\delta^{13}C$ values around -26‰. A smaller group of plants employs the C4 pathway, whereby CO_2 molecules are first fixed in a more enclosed system to reduce evapotranspiration of water. C4 plants discriminate less against ^{13}C and have average $\delta^{13}C$ values of -12‰ (B.N. Smith & Epstein, 1971). The carbon stable isotope ratio in heterotrophs is therefore an important dietary indicator discriminating between corn-based C4 diets and rice and wheat-based C3 diets (Tieszen et al., 1983). Marine algae utilize the Calvin cycle and, due to the relative ^{13}C-enrichment of dissolved CO_2 and bicarbonate ions in seawater, typically express less negative $\delta^{13}C$ values (Fry et al., 1982). Additional fractionations occur between and within each food chain member. Consequently, from primary producers, to herbivores, to carnivores, to decomposers, each successive trophic level experiences approximately 1‰ ^{13}C depletion (Ambrose & DeNiro, 1986; Schoeninger et al., 1985; Schoeninger & DeNiro, 1984).

4.3 Nitrogen isotopes

Nitrogen is incorporated into biomolecules either through direct nitrogen fixation from air (e.g., in legumes in symbiosis with nitrogen-fixing bacteria), from ammonium or nitrate in soil water, or from recycled organic nitrogen from soil (Lee-Thorp, 2008; Yakir & DeNiro, 1990). There is negligible isotope fractionation during nitrogen fixation from air (Delwiche & Steyn, 1970). Small isotope fractionations during uptake of ammonia and nitrate by autotrophs are typically dwarfed by larger isotopic variance among natural and industrial sources of ammonium and nitrate (e.g., fertilizers). Similar to carbon isotopes, the nitrogen stable isotope ratio $^{15}N/^{14}N$ in biomass increases with trophic level. The preferential retention of ^{15}N-enriched organic nitrogen in biomass is isotopically balanced by the excretion of ^{15}N-depleted waste, e.g. urea. An increase in $\delta^{15}N$ of 2 to 6‰ per trophic level has been recorded both in terrestrial and marine food webs (DeNiro & Epstein, 1981; Minagawa & Wada, 1984; Rau, 1982). Nitrogen stable isotope ratios in soils, plants and animals are also indirectly related to precipitation and atmospheric relative humidity (Pate & Janson, 2007). The highest plant $\delta^{15}N$ values are found in arid environments, and the lowest ones at high elevations and in more humid forests (Ambrose, 1991; Handley et al., 1999; Heaton, 1987).

4.4 Applications of carbon and nitrogen isotopes

Bone collagen has been used extensively in an archaeological context for paleodietary and paleoenvironmental reconstructions based on carbon and nitrogen stable isotope ratios (e.g., Ambrose, 1991; Ambrose & DeNiro, 1989; Chisholm et al., 1982; DeNiro & Epstein, 1978, 1981; Huelsemann et al., 2009; Petzke & Lemke, 2009; Walker & DeNiro, 1986; Walter & Leslie, 2009). The reliability of archaeologically recovered bone collagen for stable isotope analyses should be established by testing for contamination by allochtonous (i.e. foreign) compounds that may be adsorbed or covalently linked to collagen, and by evaluating collagen degradation with chemical methods. Collagen extraction should utilize standardized procedures to ensure comparability of results. Deteriorated collagen becomes increasingly soluble (gelatinous). Degradation has been evaluated by quantifying the loss of nitrogen from bone. More than 95% of nitrogen in fresh bone is organic nitrogen in collagen (Collins et al., 2002).

5. Physiological origin of isotopic signals in collagen

Amino acid synthesis in heterotrophs primarily relies on dietary protein rather than on lipids or carbohydrates (Schwarcz, 2000), and thus the isotopic composition of the resulting new collagen is related more to dietary protein rather than to bulk diet. Trophic recycling of dietary biomass into new live biomass is always partial and thus subject to isotope fractionation. Newly generated amino acids in heterotrophs are enriched in ^{13}C and ^{15}N relative to bulk diet (Howland et al., 2003). More specifically, glycine is enriched in ^{13}C and ^{15}N relative to bulk diet and was found to be directly integrated from food to bone collagen (Hare et al., 1991). Glycine constitutes about 1/3 of all amino acids in collagen and therefore strongly influences ^{13}C and ^{15}N- abundances in bone collagen. Carnivores rely on a protein-rich diet and produce new biomass primarily from dietary amino acids, although the enzymes required for *de novo* amino acid synthesis are present (Gannes et al., 1998). Bone collagen, muscle (meat) and apatite were analyzed for a set of modern southern African herbivores and carnivores (Lee-Thorp et al., 1989). The isotopic analyses showed ^{13}C enrichment in bone collagen, apatite and muscle, and ^{13}C depletion in lipids. Difference in $\delta^{13}C$ values between herbivores and carnivores indicates a trophic effect, which for carbon in bone collagen is 2.5-3‰ (Fig. 2).

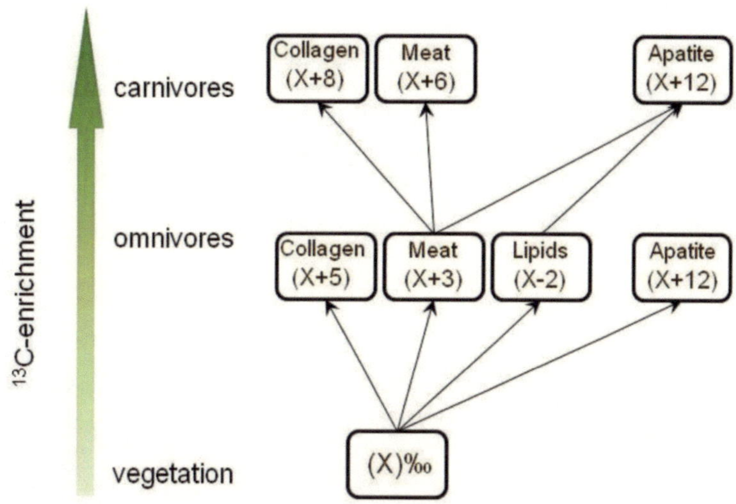

Fig. 2. Foodweb model for southern African herbivores and carnivores representing $\delta^{13}C$ enrichment/depletion from the baseline (X) $\delta^{13}C$ value for vegetation (adapted from Lee-Thorp, 1989).

Balzer et al. (1997) experimentally degraded modern bone collagen in laboratory conditions by inoculating bone with soil bacteria. Microbial biodegradation over 8 to 18 months caused isotopic shifts in $\delta^{13}C$ and $\delta^{15}N$ values by -2.9 and +5.8‰, respectively, due to altered amino acid profiles and fractionation during peptide bond cleavage (Balzer et al., 1997). The isotopic composition of bulk collagen is the weighted average of the isotopic values of individual participating amino acids that collectively span much wider isotopic ranges and contain far more detailed isotopic information than bulk stable isotope ratios. The high concentration of glycine in collagen together with glycine's systematic ^{13}C-enrichment relative to proline, hydroxyproline and glutamic acid even translates into a relative ^{13}C-enrichment of bulk collagen relative to the entire body (van Klinken, 1991). Decomposition and diagenesis of bone collagen following burial, during exposure to forest fire, or via cooking and roasting by humans may in extreme cases affect bone collagen carbon and nitrogen isotopic values (DeNiro et al., 1985). Similar limitations will need to be considered when measuring hydrogen isotopes in collagen for archaeological and forensic purposes using fossil or otherwise aged materials.

6. Hydrogen stable isotopes in bone collagen

6.1 D/H ratios in environmental water relate to D/H of organic matter

The ratio of hydrogen's two stable isotopes deuterium (^2H, traditionally abbreviated as D) and protium (^1H, here abbreviated as H) in precipitation or meteoric water shows geographic directional and altitudinal trends. Meteoric water is generally the "heaviest" (i.e. D-enriched) in equatorial regions close to warm oceans that are recharging atmospheric moisture. As air masses travel towards higher latitudes and/or gain altitude, successive loss of water during precipitation depletes the remaining atmospheric moisture in deuterium and results in progressively "lighter" (i.e. D-depleted) rain and snow (e.g., Bowen et al., 2005; van der Veer et al., 2009; Fig. 3).

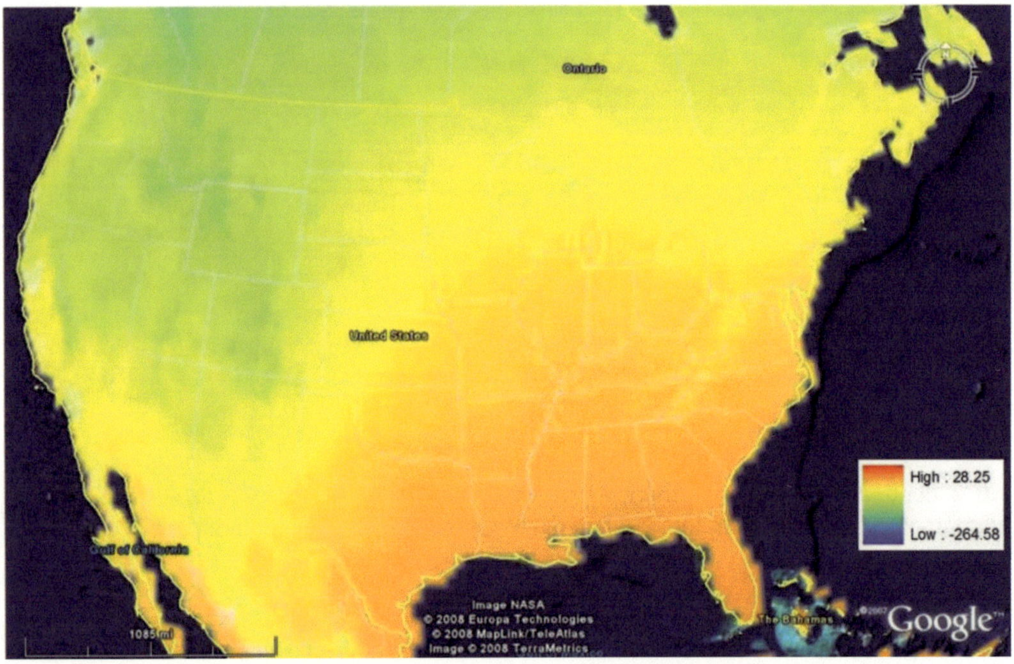

Fig. 3. Map of North America showing a general North-Northwest spatial trend in precipitation stable hydrogen values across the continent.

Hydrogen from water is the direct source of almost all organic hydrogen in tissues of autotrophs, which therefore carry the isotopic signal of environmental water, with the caveat that physical and biochemical isotope fractionations further modulate D/H ratios of individual biochemicals throughout the reaction chains of biosynthesis (e.g., Hayes, 2001). The cellulose and lipids of autotrophs (e.g., preserved in tree rings and as biomarkers in sediments) are valuable substrates in paleoclimatology for paleoenvironmental reconstruction because their organic hydrogen can be isotopically traced directly to environmental water. Heterotrophic use of autotrophic biomass and passage of organic hydrogen along the food chain progressively introduces metabolic isotope fractionations, mixing of isotopically different organic hydrogen pools, and also offers opportunities for drinking water to isotopically affect newly biosynthesized faunal biomass. Although not all contributing factors can be deconvoluted, experience has shown that careful empirical examination of the hydrogen stable isotopic signal in animal tissues can provide a sound basis for characterizing and constraining the life history and geographic origin of heterotrophs (Hobson et al., 1999; Meehan et al., 2001). In large heterotrophic animals, such as many vertebrates, the complexity of $\delta D_{organic}$ signals is enhanced because (i) animals can migrate, (ii) animals can change their diet during their life history, and (ii) different species may occupy different levels of an ecosystem's food chain (e.g., Bowen et al., 2009; Farmer et al., 2008; Solomon et al., 2009). The $\delta D_{organic}$ values of a compound or compound class within an individual animal may vary among different tissues according to when those tissues were biosynthesized. Seasonal, geographic, and other variables during the lifetime of an animal may thus leave an isotopic record (Tuross et al., 2008). For example, the collagen of tooth dentine in sheep and goats records seasonal isotopic changes in precipitation, temperature, and diet because dentine is deposited incrementally throughout the animal's life. $\delta D_{dentine}$ values from successive growth layers in teeth, when paired with tooth mineral-based $\delta^{18}O$ values from the same layers, can resolve seasonal climatic changes in temperature and humidity (Kirsanow et al., 2008). Outstanding ecological insight has been gained from $\delta D_{keratin}$ values in feathers from migrating and non-migrating birds in terms of migration patterns (e.g., Chamberlain et al., 1997; Hobson, 1999) and bird habitats (e.g., Hobson et al., 2003). Similar to collagen, keratin also is a proteinaceous biopolymer, but with prominent cross-linking by sulfur bridges to increase rigidity.

In contrast to lipids where almost all organic hydrogen is chemically strongly bound directly to carbon atoms, the analytical access to precise hydrogen stable isotope ratios in collagen has been made difficult by the fact that some organic hydrogen atoms are weakly bonded and can rapidly exchange when in contact with hydrogen from body and environmental water. As a consequence, the pool of isotopically exchangeable organic hydrogen is unable to preserve a memory of biosynthesis. Exchangeable hydrogen is also called "labile hydrogen" and is primarily located in functional groups like -OH, -COOH, -NH_2, and some specific carbon-linked positions (Schimmelmann et al., 2006). On average ~20% of total organic hydrogen in collagen is exchangeable (Cormie et al., 1994b). Collagens and other biopolymers also contain some potentially exchangeable hydrogen that is deeply embedded in macromolecules' three-dimensional matrix. Shielding from water and other chemical hydrogen donors makes this pool of organic hydrogen essentially non-exchangeable (Fig. 4). However, solubilization of collagen during sample preparation may partially unravel the tertiary polyproteinaceous structure and expose previously shielded hydrogen positions to water.

Exchangeable organic hydrogen can be chemically eliminated from carbohydrates, such as cellulose, by esterification of hydroxy groups with nitric acid (i.e. "nitration"). However, collagen and other chemically more complex organic compounds and bulk tissue require a different approach whereby exchangeable hydrogen is equilibrated, and thus isotopically controlled, with water vapors of known isotopic compositions (Sauer et al., 2009; Schimmelmann, 1991; Wassenaar & Hobson, 2000).

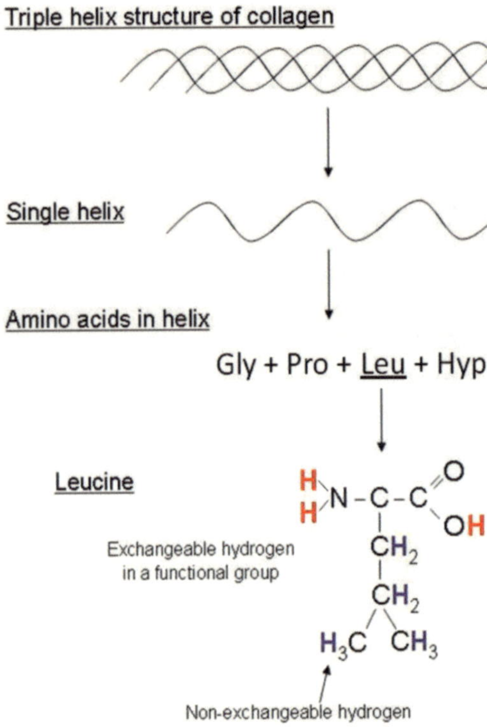

Fig. 4. Helical structure of collagen, typical amino acid sequence within a collagen strand, and exchangeable versus non-exchangeable hydrogen atoms in an individual leucine molecule (Gly - glycine, Pro - Proline, Leu - leucine, Hyp - hydroxyproline.

6.2 Hydrogen isotopic relationship between modern bone collagen and precipitation

The hydrogen stable isotope ratio of precipitation influences the δD values of bulk hydrogen in collagen more strongly in some species than in others. Cormie et al. (1994a) reported an exceptionally strong linear correlation with $R^2 = 0.92$ between both parameters (in a study of 62 white tailed and mule deer from 48 locations in North America when the effect of relative humidity was taken into account using a correction based on collagen $\delta^{15}N$. Our recent preliminary δD_n values of non-exchangeable hydrogen in bone collagen from white-tailed deer and mule deer from locations across the United States suggest a less straightforward relationship, although the data confirm that precipitation is a major factor in determining δD_n in deer (Fig. 5).

However, some of our deer individuals from the arid Joshua Tree National Park in California indicate unusual D-enrichment. This may derive from evapotranspiration in local plants that were part of the diet of the deer and/or in the body fluids of the animals themselves, as is expected in extremely dry environments (Cormie et al., 1994c; Bowen et al., 2005). Deer occupy an ecological niche that is relatively simple from the perspective of hydrogen, as their diet consists of leafy vegetation and their water is obtained from surface waters (lakes and streams) that in many cases have δD values closely representing mean annual precipitation. In contrast, omnivorous and carnivorous animals consume more diverse diets with more widely varying

δD values. Smaller animals and especially those occupying particular niches (e.g., shrews, bats) have higher turnover rates of hydrogen in their bodies. They also exploit smaller pools of water and sources of food which can have isotopic ratios that differ strongly from mean annual precipitation. Accordingly, the hydrogen isotopic relationship of bone collagen and environmental water is strongly species-specific. Using samples from freshly killed animals collected in the field, isotopic calibration studies pose a technical challenge because body water (e.g., from blood) can be collected only after death, but the δD of body water may have varied through the animal's lifespan.

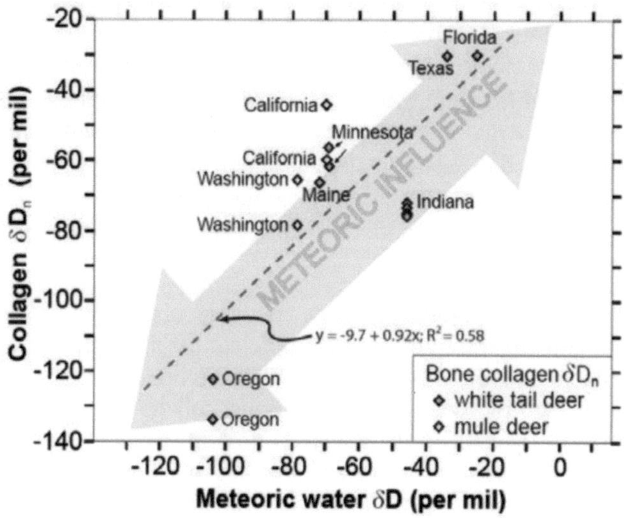

Fig. 5. δD_n of non-exchangeable hydrogen from bone collagen: Pilot data from field-collections of white tail and mule deer collected across the United States.

6.3 D/H comparison between modern and fossil bone collagen

The hydrogen isotope ratio of geochemically preserved and dated fossil bone collagens has value for paleoclimatic reconstructions (Cormie et al., 1994c; Hoppe, 2009). However, as previously stated for carbon and nitrogen, meaningful hydrogen isotopic analysis of bone collagen hinges on the prerequisite that collagen has not suffered any significant diagenetic isotopic changes (Balzer et al., 1997; Lee-Thorp, 2008). Leyden et al. (2006) compared $\delta D_{collagen}$ values from 7 modern and 52 fossil bison bones from Canada with ages of up to 10,000 years. The problem of exchangeable hydrogen in collagen was addressed by equilibration in water vapors (Wassenaar & Hobson, 2000). Leyden et al. (2006) found that $\delta D_{collagen}$ in modern and fossil bones followed the same geographic gradient as modern δD_{water} in Canadian precipitation. Furthermore, consistency between changes in $\delta D_{collagen}$ through time and independent evidence for regional Holocene climatic changes corroborates the value of $\delta D_{collagen}$ as a paleoenvironmental proxy.

6.4 Effect of trophic level on stable isotope ratios of bone collagen

δD values of amino acids from autotrophs are directly related to the δD_{water} in the cellular and intercellular water. As food is metabolized, part of its organic hydrogen has an opportunity to exchange isotopically with body water before organic components are incorporated into new biomass. Trophically "recycled" amino acids from ingested food do not exchange all of their hydrogen. Most of the hydrogen that is bonded covalently to carbon does not readily exchange with water as long as the biomolecule is chemically unaltered. Energetic transitions may foster limited isotopic exchange, e.g. during *post-mortem* racemization, but racemization rates are sufficiently slow that negligible effects on δD are expected on young bone samples that have never experienced elevated temperatures. Most of the carbon-bound hydrogen in collagen is essentially "non-exchangeable" over thousands of years at low temperature (Sessions et al., 2004; Schimmelmann et al., 2006) and is able to maintain an isotopic memory of trophic input. Because all animals are heterotrophic with a limited ability to synthesize amino acids, their collagen (and therefore the pool of H contained in collagen) represents a varying mixture of dietary biomass from lower levels in the food chain and newly synthesized amino acids. This upward-cascading flow of organic hydrogen from one trophic level to the next introduces successive isotopic patterns that relate to specific ecological and trophic situations. Although the resulting isotopic relationships can be complex, the analytically accessible isotope patterns can be empirically calibrated with modern faunal analogs and thus be used to constrain paleodiet and paleoenviroments.

This suggests (i) that isotopic patterns of bone collagens of carnivores and herbivores may differ systematically and (ii) the hydrogen isotopic composition of drinking water may be less important for collagen in carnivores than in herbivores (Pietch et al., 2011). In other words, the hydrogen isotopic composition of bone collagen should be sensitive to the amount and dietary importance of ingested protein, which should offer analytical opportunities to reconstruct aspects of diet and trophic structure.

Consumption of lower-trophic level biomass by a higher-level heterotrophic organism offers an opportunity to isotopically fractionate the pools of atoms that are utilized to build new biomass. Enrichment in ^{13}C and ^{15}N with increasing trophic level in bone collagen has been documented extensively. Similarly, trophic D-enrichment has been found in chitin (i.e. an aminosugar-based biopolymer found in arthropod exoskeletons; Schimmelmann & DeNiro, 1986). More recently, a systematic increase in δD values of vertebrate bone collagen with higher trophic levels has been documented (Birchall et al., 2005; Reynard & Hedges, 2008; Topalov et al., 2009). Non-exchangeable hydrogen in collagen from herbivores is generally D-depleted relative to environmental water, and the observed range of δD_n is smallest. Carnivores show the highest and most variable δD_n values, whereas omnivores fall into a range between herbivores and carnivores (Fig. 6). Exceptions to this trend, such small insectivores with occasional large D-enrichment (e.g., bats and shrews) may reflect special cases related to small body sizes and/or unusually high metabolic rates.

The trophic structure of ecosystems has traditionally been explored with nitrogen stable isotopes. A correlation between $\delta^{15}N$ and δD values of bone collagens from 19 aquatic and terrestrial species, including fish, birds and mammals, corroborates the value of collagen δD for trophic structural investigations (Birchall et al., 2005). δD values of collagens from both terrestrial and marine organisms at similar trophic levels were comparable, whereas the trophic signal in marine $\delta^{15}N$ values was relatively weak. Collagens from carnivores and piscivores were D-enriched by ca. 90‰ relative to herbivores and omnivores. An increase in bone collagen δD values by 10 to 30‰ from herbivores to omnivores to humans was observed by Reynard & Hedges (2008) (Fig. 7). The authors reduced the isotopic variability among faunal individuals by averaging data from 10 to 15 samples per species. Our recent preliminary data corroborate Birchall et al.'s (2005) conclusions by demonstrating a strong trophic signal with generally good separation between herbivores, omnivores, and carnivores.

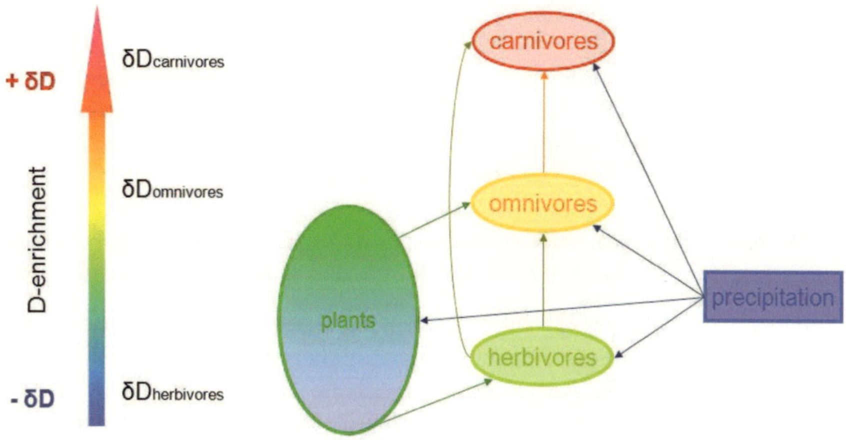

Fig. 6. Deuterium enrichment through a conceptual trophic chain.

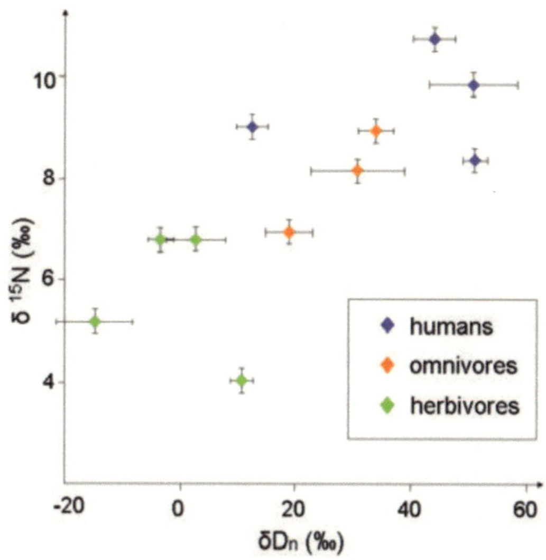

Fig. 7. $\delta^{15}N$ vs. δD mean values for bone collagens of herbivores and omnivores including humans from Great Britain, Hungary, Peru and Canada dating from the Neolithic to the mid-15th century AD (adapted from Reynard & Hedges 2008). Error bars indicate one standard error of the mean of 10 to 15 samples per species.

Terrestrial biota is exposed to potentially large seasonal and geographic hydrogen isotopic variability in meteoric and environmental waters that complicates the isotopic influence of both drinking water and diet. In contrast, the global mixing of open ocean waters results in seawaters with a narrow hydrogen isotopic range of only a few permil (Dansgaard, 1964). Consequently, the hydrogen isotopic signal of bone collagen from marine animals should predominantly reflect diet and be less sensitive to drinking water (Fetcher, 1939; Ortiz, 2001). Evaporative D-enrichment of body fluid during respiration may be limited because air near the water surface has a relatively high humidity. Marine mammals are predators at the top of a long food chain that contains multiple trophic levels of zooplankton, fish, and may even include other marine mammals. Our preliminary data from two populations of California sea lions from the Channel Islands (California, USA) indicate that the variability of collagen δD in adults is comparable to that of terrestrial carnivores from Bloomington, Indiana. This suggests that δD variability in populations is based more on individuality of animals rather than on environmental variability. Bone collagen δD of California sea lions strongly discriminates between D-depleted young pups and relatively D-enriched adult sea lions. Nursing sea lion pups (aged < 12 months), δD_n values typically range around δD =+25‰, while δD_n values for adults are substantially higher (δD =+70‰). We interpret this to reflect weaning from a D-depleted, lipid-rich milk diet to a prey-based adult diet.

Our preliminary bone collagen δD values on a wide range of animal species from a small region around Bloomington, Indiana, suggest that higher and more variable δD values are related to small body size and/or the rapid metabolic rate in some small mammals. For example, rapid metabolism and intense respiration likely entails evaporative D-enrichment in body water and greater reliance on drinking water with warm-seasonal bias towards less negative δD. In addition, seasonal changes in diet should more rapidly affect biomass of smaller animals that express rapid turnover of their biomass, including bone collagen. Even though small mammals are homeotherm organisms, their larger ratio of surface to body mass makes them more susceptible to environmental temperature. Smaller warm-blooded animals in cool climates must maintain a high metabolic rate to stay warm. Another consequence of a fast metabolism is a short life span. For example, the Eastern short-tailed shrew can live up to two years, but approximately 80% of the young will not live to the winter. Juvenile bone collagen δD will therefore be in response to summer to fall hydrological and ecological conditions rather than influenced by annual average conditions. Moreover, juveniles may still partially carry a prenatal and lactation-derived δD signal.

7. Methods of stable isotope collagen research

The use of hydrogen isotopes in biomass faces the difficulty that some organic hydrogen atoms are weakly bonded in biomolecules, exchange with hydrogen from water, and thus may lose their original biogenic isotopic information. Exchangeable hydrogen is also called labile hydrogen and occupies chemically functional groups like -OH, -COOH, -NH_2, and some specific carbon-linked positions (Schimmelmann & al., 2006). Exchangeable hydrogen can be chemically eliminated from a few organic substances, such as cellulose, through the process of 'nitration'. Most chemically more complex biochemicals and bulk tissue require a different approach whereby exchangeable hydrogen is equilibrated, and thus isotopically controlled, with waters of known isotopic compositions. Such equilibrations were traditionally performed off-line in chambers with stationary water vapor (Wassenaar & Hobson, 2000) or in dynamic flow-through conditions (Schimmelmann, 1991). A remaining problem was the labor-intensive nature of equilibrating organic samples in individual quartz tubes. Here we present a more efficient equilibration approach. Simultaneous steam-equilibration of dozens of collagen and/or other organic samples in an EA carousel with subsequent on-line continuous-flow D/H measurements via TC/EA is far more efficient than previously available methods. A variety of collegen extraction procedures have been developed following a similar general outline (e.g., Tuross et al., 1988, DeNiro & Weiner, 1988).

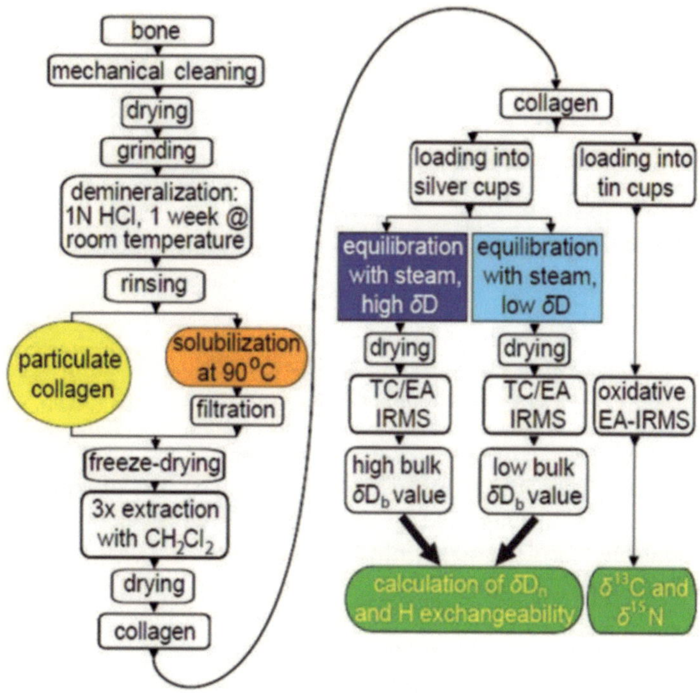

Fig. 8. The overall procedures are separated into (left) preparation of bone collagen and (right) determination of stable isotope ratios using isotope ratio mass spectrometry (IRMS). Yellow and orange fields indicate two fundamentally different approaches that need to be tested for hydrogen isotopic repercussions using collagens with different degrees of preservation and diagenetic overprinting. Light blue and dark blue fields indicate key elements of dual water vapor (steam) isotopic equilibration of exchangeable hydrogen in twin aliquots of every collagen.7.2. Isotopic δD_n, $\delta^{13}C$ and $\delta^{15}N$ characterization of collagen

7.1 Water vapor isotopic equilibration of exchangeable hydrogen in collagen

Recently we developed a fast and economical analytical procedure to isotopically equilibrate and thus control the isotopic composition of exchangeable hydrogen in bone collagen with isotopically known water vapors (Sauer et al., 2009). In brief, our method allows for reduced sample sizes of 0.3 to 1mg, depending on the hydrogen content of organic substrates. The isotopic uncertainty from exchangeable hydrogen was reduced via equilibration with isotopically known water vapors and subsequent mass-balance calculations arriving at the δD_n of non-exchangeable hydrogen in collagen. The samples in silver capsules are loaded into an EA carousel and closed off into an aluminum equilibration chamber (Fig. 9). Samples are allowed to equilibrate with steam for at least 6 hours, dried in a flow of dry N_2 and quickly transferred to the autosampler, followed by immediate flushing of the loaded autosampler with helium.

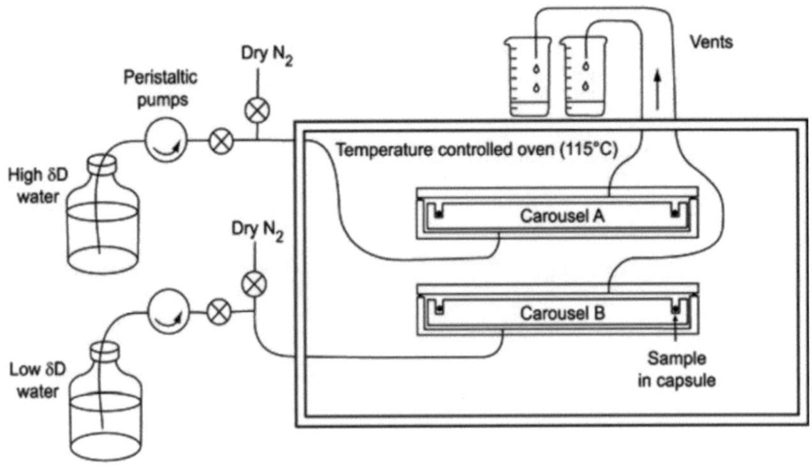

Fig. 9. Schematic of water vapor equilibration apparatus featuring carousels that hold up to 49 samples in crimped silver cups (from Sauer et al., 2009).

A reductive TC/EA (i.e. thermal conversion elemental analyzer) is used for D/H determination. The resulting δD isotopic difference between pairs of equilibrated collagens is used to calculate the percentage H_{ex} of exchangeable hydrogen in total hydrogen. All collagens are chemically similar and should have comparable H_{ex} values, which serves as quality control. In case of collagens prepared in our lab, the hydrogen exchangeability averages 22% ± 2%. Finally, collagen $\delta^{13}C$ and $\delta^{15}N$ is determined to better constrain the ecology and trophic positions of animals. For additional quality control (e.g., Lee-Thorp, 2008), the atomic C/N ratio of collagens is calculated.

8. Outlook to future isotopic research on bone collagen

The recording of environmental conditions by stable isotope ratios in biological substrates is a proven concept, but the details are more complex than anticipated by the simplistic expression "you are what you eat" (DeNiro & Epstein, 1978) "plus or minus a few permil" (Rundel et al., 1988). More laboratory-constrained experiments are needed to deconvolute and constrain the complex pathways of hydrogen, carbon and nitrogen isotopes in biochemical substrates. Analytical methods now exist to add oxygen stable isotopes to the arsenal. There is a great need for field-based research of modern non-migrating animal populations that can serve as analogs for animals in archaeological and forensic investigations. We need to assess individual isotopic variations within populations that are caused by diet, home range size, and migratory habits, as well as the causes of isotopic variations within an individual.

A promising new frontier in collagen stable isotope research is the determination of compound-specific isotope ratios of individual amino acids after hydrolysis of the biopolymer. The underlying analytical methods have long been established for carbon and nitrogen stable isotope ratios. Following hydrolytic depolymerization of collagen in acid, the free amino acids must be chromatographically separated via gas or liquid chromatography prior to combustion to carbon dioxide and nitrogen that can be measured mass-spectrometrically to yield $\delta^{13}C$ and $\delta^{15}N$ values (Chikaraishi et al., 2007). Gas chromatography (GC) requires that polar functional groups of free zwitterionic amino acids are first derivatized to yield less polar compounds that can move together with helium carrier gas through a capillary GC column. Derivatizing agents typically contain additional carbon that adds its isotopic influence to the measured analytes and must be accounted for by applying mass-balance corrections. Current analytical efforts seem to have succeeded in applying the same principle toward the measurement of compound-specific amino acid D/H ratios (e.g., Chikaraishi et al., 2003).

The stable isotope ratios of bulk collagen represent weighted averages of the stable isotope ratios of participating amino acids. Much additional isotopic information can be obtained from individual amino acids than from bulk collagen. Amino acids collectively express far larger isotopic ranges than bulk collagen. Greatly enhanced trophic information can be gleaned from the isotopic differences between essential and non-essential amino acids in collagen from heterotrophs, since only essential amino acids express a direct and unambiguous heritage from diet. We can expect that many of the shortcomings and ambiguities of stable isotope ratios in bulk bone collagen will be mitigated or overcome by novel approaches that yield compound-specific isotope ratios of individual amino acids from collagen.

9. References

Ambrose, S.H. (1991). Effects of diet, climate and physiology on nitrogen isotope abundances in terrestrial foodwebs. Journal of Archaeological Science, Vol. 18, 293-317. (http://dx.doi.org/10.1016/0305-4403(91)90067-Y)

Ambrose, S. H., & DeNiro, M. J. (1986). The isotopic ecology of East African mammals. Oecologia, 69: 395-406.

Ambrose, S.H. & DeNiro, M.J. (1989). Climate and habitat reconstruction using stable carbon and nitrogen isotope ratios of collagen in prehistoric herbivore teeth from Kenya. Quaternary Research 31, 407-422.

Balzer, A., Gleixner, G., Grupe, G., Schmidt, H.-L., Schramm, S., & Turban-Just, S. (1997). In vitro decomposition of bone collagen by soil bacteria: the implications for stable isotope analysis in archaeometry. Archaeometry 39 (2), 415-429. (http://dx.doi.org/10.1111/j.1475-4754.1997.tb00817.x)

Babraj, J., Cuthbertson, D.J., Rickhuss, P., Meier-Augenstein, W., Smith, K., Bohé, J., Wolfe, R.R., Gibson, J.N., Adams, C. & Rennie, M.J. (2002). Sequential extracts of human bone show differing collagen synthetic rates. Biochemical Society Transactions, Vol. 30, pp. 61–65.

Bell, L.S., Skinner, M.F., & Jones, S.J. (1996). The speed of post mortem change to the human skeleton and its taphonomic significance. Forensic Science International, Vol. 82, pp. 129–140.

Birchall J., O'Connell, T.C., Heaton, T.H.E., & Hedges, R.E.M. (2005). Hydrogen isotope ratios in animal body protein reflect trophic level. Journal of Animal Ecology, 74, 877-881. (http://dx.doi.org/10.1111/j.1365-2656.2005.00979.x)

Bocherens, H., Billiou, D., Mariotti, A., Patou-Mathis, M., Otte, M., Bonjean, D., & Toussaint, M. (1999). Palaeoenvironmental and paleodietary implications of isotopic biogeochemistry of Last Interglacial Neanderthal and mammal bones in Scladina Cave (Belgium). Journal of Archaeological Science, Vol. 26, pp.599-607.

Bowen, G.J., L.I. Wassenaar, & Hobson, K.A. (2005). Global application of stable hydrogen and oxygen isotopes to wildlife forensics. Oecologia, Vol. 143, pp. 337-348. (http://dx.doi.org/10.1007/s00442-004-1813-y)

Chamberlain, C.P., Blum, J.D., Holmes, R.T., Feng, X., Sherry, T.W. & Graves, G.R. (1997). The use of isotope tracers for identifying populations of migratory birds. Oecologia, Vol. 109, No. 1, pp. 132-141. (http://dx.doi.org/10.1007/s004420050067)

Chikaraishi, Y. & Naraoka, H. (2003). Compound-specific _D-_13C analyses of n-alkanes extracted from terrestrial and aquatic plants. Phytochemistry, Vol. 63, pp. 361–371.

Chikaraishi, Y., Kashiyama, Y., Ogawa, N.O., Kitazato, H., & Ohkouchi, N. (2007). Metabolic control of nitrogen isotope composition of amino acids in macroalgae and gastropods: Implications for aquatic food web studies. Marine Ecology Progress Series, Vol. 342, pp. 85–90.

Child, A.M. (1995). Microbial taphonomy of archaeological bone. Studies in Conservation, Vol.40, pp.19–30.

Chisholm, B.S., Nelson, D.E., & Schwarcz, H.P. (1982). Stable-carbon isotope ratios as a measure of marine versus terrestrial protein in ancient diets. Science, Vol. 216 No. 4550, pp. 1131-1132. (http://dx.doi.org/10.1126/science.216.4550.1131)

Collins, M. J., Riley, M., Child, A. M., & Turner-Walker, G. (1995) A basic mathematical simulation of the chemical degradation of ancient collagen, Journal of Archaeological Science, Vol.22, pp. 175–83.

Collins, M.J., Nielsen-Marsh, C.M., Hiller, J., Smith, C.I., Roberts, J.P., Prigodich, R.V., Weiss, T.J., Csapò, J., Millard, A.R., & Turner-Walker, G. (2002). The survival of organic matter in bone: a review. Archaeometry, Vol. 44, No. 3, pp. 383-394. (http://dx.doi.org/10.1111/1475-4754.t01-1-00071)

Coplen, T.B. (1996). New guidelines for reporting stable hydrogen, carbon, and oxygen isotope-ratio data. Geochimica et Cosmochimica Acta Vol. 60, No. 17, pp. 3359-3360. (http://dx.doi.org/10.1016/0016-7037(96)00263-3)

Coplen, T.B. (2011). Guidelines and recommended terms for expression of stableisotope-ratio and gas-ratio measurement results. Rapid Communications in Mass Spectrometry, 25, 2538–2560. DOI: 10.1002/rcm.5129

Cormie, A.B., Luz, B., & Schwarcz, H.P. (1994a). Relationship between the hydrogen and oxygen isotopes of deer bone and their use in the estimation of relative humidity. Geochimica et Cosmochimica Acta, 58, pp. 3439-49. (http://dx.doi.org/10.1016/0016-7037(94)90097-3)

Cormie, A.B., Schwartz, H.P., & Gray, J. (1994b). Determination of the hydrogen isotopic composition of bone collagen and correction for hydrogen exchange. Geochimica et Cosmochimica Acta, Vol. 58, pp. 365-375.

Cormie, A.B., Schwartz, H.P., & Gray, J. (1994c). Relation between hydrogen isotopic ratios of bone collagen and the rain. Geochimica et Cosmochimica Acta 58, pp. 377-391. (http://dx.doi.org/10.1016/0016-7037(94)90471-5)

Dansgaard, W. (1964). Stable isotopes in precipitation, Tellus, Vol. 16, pp. 436-468.

Delwiche, C.C. & Steyn, P.L. (1970). Nitrogen isotope fractionation in soils and microbial reactions. Environmental Science and Technology, Vol. 4, pp. 929-935.

DeNiro, M.J., & Epstein, S. (1978). Influence of diet on the distribution of carbon isotopes in animals. Geochimica et Cosmochimica Acta, 42, pp. 495-506. (http://dx.doi.org/10.1016/0016-7037(78)90199-0)

DeNiro, M.J., & Epstein, S. (1981) Influence of diet on the distribution of nitrogen isotopes in animals. Geochimica et Cosmochimica Acta, 45, pp. 341-351. (http://dx.doi.org/10.1016/0016-7037(81)90244-1)

DeNiro, M.J., & Weiner, S. (1988). Chemical, enzymatic and spectroscopic characterization of "collagen" and other organic fractions from prehistoric bones. Geochimica et Cosmochimica Acta Vol. 52, No. 9, pp. 2197-2206. (http://dx.doi.org/10.1016/0016-7037(88)90122-6)

DeNiro, M.J., & Weiner, S. (1988). Chemical, enzymatic and spectroscopic characterization of "collagen" and other organic fractions from prehistoric bones. Geochimica et Cosmochimica Acta, Vol.52, pp. 2197–2206.

Eyre, D.R. (1980). Collagen: molecular diversity in the body's protein scaffold. Science, Vol. 207, No. 4437, (March 21, 1980), pp. 1315–1322.

Farmer, A., Cade, B.S., & Torres-Dowdall, J. (2008). Fundamental limits to the accuracy of deuterium isotopes for identifying the spatial origin of migratory animals. Oecologia, Vol. 158, pp. 183-192. (http://dx.doi.org/10.1007/s00442-008-1143-6)

Faure, G. & Mensing, T.M. (2005). Isotopes: Principles and Applications.(3rd edition), Wiley, ISBN-10: 0471384372, Hoboken, NJ, USA.

Fetcher, E.S., Jr (1939). The water balance in marine mammals. The Quarterly Review of Biology, Vol. 14, pp. 451–459.

Fry, B., Lutes, R., Northam, M., Parker, P.L., & Ogden, J. (1982). A 13C/12C comparison of food webs in Caribbean seagrass meadows and coral reefs, Aquatic Botany, Vol. 14, pp. 389–398.

Gannes, L.Z., Martinez del Rio, C., & Koch, P. (1998). Natural abundance variations in stable isotopes and their potential uses in animal physiological ecology. Comparative Biochemistry and Physiology - Part A: Molecular & Integrative Physiology, Vol. 119, No. 3, pp. 725-737. (http://dx.doi.org/10.1016/S1095-6433(98)01016-2)

Gelse, K., Poschl, E., & Aigner, T. (2003). Collagens - structure, function, and biosynthesis. Advanced Drug Delivery Reviews, Vol. 55, pp. 1531-1546. (http://dx.doi.org/10.1016/j.addr.2003.08.00)

Handley, L.L., Austin, A.T., Robinson, D., Scrimgeour, C.M., Raven, J.A., Heaton, T.H.E., Schmidt, S., & Stewart, G.R. (1999). The 15N natural abundance(δ15N) of ecosystem samples reflects measures of water availability. Australian Journal of Plant Physiology, Vol. 26, pp.185–199.

Hare, P.E., Fogel, M.L., Stafford Jr., T.W., Mitchell, A.D., & Hoering, T.C. (1991). The isotopic composition of carbon and nitrogen in individual amino acids isolated from modern and fossil proteins. Journal of Archaeological Science, Vol. 18, pp. 211-292. (http://dx.doi.org/10.1016/0305-4403(91)90066-X)

Hayes, J.M. (2001). Fractionation of carbon and hydrogen isotopes in biosynthetic processes. In: Stable Isotope Geochemistry (J.W. Valley, D.R. Cole, Eds.), Reviews in Mineralogy and Geochemistry 43, 225-277. Washington DC: Mineralogical Society of America, 662 pp.

Heaton, T.H.E. (1987). Isotopic studies of nitrogen in the hydrosphere and atmosphere: a review. Chemical Geology (Isotope Geoscience), Vol. 59, pp. 87–102.

Hedges, R. E. M., Stevens, R. E., & Koch, P. L. (2006). Isotopes in bone and teeth. In Leng, M. J. (ed.), Isotopes in Palaeoenvironmental Research, Springer, pp. 117–146.

Hedges, R. E. M., Clement, J. G., Thomas, C. D. L., & O'Connell, T. C. (2007). Collagen turnover in the adult femoral mid-shaft: Modeled from anthropogenic radiocarbon tracer measurements. American Journal of Physical Anthropology, 133, pp. 808–816.

Hedges, R. E. M. & Law, I. H. (1989). The radiocarbon dating of bone, Applied Geochemistry, Vol. 4, pp. 249–53.

Hobson, K. A. & Clark, R. G. Assessing Avian Diets Using Stable Isotopes I: Turnover of 13 C in Tissues. The Condor, Vol. 94, No. 1. (Feb., 1992), pp. 181-188.

Hobson, K.A., Atwell, L., & Wassenaar, L.I. (1999). Influence of drinking water and diet on the stable- hydrogen isotope ratios of animal tissues. Proceedings of the National Academy of Sciences, Vol. 96, 8003-8006.

Hobson, K.A., Wassenaar, L.I., Mila, B., Lovette, I., Dingle, C., & Smith, T.B. (2003). Stable isotopes as indicators of altitudinal distributions and movements in an Ecuadorian hummingbird community. Oecologia, Vol. 136, pp. 302-308. (http://dx.doi.org/10.1007/s00442-003-1271-y)

Hoppe, K.A. (2009). How well do the stable hydrogen isotope ratios of bone collagen from modern herbivores track climatic patterns? 2009 Portland GSA Annual Meeting (18-21 October 2009), GSA Abstracts with Programs 41 (7), pp. 311. (http://gsa.confex.com/gsa/2009AM/finalprogram/abstract_166384.htm)

Howland, M.R., Corr, L.T., Young, S.M.M., Jones, V., Jim, S., Van Der Merwe, N.J., Mitchell, A.D., & Evershed, R.P. (2003). Expression of the dietary isotope signal in the compound-specific $\delta^{13}C$ values of pig bone lipids and amino acids. International Journal of Osteoarchaeology, 13, pp. 54-65. (http://dx.doi.org/10.1002/oa.658)

Huelsemann F., Flenker, U., Koehler, K., & Schaenzer, W. (2009). Effect of a controlled dietary change on carbon and nitrogen stable isotope ratios of human hair. Rapid Communications in Mass Spectrometry Vol. 23, pp. 2448-2454. (http://dx.doi.org/10.1002/rcm.4039)

Kirsanow K., Makarewicz, C., & Tuross, N. (2008). Stable oxygen ($\delta^{18}O$) and hydrogen (δD) isotopes in ovicaprid dentinal collagen record seasonal variation. Journal of Archaeological Science Vol. 35, No. 12, pp. 3159-3167. (http://dx.doi.org/10.1016/j.jas.2008.06.025)

Lee-Thorp, J. A., Sealey, J. C. & van der Merwe, N. J. (1989). Stable carbon isotope ratio differences between bone collagen and bone apatite, and their relationship to diet. Journal of Archaeological Science, Vol. 16, pp. 585–599.

Lee-Thorp, J.A. (2008). On isotopes and old bones. Archaeometry, Vol. 50, No. 6, pp. 925-950. (http://dx.doi.org/10.1111/j.1475-4754.2008.00441.x)

Leyden, J.J., Wassenaar, L.I., Hobson, K.A., & Walker, E.G. (2006). Stable hydrogen isotopes of bison bone collagen as a proxy for Holocene climate on the Northern Great Plains. Palaeogeography, Palaeoclimatology, Palaeoecology, Vol. 239, pp. 87-99. (http://dx.doi.org/10.1016/j.palaeo.2006.01.009)

Lis, G., Wassenaar, L.I., & Hendry, M.J. (2008). High-precision laser spectroscopy D/H and $^{18}O/^{16}O$ measurements of microliter natural water samples. Analytical Chemistry, Vol. 80, No. 1, pp. 287–293. (http://dx.doi.org/10.1021/ac701716q)

Meehan, T.D., Lott, C.A., Sharp, Z.D., Smith, R.B., Rosenfield, R.N., Stewart, A.C., & Murphy, R.K. (2001). Using hydrogen isotope geochemistry to estimate the natal latitudes of immature Cooper's hawks migrating through the Florida keys. Condor, Vol. 103, 11-20. (http://dx.doi.org/10.1650/0010-5422(2001)103[0011:UHIGTE]2.0.CO;2)

Minagawa, M., & Wada, E. (1984). Stepwise enrichment of 15N along food chains: further evidence and the relation between 15N and animal age. Geochimica et Cosmochimica Acta, Vol. 48, pp.1135–1140.

Ortiz, R.M. (2001). Osmoregulation in marine mammals. The Journal of Experimental Biology, Vol. 204, 1831–1844.

Pate, F.D., & Anson, T.J. (2007). Stable nitrogen isotope values in arid-land kangaroos correlated with mean annual rainfall: potential as a palaeoclimatic indicator. International Journal of Osteoarchaeology, Vol. 18, 317–326.

Petzke, K.J. & Lemke, S. (2009). Hair protein and amino acid ^{13}C and ^{15}N abundances take more than 4 weeks to clearly prove influences of animal protein intake in young women with a habitual daily protein consumption of more than 1g per kg body weight. Rapid Communications in Mass Spectrometry, Vol. 23, 2411-2420. (http://dx.doi.org/10.1002/rcm.4025)

Pietsch, S.J., Hobson, K.A., Wassenaar, L.I., & Tütken, T. (2011). Tracking Cats: Problems with Placing Feline Carnivores on $\delta^{18}O$, δD Isoscapes. Public Library of Science, ONE Vol. 6, No. 9, e24601.

Price, T.D., Schoeninger, M.J., & Armelagos, G.J. (1985). Bone chemistry and past behavior: an overview. Journal of Human Evolution Vol. 14, pp. 419-447. (http://dx.doi.org/10.1016/S0047-2484(85)80022-1)

Rau, G.H. (1982). The relationship between trophic level and stable isotopes of carbon and nitrogen. In Coastal water research project biennial report for the years 1981–1982. Edited by W. Bascom. Southern California Water Research Project, Long Beach, Calif. pp. 143–148.

Reynard, L.M., & Hedges, R.E.M. (2008). Stable hydrogen isotopes of bone collagen in palaeodietary and palaeoenvironmental reconstruction. Journal of Archaeological Science, Vol. 35, pp. 1934-1942. (http://dx.doi.org/10.1016/j.jas.2007.12.004)

Rundel, P.W., Ehleringer, J.R, & Nagy, K..A eds. (1988). Stable Isotopes in Ecological Research. Ecollogical Studies. ISBN 0-387-96712-5. Springer-Verlag, New York.

Sauer, P.E., Schimmelmann, A., Sessions, A.L., & Topalov, K. (2009). Simplified batch equilibration for D/H determination of non-exchangeable hydrogen in solid organic material. Rapid Communications in Mass Spectrometry, Vol. 23, No. 7, pp. 949-956. (http://dx.doi.org/10.1002/rcm.3954)

Schimmelmann, A. (1991). Determination of the concentration and stable isotopic composition of nonexchangeable hydrogen in organic matter. Analytical Chemistry, Vol. 63, pp. 2456-2459. (http://dx.doi.org/10.1021/ac00021a013)

Schimmelmann, A., & DeNiro, M.J. (1986). Stable isotopic studies on chitin III. The $^{18}O/^{16}O$ and D/H ratios in arthropod chitin. Geochimica et Cosmochimica Acta, Vol. 50, pp. 1485-1496. (http://dx.doi.org/10.1016/0016-7037(86)90322-4)

Schimmelmann, A., Sessions, A.L., & Mastalerz, M. (2006). Hydrogen isotopic (D/H) composition of organic matter during diagenesis and thermal maturation. Annual Review of Earth and Planetary Sciences, Vol. 34, pp. 501-533. (http://dx.doi.org/10.1146/annurev.earth.34.031405.125011)

Schoeninger, M.J. (1985). Trophic level effects on 15N/14N and 13C/12C ratios in bone collagen and strontium levels in bone mineral. Journal of Human Evolution, Vol.14, pp. 515-525.

Schoeninger, M.J. & DeNiro, M.J. (1984). Nitrogen and carbon isotopic composition of bone collagen from marine and terrestrial animals. Geochimica et Cosmochimica Acta, Vol. 48, pp. 625–639.

Schwarcz, H.P. (2000). Some biochemical aspects of carbon isotopic paleodiet studies. In: Biogeochemical Approaches to Paleodietary Analysis (S.H. Ambrose, M.A. Katzenberg, Eds.), Kluwer Academic/Plenum Publishers New York, pp. 189-209. (http://dx.doi.org/10.1007/0-306-47194-9_10)

Sessions, A.L., Burgoyne, T.W, Schimmelmann, A. & Hayes, J.M. (1999). Fractionation of hydrogen isotopes in lipid biosynthesis. Organic Geochemistry, Vol. 30, pp. 1193-1200. (http://dx.doi.org/10.1016/S0146-6380(99)00094-7)

Smith, B. N. & Epstein, S. (1971). Two categories of ~C/~2C ratios for higher plants. Plant Physiology, Vol. 47, pp. 380-384.

Smith, C. I., Craig, O. E., Prigodich, R. V., Nielsen-Marsh, C. M., Jans, M. M. E., Vermeer, C., Collins, M. J. (2005). Diagenesis and survival of osteocalcin in archaeological bone. Journal of Archaeological Science, Vol. 32, pp. 105–113.

Solomon, C.T., Cole, J.J., Doucett, R.R., Pace, M.L., Preston, N.D., Smith, L.E., & Weidel, B.C. (2009). The influence of environmental water on the hydrogen stable isotope ratio in aquatic consumers. Oecologia, Vol. 161, pp. 313-324. (http://dx.doi.org/10.1007/s00442-009-1370-5)

Tieszen, L.L., Boutton, T.W., Tesdahl, K.G., & Slade, N.A. (1983). Fractionation and turnover of stable carbon isotopes in animal tissues: implications for d13C analysis of diet. Oecologia, Vol. 57, pp. 32-37.

Topalov, K., Schimmelmann, A., Polly, P.D., Sauer, P.E., & Lowry, M. (2009). D/H of bone collagen as environmental and trophic indicator. Goldschmidt Conference Abstracts 2009, Davos, Switzerland (http://goldschmidt.info/2009/abstracts/T.pdf)

Tuross, N. (2002). Alterations in fossil collagen. Archaeometry, Vol. 44, pp. 427–434.

Tuross, N., Fogel, M.L., & Hare, P.E. (1988). Variability in the preservation of the isotopic composition of collagen from fossil bone. Geochimica et Cosmochimica Acta, Vol. 52, No.4, pp. 929-935. (http://dx.doi.org/10.1016/0016-7037(88)90364-X)

Tuross, N., Warinner, C., Kirsanow, K., & Kester, C. (2008). Organic oxygen and hydrogen isotopes in a porcine controlled dietary study. Rapid Communications in Mass Spectrometry, Vol. 22, pp. 1741-1745. (http://dx.doi.org/10.1002/rcm.3556)

van der Veer, G., Voerkelius, S., Lorentz, G., Heiss, G., & Hoogewerff, J.A. (2009). Spatial interpolation of the deuterium and oxygen-18 compositions of global precipitation using temperature as ancillary variable. Journal of Geochemical Exploration, Vol. 101, pp. 175-184. (http://dx.doi.org/10.1016/j.gexplo.2008.06.008)

van Klinken, G.J. (1991). Dating and dietary reconstruction by isotopic analysis of amino acids in fossil bone collagen - with special reference to the Caribbean. PhD thesis, University of Groningen.

Walker, P.L. & DeNiro, M.J. (1986). Stable nitrogen and carbon isotope ratios in bone collagen as indices of prehistoric dietary dependence on marine and terrestrial resources in Southern California. American Journal of Physical Anthropology, Vol. 71, pp. 51-61. (http://dx.doi.org/10.1002/ajpa.1330710107)

Walter, W.D. & Leslie, Jr., D.M. (2009). Stable isotope ratio analysi to differentiate temporal diets of a free-ranging herbivore. Rapid Communications in Mass Spectrometry, Vol. 23, pp. 2190-2194. (http://dx.doi.org/10.1002/rcm.4135)

Wassenaar, L.I. & Hobson, K.A. (2000). Improved method for determining the stable-hydrogen isotopic composition (δD) of complex organic materials of environmental interest. Environmental Science and Technology, Vol. 34, pp. 2354-2360. (http://dx.doi.org/10.1021/es990804i)

Waterlow, J.C. (2006). Protein Turnover. ISBN 0851996132. CABI Publishing, Wallingford, UK.

Yakir, D. & DeNiro, M.J. (1990). Oxygen and hydrogen isotope fractionation during cellulose metabolism in Lemna gibba L. Plant Physiology, 93, pp. 325-332.

8

A Facile One-Pot Synthesis of MSe (M = Cd or Zn) Nanoparticles Using Biopolymer as Passivating Agent

Oluwatobi S. Oluwafemi[1,*] and Sandile P. Songca[2]
[1]*Department of Chemistry and Chemical Technology, Walter Sisulu University Mthatha Campus, Private Bag XI*
[2]*Executive Dean, Faculty of Science, Engineering and Technology Walter Sisulu University, Tecoma, East London South Africa*

1. Introduction

In recent years, semiconductor quantum dots (QDs) sometimes also referred to as nanoparticles (NPs) or nanocrystals (NCs) have attracted much attention for many potential applications due to their unique physical and chemical properties such as size-dependent band gap, size dependent excitonic emission, enhanced nonlinear optical properties and size-dependent electronic properties attributed to quantum size-effect and enormously high specific surface area. Hence they are different from those of the corresponding bulk materials and contain a relatively large percentage of surface atoms which makes them extremely active (Alivastos, 1996; Bruchez et al., 1998). They have been extensively studied over the past decade and are useful in many wider areas of applications hence, they have become an important class of material for the photonic, electronic, biological and other technological industries in the 21st century (Bruchez et al., 1998; Chan & Nie, 1998; Coe et al., 2003; Murphy, 2002). The applicative utility of QDs as seen in various field are: medicine - for diagnostics, drug delivery and tissue engineering, chemistry and environment -for catalysis and filtration, energy - for reduction of energy consumption, increasing the efficiency of energy production, the use of more environmentally friendly energy systems, recycling of batteries, information and communication - novel semiconductor devices, novel optoelectronic devices, displays, quantum computers, heavy industry- aerospace, refineries, vehicle manufacturers to mention a few. Compared with conventional organic fluorophores, QDs exhibit bright fluorescence, narrow emission, broad UV excitation, high quantum yield and high photostability (Coastal –Fernandez et al., 2006; Ozkan, 2004; Derfus et al., 2004).

Among all semiconductor NPs, metal selenides have been the focus of great attention due to their importance in various applications such as thermoelectric cooling materials, optical filters and sensors, optical recording materials, solar cells, superionic materials, laser materials and biological labels. Many synthetic methods have been developed for the preparation of relatively monodispersed selenide nanoparticles (Murray et al., 1993; Korgel

* Corresponding Author

and Monbouquette, 1996; Trindade and O'Brien, 1996; Pileni, 1997). These methods have proved to be effective in the preparation of high quality selenide nanoparticles, using low molecular weight stabilizers such as thiols (Rockenberger et al., 1997), ethylhexanoate (Diaz et al., 1999), polyphosphate (Spanhel et al., 1987) and trioctylphosphine oxide (Murray et al., 1993; Trindade and O'Brien, 1996, 1997; Malik et al., 2001). However the synthesis usually involve high temperature, sophisticated equipment, complicated processes or complex reagents which are not environmentally friendly and rigorous conditions such as injection of hazardous metal alkyls, which are toxic, volatile, low boiling point materials, explosive at elevated temperature and pyrophoric. Therefore standard airless techniques are required to protect reaction reagents and the as-synthesised nanoparticles from oxidation, a major source of their cytotoxicity (Guo et al., 2007; Austin et al., 2004; Gaunt et al., 2005). As a result of these concerns on toxicity coupled with rigorous conditions involved in the preparation of these materials, the search for greener, sustainable and environmentally benign methods is still ongoing.

Apart from those that utilized highly toxic H_2Se (Trindade et al., 2001) or N, N-dimethylselenourea (Rogach et al., 2000) as selenium source, the solution reaction growth techniques appear to be the cheapest and most convenient approaches (Wang et al., 1999; Li et al., 2007; Ma et al., 2002a, 2002b; Yang and Xiang, 2005). However in solution, bare nanoparticles are thermodynamically unstable and always tend to aggregate, losing their peculiar properties. As a result of this, passivation of the surface is necessary in order to prevent uncontrolled growth and agglomeration of the nanoparticles. Most of the suitable capping agents used as stabilisers to modify the surface include thiols, ethylhexanoate, polyphosphate, trioctylphosphine oxide (TOPO), polyesters, starburst dendrimers and amino-derivatised polysaccharides. Among all these materials used as stabilisers, polymers, by far usually provide an excellent steric hindrance effect with robust stability against the environmental variation on the NPs. In addition, through hybridization with polymer, NPs are envisaged to inherit good compatibility, excellent capability and the high engineering performance which are sought after in most technical applications of the NPs (Abouraddy et al., 2007). Therefore polymer-inorganic nanocomposites have attracted much attention. It has been well established that solutions of polymeric materials contain size-confined, nanosized pools of inter - and intramolecular origin, which can be used for the synthesis of nanoparticles (Raveendran et al., 2003). In nanoparticles synthesis, linear as well as dendritic polymers have been used successfully. The Polyhydroxylated macromolecules in the polymer present interesting dynamic supramolecular associations facilitated by inter and intramolecular hydrogen bonding resulting in molecular level capsules, which can act as templates for nanoparticle growth. Thus these nanocomposites combine advantageous properties of polymers with properties of semiconductor nanoparticles. In general, the role of the polymers is to encapsulate the nanoparticles and enable better exploitation of their characteristic properties. However, it should be noted that polymers do not only serve as good host materials, they can also be used to modify the surface and/or to control the growth of nanoparticles. Surface modification could be of great importance for possible use of semiconductor nanoparticles in biomedical applications and diagnostics. In addition, biopolymers have been proved to be good controlled environments for the growth of metallic and semiconductor nanoparticles (Bozanic et al., 2009). A biopolymer such as starch is inexpensive, hydrophilic, nontoxic, biocompatible, biodegradable, is readily available from agriculture as the major component of carbohydrates and can be easily modified and transformed into other products (Mondal et al., 2004; Dragunki &Pawlicka 2001; Fang et al., 2005; Patel et al., 2010; Dzulkefly et al., 2007). It has a wide range of potential applications because it is abundant, renewable, safe and economic. Starch based materials have been used as substituent for petroleum-based plastic material especially in packaging industries and thus offer an alternative solution to the disposal and biodegradability problem of petroleum based materials (Fang et al., 2005; Patel et al., 2010; Dzulkefly et al., 2007). In addition, its biodegradable nature may also help in reducing cytotoxicity problems of QDs materials and hence extend their application to food and pharmaceutical products. Other polymers such as polyvinlylpyrolindone (PVP) and polyvinyl alcohol (PVA) are good choices as stabilizers because they can interact with metal ions by complex or ion-pair formation and can be designed to improve certain physical properties of semiconductor nanoparticles (Colvin et al., 1994; Dabbousi et al., 1996; Selim et al., 2005). In addition, the high viscosity of the polymer solution would be helpful in controlling the growth of selenide nanoparticles and thus prevent particles from aggregating hence no additional stabilizer would be needed. Furthermore, these polymers have high aqueous solubility and from an applications point of view, the polymer matrix would protect the selenide against photo-oxidation, a major factor responsible for the cytotoxicity of the quantum dots.

Most of the works on polymer-inorganic nanocomposites containing chalcogenide quantum dots have been concentrated on sulphide because of the accessibility of sulphide sources in wet chemical control synthetic methods, while for the preparation of polymer selenide nanocomposites there were some difficulties. Polymer-selenide nanocomposites are usually prepared via a two-stage process. This involves the synthesis of selenide nanocrystals and polymer separately, followed by the dispersion of the selenide nanoparticles into the polymer matrix. Though well dispersed polymer nanocomposites have been successfully prepared using these methods, as mentioned earlier, the preparation of the selenide nanocrystals usually involves sophisticated equipment, complex or toxic reagents which are not environmentally friendly and rigorous conditions in order to protect reaction reagents and the as-prepared material from oxidation. Therefore in order to develop technologies that can be used to improve or protect the environment, it is desirable to design and use greener methods to synthesise nanomaterials. As described by Murphy and Dahi (Murphy, 2008; Dahi et al., 2007) "green" synthesis involves the use of less toxic starting materials, limited reagents, fewer synthetic steps, reduced amounts of by-products and waste, low reaction temperature and if possible, the use of water as a solvent. Another drawback in the preparation of selenide nanocomposites is the dispersion process which is tedious, time-consuming, involving wastage of ancillary materials used in the reactions and reduction in the optical properties of the final nanocomposite hence, direct synthesis of the polymer selenide nanocomposites using a one – pot synthesis without the use of additional stabiliser will be a better alternative route.

Based on these two considerations, there has been few reports on the synthesis of polymer selenide nanocomposites. Yang et al., (2002) used the redox reaction of selenite and tellurite salts with cadmium nitrate to produce CdSe and CdTe nanowires in an autoclave, through a poly vinyl alcohol (PVA) assisted ethylenediamine solvothermal method at 160 - 180 ^{o}C. The PVA used in the process was favourable for the formation of nanowires by promoting the oriented attachment growth under solvothermal conditions. The as-synthesised CdSe and CdTe nanowires are mostly in cubic zinc - blende phase, growing along (111) zone axis direction. By varying the current density and temperature of the solution, Sarangi and Sahu

(2004) used cathodic electrodeposition techniques to produce nanocrystalline CdSe semiconductor thin films of different crystalline sizes. Using sodium selenosulphate as the selenium source, Li et al. (2007) reported the synthesis of spherical cubic structured CdSe nanoparticles with starch as capping agent. Ma and co-workers also reported a series of polyvinyl alcohol (PVA)-selenide nanocomposites via a one step solution growth technique at room temperature and ambient pressure (Ma et al., 2002a, 2002b). Though this technique is simple and its flexibility allows the use of different passivating agents such as starch, PVA, polyvinyl pyrolindone (PVP) (Yang and Xiang, 2005) and tartaric acid (Behboudnia and Azizianekalandaragh, 2007), the preparation of sodium selenosulphate used as the selenide source requires a longer reaction time and must be stored in the dark at 60 °C due to its instability at room temperature. Recently we have reported a facile, safe and inexpensive synthesis of nearly monodispersed organically soluble CdSe nanoparticles using selenide ion produced via reduction of selenium powder in water as the selenium source (Oluwafemi & Revaprasadu, 2008, 2009; Oluwafemi et al., 2010). Based on the modification of this technique, we have also reported for the first time, a series of water soluble selenide nanoparticles using cysteine, ascorbic acid, methionine and starch as passivating agents (Oluwafemi & Revaprasadu, 2007, 2009; Oluwafemi et al., 2008, 2010; Oluwafemi, 2009; Oluwafemi and Adeyemi, 2010). In this chapter, we will give a review of the starch work and adaptation of this synthetic route to other polymers. An entire 'green' chemistry is explored in this synthetic procedure without further purification steps or adjustments of the reaction environment. The hydroxyl groups of the polymers are expected to facilitate the complexation of the metal ion, solubilisation in water and conjugation site for further functionlisation of the as-synthesised nanoparticles hence no additional stabilisers are required. By varying the reaction time and Cd:Se ratio, the temporal evolution of the optical properties, shape and growth of the as-synthesised NPs were monitored. It is believed that the insight gained from this green synthetic approach will enable an economically viable and environmentally benign method for the synthesis of functionalised water soluble nanoparticles for large scale production and commercialisation.

This chapter is organised as follows: Following this introduction as section 1, a brief description of the synthesis and characterisation techniques used for the as-synthesised polymer capped selenide nanoparticles is given as section 2. In section 3, the mechanism of the reaction, results and discussion of the different selenide nanocomposites obtained using different polymers are given. Section 4, the last section gives a summary of the whole process, followed by references. Acknowledgements are cited before references.

2. Methodology

2.1 Synthesis

2.1.1 Synthesis of the precursor solution

Selenium precursor stock solution was prepared by adding 0.32 mmol of selenium powder to 20 mL deionised water in a three-necked flask. Sodium borohydride (0.81 mmol) was carefully added to this mixture and the flask was immediately purged with nitrogen gas to create an inert environment. The mixture was then stirred for 2 h, at room temperature. The entire selenium dissolves in water giving rise to a colourless selenium solution. The cadmium solution was prepared by adding 0.32 mmol of $CdCl_2$ powder in 20 mL of deionised water. The zinc solution was prepared by dissolving 0.32 mmol $ZnCl_2$ in 20 mL of deionised water. The starch solution was prepared by dissolving 0.05 % wt of soluble starch in 20 mL of deionised water under constant stirring at room temperature. The PVA solution was prepared by adding 6.0 g of PVA into 50 mL of deionised water and heated at 90 °C for 1 h under constant stirring to obtain a viscous transparent solution. The PVP solution was prepared by dissolving 1.0 g of PVP in 20 mL of deionised water under constant stirring at room temperature.

2.1.2 Synthesis of polymer capped MSe nanoparticles

The synthetic approach is very simple and does not require any special set up. In a typical room temperature reaction, 1.0 mL aqueous solution of cadmium chloride was added to 20 mL aqueous solution of soluble starch in a 50 mL one-necked round-bottom flask with constant stirring at room temperature. The pH of the solution was adjusted from 6 to 11 using 0.1 M ammonia solution. This was followed by a slow addition of 1.0 mL colourless selenide ion stock solution. The mixture was further stirred for 2 h and aged for 18 h. The resultant solution was filtered and extracted with acetone to obtain a red precipitate of CdSe nanoaprticles. The precipitate was washed several times and dried at room temperature to give a material which readily dispersed in water. The same procedure was repeated for the synthesis of PVA and PVP - capped CdSe nanoparticles by replacing the starch solution with the PVA and PVP polymers while the synthesis of elongated nanoparticles was achieved by changing the Cd:Se precursor ratio from 1:1 to 1:2. The synthesis of polymer capped ZnSe nanoparticles also follows the same procedure except that $ZnCl_2$ solution was used instead of $CdCl_2$ solution.

2.2 Characterisations

A Perkin Elmer Lamda 20 UV-vis Spectrophotometer was used to carry out optical measurements in the 200-1100 nm wavelength range at room temperature. Samples were placed in quartz cuvettes of 1 cm path length. Room temperature photoluminescence (PL) spectra were recorded on a Perkin Elmer LS 55 luminescence spectrometer with Xenon lamp over a range 400 – 800 nm. The samples were placed in quartz cuvettes of 1 cm path length. Fourier transform infrared (FT-IR) analysis was carried out using a Perkin Elmer spectrum FTIR Spectrometer with the universal ATR sampling accessory. A Philips CM120 BIOTWIW transmission electroscope sample viewed at 80k was used for the transmission emission microscopic (TEM) analysis while a JEOL 2100 TEM operating at 200 KV was used for the high resolution transmission electron microscopic (HRTEM) measurements. Samples for the analysis were prepared by putting an aliquot solution of the water soluble nanocrystalline material onto an amorphous carbon substrate supported on a copper grid and then allowing the solvent to evaporate at room temperature. XRD measurements of the samples were performed using a Bruker aXS D8 Advance diffractometer with Cu Kα radiation (λ = 1.5406 Å) operated at 40 kV and 40 mA.

3. Results and discussion

3.1 Starch-capped MSe nanoparticles

Starch is an inexpensive, hydrophilic, nontoxic, biocompatible and totally biodegradable polymer. It is a mixture of two main components: amylose formed by the α-1,4 glycosidic bonds between D-glucose units and amylopectin, a highly branched water-soluble macromolecule consisting of both α-1,4 and α-1,6 glycosidic bonds (Raveedran *et al.*, 2003). The large number of hydroxyl groups present in glucose monomers can facilitate the complexation of metallic ions to the starch matrix, while nano-supramolecular structures formed by intermolecular and intramolecular hydrogen bonding can act as templates for nanoparticle growth. Thus making starch a good candidate for passivating QDs. Using starch as passivating agent also comes with other advantages which can help in achieving green synthesis of nanoparticles. These includes (i) the high dispersity of starch in water hence, one can completely avoid the use of organic solvents, (ii) the relatively weak binding interaction between starch and nanoparticles compared to the interaction between the nanoparticles and typical thiol-based protecting groups which makes the separation of these particles feasible and (iii) integration of starch-protected nanoparticles readily into systems relevant for pharmaceutical and biomedical applications (Mondal et al., 2004; Dragunki & Pawlicka, 2001; Fang et al., 2005; Patel et al., 2010; Dzulkefly et al., 2007 ; Raveedran *et al.*, 2003). Few authors (Li et al., 2007, Taubert and Wegner, 2002; Raveendran et al., 2003; Wei et al., 2004) have reported the synthesis of inorganic nanoparticles using starch as capping agent. Most of these syntheses are either based on surface modification of the organically soluble nanoparticles prepared using a selenium source that is usually complexed to toxic organic material like trioctylphosphine oxide (TOP) before hot injection or the use of selenosulphate for direct synthesis in aqueous solution. Our work was the first reported 'green' room temperature one – pot synthesis of MSe (M= Cd or Zn) nanoparticles under a mild and environmentally benign conditions using starch as a passivating agent at room temperature (Oluwafemi, 2009; Oluwafemi and Adeyemi, 2010).

3.1.1 Reaction mechanisms

The overall chemical reaction involved in the process is represented by the following equations:

$$4NaBH_4 + 2Se + 7H_2O \longrightarrow 2NaHSe + Na_2B_4O_7 + 14H_2 \quad (1)$$
$$MCl_2 + \text{Starch} \longrightarrow \text{M-starch complex} \quad (2)$$
$$\text{M-starch complex} + HSe^- + OH^- \longrightarrow \text{starch-MSe} + H_2O \quad (3)$$
$$HSe^- + OH^- \longrightarrow Se^{2-} + H_2O \quad (4)$$

Scheme 1. Proposed chemical reactions involved in the formation of starch capped MSe nanoparticles

We propose a series of equations to represent the whole process of the formation of starch capped MSe nanoparticles at room temperature based on this synthetic route. The whole process is a redox reaction with selenium acting as the oxidant and MSe as the reduction product, while $NaBH_4$ acts as the reducing agent and $Na_2B_4O_7$ should be the oxidation product. In the experiments, at room temperature, Se is reduced to NaHSe, while the $NaBH_4$ is oxidised to $Na_2B_4O_7$. MCl_2 reacts with the starch to form the starch-metal complex. Finally the starch–metal complex reacts with Se^{2-} to form starch-capped MSe NPs. Scheme 1 shows the proposed chemical reactions involved in the formation of starch-capped MSe nanoparticles. Equation 1 is the reduction process of selenium powder in aqueous solution at room temperature under an inert atmosphere, for the formation of highly active hydrogen selenide ions. Equation 2 demonstrates that the metal chloride reacts with the starch to generate starch–

metal ion solution. In aqueous solution, starch adopts a right-handed helical conformation in which the ubiquitous hydroxyl groups are expected to facilitate the complexation of metal ions to the molecular matrix. Equation 3 is the reaction of starch–metal ion complex with the active HSe^- ions under alkaline medium to produce starch capped metal selenide nanoparticles. When hydrogen selenide ion solution was added to the starch - metal ion solution, it gradually released selenide ions (Se^{2-}) upon hydrolytic decomposition in alkaline media (Equation 4). The released selenide ions then react with metal ion to form seed particles (nucleation). The hydroxyl groups of the starch act as the coordination site for the release of the metal ions and hence control the chemical reaction rate of M^{2+} and Se^{2-} to produce MSe. These hydroxyl groups also act as passivating centres for stabilisation and solubility of the as-synthesized nanoparticles in water, while the free aldehyde group on one end of a starch polymer could also act as conjugation site for further functionalization which makes it easier for its biological applications (Raveendran et al., 2003;, Mishra et al., 2009; Rodriguez et al., 2008).

3.1.2 Starch capped CdSe nanoparticles

3.1.2.1 Structural analysis

The TEM image of the as-synthesised material at 1:1 precursor molar ratio (Figure 1) showed the presence of monodispersed spherical particles together with elongated particles of low aspect ratios [Oluwafemi, 2009]. The mean particle diameter was calculated to be 4.3 nm with standard deviation (σ) of 0.414 nm. The image also showed different patterns of how the dot particles arranged themselves into elongated ones. Different patterns such as tripods (I), sinusoidal (II) and S-shaped (III) were clearly visible. These different arrangement patterns clearly indicate the mechanism for the formation of one dimensional (1D) nanostructures i.e self-reorganisation occurring via adhesions between the spherical nanoparticles as a result of dipole-dipole interactions between the highly charged surfaces of II-VI semiconductor nanocrystals (Adam and Peng, 2001, 2002; Peng, 2003, Lee et al.,2002, 2003; Wang et al.,2006). The TEM image of the as-synthesised material at 1:2 precursor molar ratio consists of nanorods and nanowires of very high aspect ratio, together with interspersed monodispersed spherical particles (Figure 2). The presence of interspersed spherical particles supports the fact that the formation of elongated nanocrystals, is possibly due to self assembly between the spherical nanoparticles. The increase in the diameter of the interspersed particles at 1:2 (5.5 nm) and its broad size distribution (σ = 1.629 nm) further confirmed adhesion between the nanoparticles with different sizes. This morphology mutation indicates that, under this synthetic route, the final shape of the CdSe nanocrystals can be determined by controlling the precursor ratio. We proposed that at 0.5:1 (Cd:Se) molar ratio, high selenium monomer concentration in the precursor solution promote anisotropic growth of the pre-exiting particles by supplying the reaction with high kinetic drive and the slowest growth rate. The high kinetic drive in conjunction with the reaction temperature, which also provides the reaction system with low thermal energy (KT) and slower reaction rate, enhanced interaction between the existing spherical particles in the product solution, resulting in the formation of elongated particles. In addition, the presence of excess Se increases the Se dangling sites on the surface of CdSe nanoparticles. This makes the spherical particles thermodynamically less stable and increases the energy of the dipole-dipole attraction between the nanocrystals thus, enables the formation of 1D CdSe nanostructures. That is, under this condition, some of the nanocrystals can act as "adhesives" for the adhesion of two or more quantum dots to give the elongated particles. A similar observation has been reported for organically-capped nanocrystals (Tang et al., 2002; Wang et al., 2006; Deng et al., 2006). At lower initial precursor concentration (1:1), thermodynamically stable, isotropic quasi-spherical growth was observed. These results indicate that, the molar ratio of precursor plays an important role in the shape-control of CdSe nanocrystals and, the shape evolution from spherical to rod was the result of the reaction system shifting from thermodynamically-driven to kinetics-driven control.

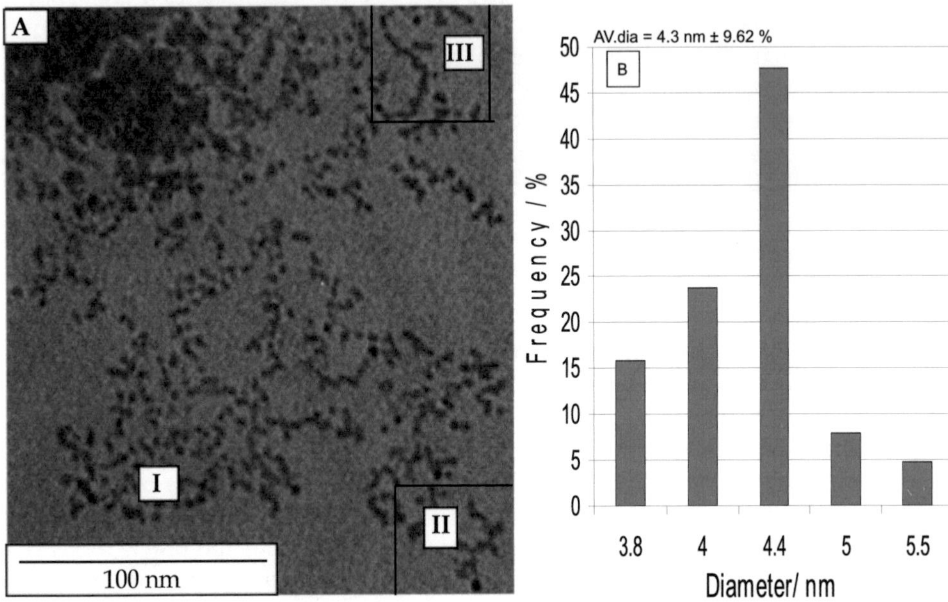

Fig. 1. (a) The TEM image and (b) particle size distribution of starch-capped CdSe nanoparticles at 1:1 precursor molar ratio.

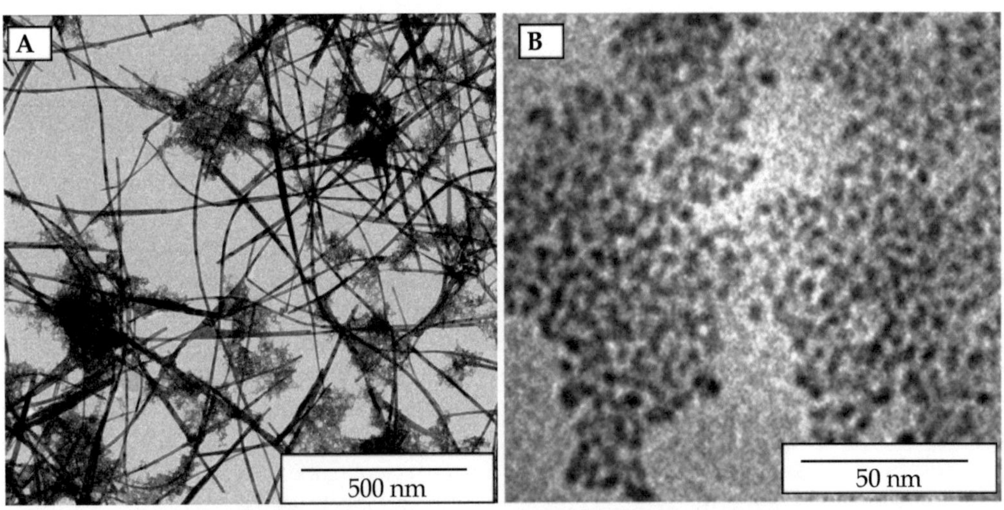

Fig. 2. The TEM images of starch-capped CdSe nanoparticles at 0.5:1 precursor molar ratio showing (A) nanowires with interspersed spherical particles and (B) a fraction of the interspersed spherical particles at low magnification.

The XRD patterns of the starch-capped CdSe nanoparticles at the two molar ratios exhibited predominantly wurtzite crystal structure with distinct diffraction peaks corresponding to the (002), (101), (102), (110), (103), and (112) crystalline planes of hexagonal CdSe (International centre for data diffraction PDF 2). Although the cubic zinc-blende and the hexagonal wurtzite phases could produce similar X-ray powder diffraction (XRD) results in nanocrystals (NCs), the XRD pattern showed unambiguous evidence that the particles are of a wurtzite structure: the (102) and (103) diffraction peaks characteristic of hexagonal-wurtzite CdSe structure, which are usually observed between 35° - 37° and 45° - 47° 2θ values respectively, were clearly seen (Oluwafemi, 2009). The sharp X-ray diffraction patterns suggest that the diffraction is predominantly due to the presence of elongated particles of a very high aspect ratio. This also accounts for the slight shift of the diffraction peaks to higher 2θ values. The sharp diffraction at 1:1 molar ratio was attributed to the self-assembly of the dot particles as indicated by the different arrangement pattern. Generally in thermodynamic terms, it is notable that cubic zinc - blende is the most stable form at lower temperatures (Li et al., 2007; Ma et al., 2002a, 2000b; Yang & Xiang, 2005), while wurtzite is more stable at high temperatures (Murray et al., 1993; Trindade & O'Brien, 1996, 1997; Malik et al., 2001). In the present study, hexagonal wurtzite phase was obtained at room temperature. The invariability of the crystal structure at the two molar ratios indicates that wurtzite is the sole stable resultant phase for this aqueous phase synthesis at room temperature. This shows that temperature is not the factor responsible for the crystal phase in the present case. The high Se:Cd ratio and the nature of the capping agent might induce the change in shape and phase of the CdSe nanocrystals under this synthetic method.

3.1.2.2 Optical analysis

The absorption spectra at 1:2 and 1:1 molar ratios showed an absorption shoulder at 602 nm and 558 nm with an absorption edge at 670 nm (1.85 eV) and 652 nm (1.90 eV) respectively. The broadness of the absorption shoulder was attributed to the self-assembly of the dot particles and the formation of elongated particles. Estimation of the particle diameters using the effective mass approximation (EMA) model (Brus, 1984): 5.59 nm (1:2) and 4.70 nm (1:1) are in agreement with the particle diameter calculated using the TEM. The slight red-shift observed in the absorption spectrum of the elongated particles, as compared to the spherical particles was attributed to the fact that, during the formation of elongated particles through adhesion of the dot particles, the band-gap of elongated particles depended more sensitively on their width than on their length although both parameters were responsible for the variation of band-gap (Wang et al., 2006). These particles emitted in the blue region. The position of the absorption and emission spectra of these particles appeared atypical, as the emission peaks were blue-shifted from the absorption band edges. This suggested that the emission originated from a higher energy state than the band gap, in total contrast to other reports of CdSe nanomaterials. However similar observation has been reported recently for aqueous synthesis of cysteine-capped CdTe by Green *et al.*, (2007).

3.1.3 Starch capped ZnSe nanoparticles

3.1.3.1 Structural analysis

The XRD pattern of the as-prepared ZnSe nanoparticles (Figure 3) exhibited predominantly wurtzite crystal structure with distinct diffraction peaks corresponding to the crystalline planes of hexagonal ZnSe and lattice constant of 0.397 nm, which was consistent with the JCPDS card for ZnSe (JCPDS no 89-2840). The mean particle diameter as calculated from the line width of the diffraction peak using Debye-Scherrer equation was 3.50 nm. As expected, the XRD peaks of the ZnSe NCs were considerably broadened compared to those of the bulk CdSe due to the nanocrystalline nature of the particles. The stronger and narrower (002) peak in the diffraction pattern indicated that the nanoparticles were elongated along the *c*-axis. The formation of hexagonal ZnSe NCs at room temperature further showed that this synthetic route and the capping agent might be responsible for the formation of hexagonal-wurtzite phase. Similar observation was made for the organically soluble CdSe NCs produced using this synthetic route and hexadecylamine as the capping agent (Oluwafemi and Revaprasadu, 2008; Oluwafemi et al., 2010). The mechanism behind this still needs to be elucidated.

The TEM images of the starch-capped ZnSe nanoparticles showed that the particles were well dispersed, small and spherical in shape. The particles are in the range 2.3 to 4.2 nm with mean particle diameter of 3.3 nm and standard deviation (σ) of 0.562 nm indicating broad size distribution. The TEM diameter is in accordance with the XRD result. The broad size distribution was attributed to the aggregation of the smaller particles which are thermodynamically unstable in the solution as a result of their high surface energy which makes them very reactive (Wei et al., 2004; Lee at al., 2002, 2003).

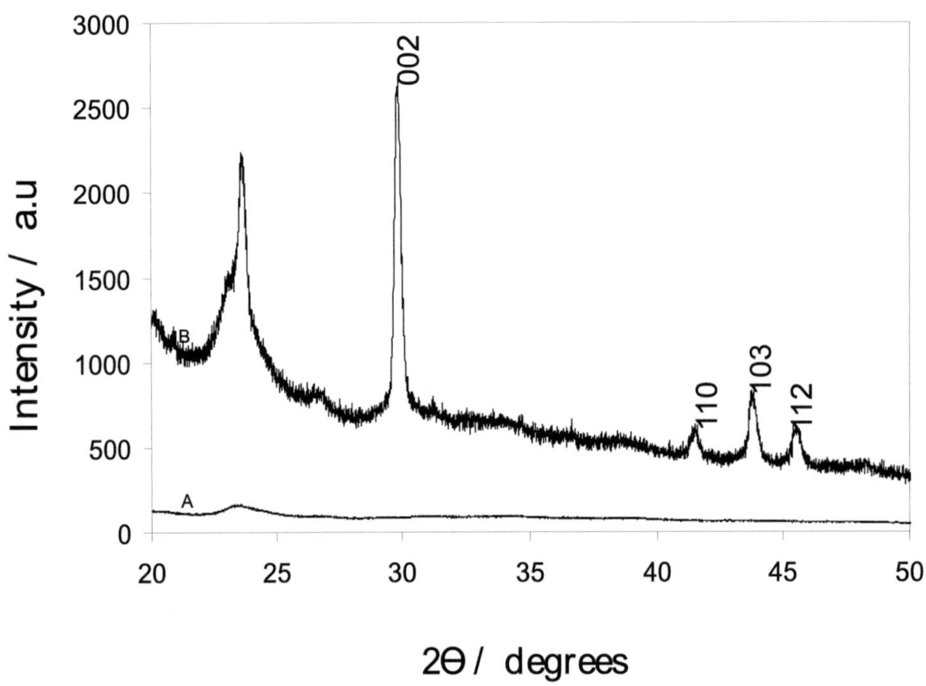

Fig. 3. XRD patterns of (a) pure starch and (b) starch -capped ZnSe nanoparticles.

3.1.3.2 Optical analysis

The absorption spectrum consists of an absorption shoulder at 282 nm (4.40 eV) signifying the monodispersity of the particles and an absorption edge at 303 nm (4.09 eV) which was blue shifted from the bulk band gap of 460 nm (2.70 eV) for bulk ZnSe. This shifting from the bulk band gap was attributed to strong quantum confinement and also signified small particle size. The emission spectrum was broad and showed a trap state emission with maximum at 408 nm which was red-shifted in relation to its absorption maximum. The emission curve was not smooth/symmetrical suggesting the presence of surface defects which was attributed to the smaller size of the particles. As the particle size decreases, the surface to volume ratio increases, thereby increasing the number of surface atoms and energy. This increased the density of the surface defects and resulted in ineffective passivation of the surface states (Xiong et al., 2007; Qu & Peng, 2002). The surface chemistry of the as-synthesised material was investigated using FT-IR spectroscopy. The spectra showed a strong band at 3298 cm^{-1} attributed to the O-H stretching of starch and its width was ascribed to the formation of inter and intramolecular hydrogen bond. The asymmetric stretching of C-H band was at 2921 cm^{-1} while the band at 1630 cm^{-1} was attributed to tightly bound water present in the starch. The two characteristic bands at the fingerprint region of spectra at 1146 cm^{-1} and 1075 cm^{-1} were attributed to the C-O, C-C stretching and C-O-H bending of starch respectively while the band at 1330 was attributed to the angular deformation of C-H. The prominent band at 992 was attributed to the skeletal mode vibration of α 1,4-glycosidic linkage (C-O-C) while the band at 757 was attributed C-C stretching. All these characteristic bands confirmed the capping of the particles by starch.

3.2 PVA and PVP-capped MSe nanoparticles

PVA and PVP are good choices as stabilizers because they can interact with metal ions by complex or ion-pair formation and can be designed to improve certain physical properties of semiconductor nanoparticles (Colvin et al., 1994; Dabbousi et al., 1996; Selim et al., 2005). In addition, the high viscosity of the polymer solution would be helpful in controlling the growth of selenide nanoparticles and thus prevent particles from aggregating hence no additional stabilizer would be needed. Furthermore, these polymers have high aqueous solubility and from an applications point of view, the polymer matrix would protect the selenide against photo-oxidation. Most of the direct aqueous synthesis for the polymer selenide nanocomposite usually involves the use of selenosulphate as the source of selenium. Ma and co-workers had reported a series of polyvinyl alcohol (PVA)-selenide nanocomposites using this technique (Ma et al., 2002a, 2002b). They observed that, the particle size can be determined to some extent by the experimental conditions such as pH, concentration of the free ions and reaction temperature. They indicated that under weakly basic conditions (pH ~10), sodium selenosulphate gradually released Se^{2-} upon hydrolytic decomposition, while in the acidic system, it decomposed quickly to produce selenium and no crystalline selenides were obtained. In all, their reactions complex ions play a significant role of controlling the release of free metal ions, which can then steadily combine with Se^{2-} to form nanocrystalline selenides. Otherwise, free metal ions with a high concentration would have rapidly combined with Se^{2-}, hence difficulties in controlling the nucleation and growth of the selenide nanocrystallites. In another development, Badr and Mahmoud (2005) using this technique prepared different sizes of PVA-capped CdSe NCs by varying the Cd:Se ratio. The optical properties showed that the particle size decreases with increasing cadmium ratio. The complexation between the polymer and the CdSe NPs as well as the conductivity of the PVA was also found to increase with decreasing particle size. This complexation leads to decrease in the intermolecular interaction between the PVA chains and thus decrease in the degree of PVA crystallinity. A similar observation has also been reported by Ma and co-workers (Ma et al., 2002b). Using selenosulphate, Yang and Xiang (2005) also reported the synthesis of PVP-capped CdSe nanoparticles. However, the reaction takes place in the presence of thioglycerol which acts as both complexant and catalyst for the decomposition of selenosulphate. In the absence of the complexant, no UV-vis absorption peak and PL emission peak was observed. The mechanism of the direct reaction between cadmium ion and selenosulphate in aqueous solution to form CdSe had been investigated by Yochelis and Hodes (Yochelis & Hodes, 2004). They discovered that CdSe were formed as a disordered phase surrounding the $CdSO_3$ produced via reaction between the Cd precursor and sodium selenosulphate in the presence of excess sodium sulphite. These amorphous CdSe then break off from $CdSO_3$ crystals to form nanocrystals of CdSe. In this section, we report the synthesis of PVA and PVP-capped MSe nanoparticles using selenide ion produced via reduction of selenium powder at room temperature. However as far as we know, our work will be the first reported 'green' room temperature one pot synthesis of PVA and PVP capped MSe (M= Cd or Zn) nanoparticles under a mild and environmental benign synthetic route.

3.2.1 Reaction mechanisms

The overall chemical reaction involved in the process is represented by the following equations:

$$4NaBH_4 + 2Se + 7H_2O \longrightarrow 2NaHSe + Na_2B_4O_7 + 14H_2 \quad (5)$$
$$M^{2+} + \text{Polymer} \longrightarrow \text{Polymer- metal ion complex (P-MC)} \quad (6)$$
$$\text{P-MC} + HSe^- + OH^- \longrightarrow \text{P-MSe} + H_2O \quad (7)$$
$$HSe^- + OH^- \longrightarrow Se^{2-} + H_2O \quad (8)$$

Scheme 2. Proposed chemical reactions involved in the formation of polymer capped MSe nanoparticles

As discussed earlier the whole process is a redox reaction. Selenium is reduced using sodium borohydride to give selenide ions. In the above reaction, the metal ion reacts with the polymer (PVP or PVA) solution to form the polymer-metal ion solution. Addition of the selenide ion solution to the polymer-metal ion solutions resulted in instantaneous change in the colour of the solutions from colourless to orange (PVA) and orange red (PVP). This indicates the formation of CdSe nanoparticles. The addition of the selenide solution to the polymer - metal ion solution resulted in gradual release of selenide ion (Se^{2-}) upon hydrolytic decomposition in alkaline media (equation 4). The released selenide ions then react with metal ion to form seed particles (nucleation).

$$mCd^{2+} + mSe^{2-} \rightleftharpoons (CdSe)m \quad (9)$$

The growth of the nanoparticles could have occurred either by the growth of CdSe on the seeds (growth from supersaturated solution) or by the process of Ostwald ripening whereby larger seed grow at the expense of the smaller ones.

In this reaction, the complexation of the metal ion by the polymer is expected to play a significant role in the formation of the nanocrystalline selenides. During the reaction, the complex ion controls the gradual release of the free metal ion which can then steadily combine with Se^{2-} to produce CdSe nanocrystals. This gradual release controls the nucleation and growth process. In addition, the high viscosity of the polymer solution would be helpful in preventing the particles from aggregating hence no additional stabilizer would be needed.

$$m\mathrm{Cd}^{2+} + m\mathrm{Se}^{2-} \rightleftharpoons (\mathrm{CdSe})m \qquad (10)$$

$$m\mathrm{Cd}^{2+} + m\mathrm{Se}^{2-} \rightleftharpoons (\mathrm{CdSe})m \longrightarrow (\mathrm{CdSe})m{+}1 \qquad (11)$$

$$(\mathrm{CdSe})n \longrightarrow (\mathrm{CdSe})n{-}1 + \mathrm{Cd}^{2+} + \mathrm{Se}^{2-} \quad (m > n) \qquad (12)$$

3.2.2 PVA- capped CdSe nanoparticles

3.2.2.1 Optical analysis

The absorption and emission spectra of PVA-capped CdSe nanoparticles are shown in Figure 4. The absorption band edges are all blue-shifted in relation to the bulk band gap (indicating quantum confinement effects) and appeared at the same position without any significant shift throughout the reaction time. The absorption spectra consist of an absorption band edge at 648 nm (1.91 eV) and excitonic shoulder at 299 nm and 315 nm signifying the monodispersity of the particles. The particle sizes as estimated using the effective mass approximation (EMA) model (Brus, 1984), is 4.57 nm. The appearance of the absorption features at the same position is an indication that the growth of the nanoparticles ceases after 1 hr. This shows that the generation of the CdSe particles reaches a homeostatic state after 1 hr, suggesting that the polymer matrix confines the growth of particles and prevents further growth or aggregation. The emission spectra (Figure 1 insert) show that the emission line width becomes narrower as the reaction time increases indicating that, extending the reaction further only results in better passivation of the as-synthesised CdSe nanoparticles. This shows that, there is optimal surface structure reconstruction of the nanocrystals with increasing reaction time which can be attributed to the effective passivation of the surface trapping states that are normally associated with semiconductor nanoparticles. All the absorption spectra showed the presence of two bands in the UV region located at 269 nm and 278 nm. These bands are attributed to the n-π^* excited transition of CdSe-PVA nanocomposite confirming the interaction/complexation between the polymer and the CdSe (Selim et al., 2005). All the particles emit in the blue region without any significant change in the emission maxima, which is at 406 nm. The observed blue shift in the emission spectra in relation to the absorption suggested that the emission originated from a higher energy state than the band gap (Green et al., 2007). This can be attributed to the shrinkage of the CdSe emitting core due to the interaction between the CdSe particles and the polymer. Similar observation has been reported by (Akamatsu *et al.*, (2005), for CdTe particles and was attributed to the particle reacting with the thiol transferring agent. This atypical trend in the position of the absorption and emission spectra of these particles was also reported earlier for the starch - capped CdSe nanoparticles under this synthetic route (Oluwafemi, 2009). It also suggests that the bigger particle size calculated from the absorption spectroscopy might be a contribution from both the capping group at the surface of the emitting CdSe core and the emitting CdSe core.

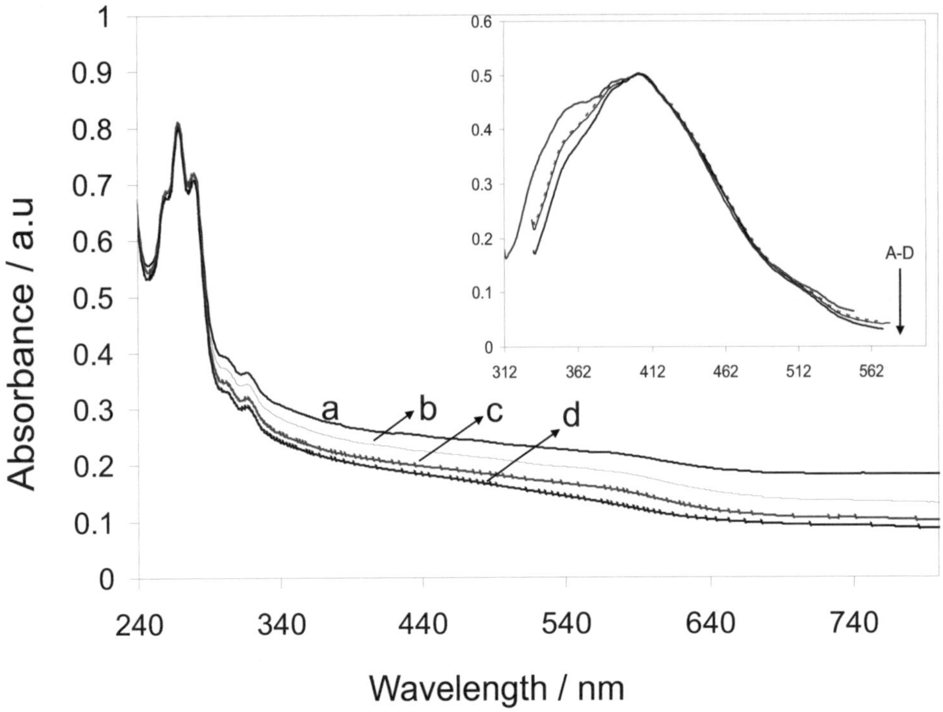

Fig. 4. Absorption and photoluminescence (inlet) spectra of PVA-capped CdSe nanoparticles at (a) 1 hr, (b) 3 hrs, (c) 5 hrs and (d) 24 hrs reaction times.

3.2.2.2 Structural analysis

A typical representative of the TEM image is shown in Figure 5A. The image shows well-defined, monodispersed, spherical particles. The particles are in the range of 3.80 to 5.0 nm with mean particle diameter of 4.51 nm. The relative standard deviation of the size of the nanoparticles shown in Figure 5A (insert) is 0.37 nm (8 %) indicating a narrow size distribution. These particle sizes suggest that, there is possibility for the adhesion of the smaller nanoparticles (seed particles) produced at the beginning of the reaction before complete passivation by the polymer. These smaller particles possesses high surface to volume ratio and are therefore very reactive. The surface chemistry of the as-synthesised materials investigated using FTIR spectroscopy confirmed the interaction of the particles with PVA as shown in figure 5B. The spectrum exhibits bands characteristic of stretching and bending vibrations of O-H, C-H, C=C and C-O groups. The strong bands observed at 3229-3365 cm^{-1} corresponding to the O-H stretching frequency indicates the presence of hydroxyl groups (Sweeting, 1968). The broad weak band observed at 2159 cm^{-1} has been assigned to the combination frequency of C-H with C-C (Raju et al., 2007). The bands observed at 1634 cm^{-1} and 1411 cm^{-1} have been assigned to the C=C stretching mode and bending modes of CH_2 group respectively. The weak broad band at 1103 cm^{-1} is attributed to the stretching mode of C=O group. The absence of non bonded -OH stretching band at $v = 3600\text{-}3650$ cm^{-1} and the decrease in the intensity of the O-H stretching vibration peak at $v = 3229\text{-}3365$ cm^{-1} in the PVA - CdSe spectrum further confirm the complexation of PVA chains CdSe nanoparticles.

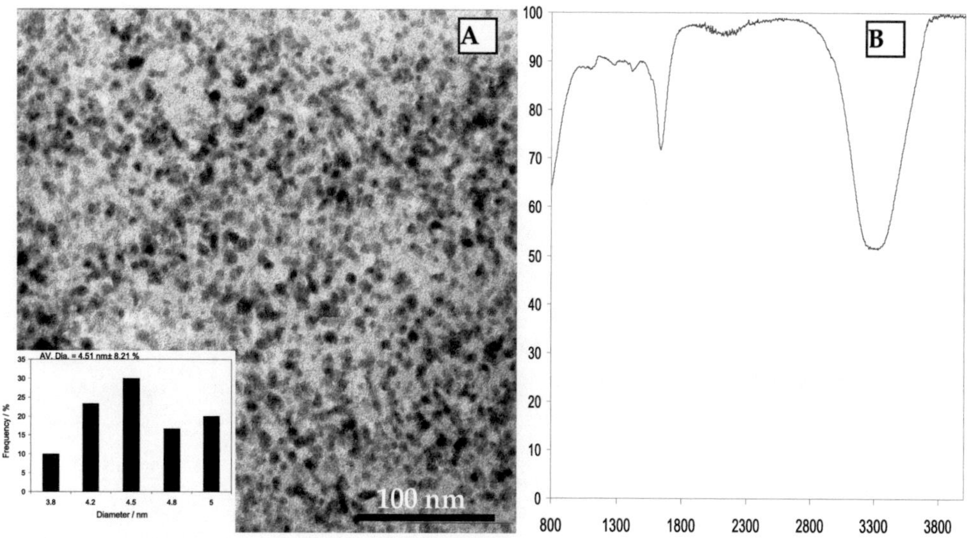

Fig. 5. The TEM image (A), particle size distribution (inlet) and FTIR spectra (B) of PVA - capped nanoparticles at 5 hrs reaction time.

3.2.3 PVP- capped CdSe nanoparticles

The addition of the colourless selenide ion solution to the PVP-metal ion solutions resulted in instantaneous change in the colour of the solutions from colourless to orange red. The orange red seed particles (after 1-h reaction) showed an absorption shoulder at about 546 nm and the band gap (*E*g) calculated was 2.27 eV. This absorption edge was significantly blue-shifted (167 nm) compared to bulk CdSe (*E*g =1.74 eV) indicating quantum effect. As the reaction times increases, the samples showed an absorption band edge at about 566 nm (*E*g (556) =2.19 eV) at 3 h with increased intensity in the colour of the solution. This increase in the band edge suggests the growth process via Ostwald ripening whereby larger seeds grew at the expense of the smaller ones. The absorption band edge remained at the same position at 5 h and for the rest of the reaction indicating that the polymer matrix confined the growth of the particles and the generation of CdSe particles reached a homeostatic state. Since the absorption edge was an index of particle size, smaller particles have larger band gaps and absorb at shorter wavelengths (Brus, 1983). The particles emit in the blue region and the emission line width is narrow, signifying particles with narrow size distributions. The emission maxima appeared at the same position through out the reaction time and are blue shifted in relation to the absorption band edge. This blue shift in the emission spectra in relation to the absorption has been attributed to emission from a higher energy state than the band gap (Green et al., 2007) due to the shrinkage of the CdSe emitting as a result of the interaction between the CdSe particles and the polymer. Since this atypical trend in the position of the absorption and emission spectra was reported for all the polymer capped CdSe nanoparticles under this synthetic route, it will be good to study the reaction earlier, i.e. before 1 hr, in order to gain better understanding of the reaction process.

The TEM image of the PVP capped CdSe at 5 hr in figure 6A showed small, spherical particles with some aggregation. The aggregation is due to the oriented attachment between the spherical nanoparticles as a result of dipole-dipole interactions between the highly charged surfaces of seed particles of nanocrystals. The particle sizes are in the range 3.5 to 5.8 nm with mean particle diameter of 4.49 nm. The HRTEM image in figure 6B shows clearly the lattice fringes, indicating the crystalline nature of the as-synthesised material. There are also discontinuities in the lattice fringes which are due to dislocations and stalking faults which further confirm oriented attachment growth among the nanoparticles.

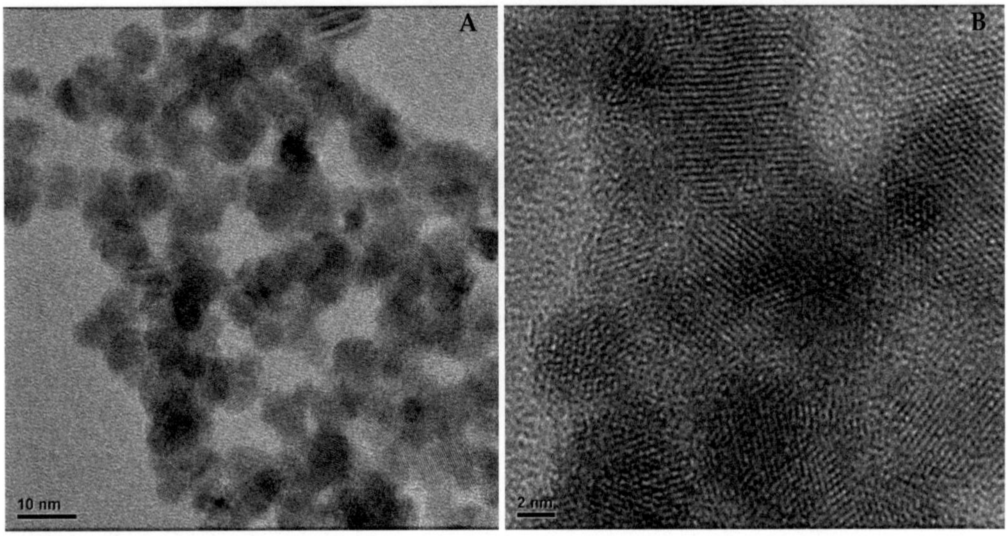

Fig. 6. Typical representative of the TEM image (A), and HRTEM image (B) of PVP -capped nanoparticles.

3.2.4 PVP-capped ZnSe nanoparticles

The absorption and emission spectra of PVP-capped ZnSe nanoparticles at different reaction times are shown in Figures 7 and 8. The absorption band-edges and emission maxima are slightly blue-shifted from the bulk band gap. The absorption spectra are characterised by tailing band-edges with no distinctive excitonic feature at lower reaction times (1–3 hrs) indicating the presence of large particles. As the reaction progressed, a distinctive excitonic shoulder appeared in the spectra, without any changes in the absorption band-edges of the particles at 5 and 24 h. The emission spectra of the particles showed band-edge luminescence. The emission maxima appeared at the same position as the reaction time increased for the first 3 hrs and decreased slightly as the reaction time increased. The emission maxima appeared at 446 nm (1 and 3 h), 435 nm (5 h) and 427 nm (24 h). The blue shift in the emission spectra and the appearance of the distinct excitonic shoulder at 5 and 24 h indicates the presence of smaller particles. This type of atypical optical spectra as the reaction time increases has been reported recently and was attributed to digestive ripening (Green et al, 2007). According to Green *et al.*, digestive ripening is essentially the opposite of Ostwald ripening. It usually involves the breaking of particles into smaller fragments or digestion of a layer at a time. Several factors such as ligands (Prasad et al.2002), solvent (Hines & Scholes, 2003) and surfactant (Green et al., 2007; Samia et al., 2005), have been reported as the driving force for this mode of particle growth. The digestive ripening reported here is possibly a contribution from both the ligand and reaction medium. Another possible explanation for this atypical optical spectrum, like the CdSe NPs discussed earlier, could be shrinkage of the ZnSe core as the reaction time increases which could be due to the interaction between the ZnSe particles and the polymer.

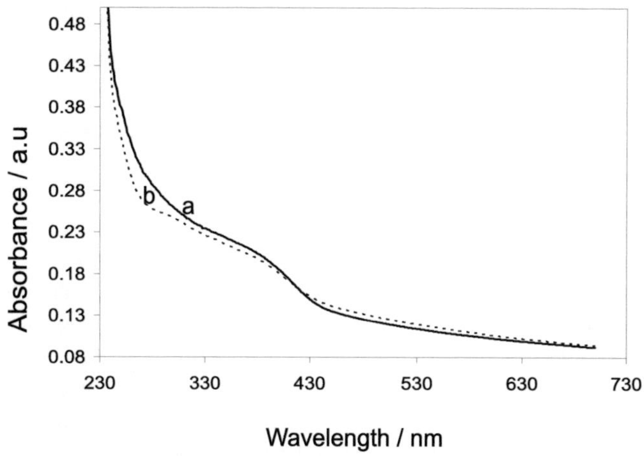

Fig. 7. Absorption spectra of PVP-capped ZnSe nanoparticles at (a) 5 h, (b) 24 h and Fig. 5 the photoluminescence (PL) at (a) 1 h, (b) 3 h, (c) 5 h and (d) 24 h reaction times

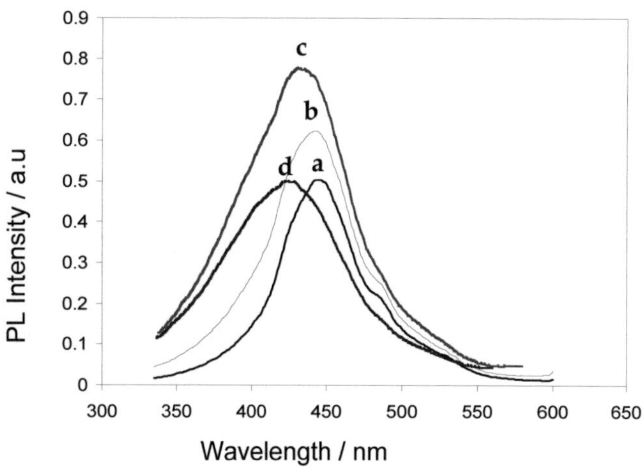

Fig. 8. The photoluminescence (PL) spectra of spectra of PVP-capped ZnSe nanoparticles at (a) 1 h, (b) 3 h, (c) 5 h and (d) 24 h reaction times

4. Conclusion

We have reported a simple, 'green', bench top, economical and environmentally benign room temperature synthesis of MSe (M=Cd or Zn) nanoparticles using starch, PVA and PVP as passivating agents. The whole process is a redox reaction with selenium acting as the oxidant and MSe as the reduction product. An entire "green" chemistry was explored in this synthetic procedure and it is reproducible. The optical spectroscopy showed that all the particles are blue shifted from the bulk band gap clearly due to quantum confinement. Starch capped CdSe nanoparticles showed the presence of monodispersed spherical particles together with elongated particles of low aspect ratios at 1.1 precursor molar ratio. Different arrangement patterns such as tripods, sinusoidal and S-shaped were also clearly visible. These arrangement patterns clearly indicate the mechanism for the formation of one dimensional (1D) nanostructure under this aqueous route. At 1:2 precursor molar ratio, nanorods and nanowires of very high aspect ratio were produced indicating that, under this synthetic route, the final shape of the CdSe nanocrystals can be determined by controlling the precursor ratio. The XRD patterns of the starch-capped CdSe nanoparticles at the two molar ratios exhibited predominantly hexagonal wurtzite crystal structure, in contrast to the cubic zinc-blende structure normally obtained at lower temperature. Well dispersed, small and spherical particles with wurtzite crystal structure were observed for the starch capped ZnSe nanoparticles. The absorption spectra of the PVA and PVP-capped CdSe nanoparticles showed that the growth of the particles reach a homeostatic state after 1 h and 3 h respectively, suggesting that, the polymer matrix confines the growth of particles. FTIR spectroscopic analysis confirmed the capping/interaction of the nanoparticles surface by/with the polymers. All the as-synthesised CdSe nanoparticles under this synthetic route emitted at lower wavelength compared to the absorption wavelength suggesting that, the emission originated from a higher energy state than the band gap. This has been attributed to the shrinkage of the CdSe emitting core due to the interaction between the CdSe particles and the polymers. More work is still required for better understanding of the whole process. Compared to the exiting methods, the solution route reported here is a relatively simple, "green", low-cost technique, one - pot and reproducible. We believe that the knowledge gained from this study will enable an easy greener approach that would be more promising for practical production and will open the way for direct synthesis of highly luminescent water-soluble CdSe nanoparticles using different biomolecules for large scale productions.

5. Acknowledgement

This work was supported by National Research Foundation (NRF), South Africa, University of Zululand research committee and Walter Sisulu University (WSU) directorate of research. The authors thank Dr I. A Oluwafemi for technical assistance and Prof. Revaprasadu for the opportunity to work in his group. We also thank Manfred Scriba (CSIR) for TEM and HRTEM measurements and Ms V. Ncapayi for the FTIR measurements.

6. References

Abouraddy, A. F.; Bayindir, M. and Benoit, G. T. (2007). Towards multimaterial multifunctional fibres that see,hear,sense and communicate. *Nat. Mater.* 6 336-347.

Adam, Z. and Peng, X. G. (2001). Mechanisms of the Shape Evolution of CdSe nanocrystals. *J. Am. Chem. Soc.*, 123, 1389-1395

Adam, Z. and Peng, X. G. (2002). Nearly Monodisperse and Shape-controlled CdSe nanocrystals via alternative Routes: Nucleation and growth. *J. Am. Chem. Soc.*, 124, 3343-3353.

Alivisatos, A. P. (1996). Semiconductor clusters, nanocrystals, and quantum dots. *Science,* 271 933-937.

Austin, M. D.; Warren, C. W. C. and Sangeeta, N. B. (2004). Probing the Cytotoxicity of Semiconductor Quantum Dots. *Nano Lett.*, 4, 11-18.

Badr, Y. and Mahmoud, M. A. (2005). Optimization and photophysics of cadmium selenide nanoparticles. *Physica* B: *condensed matter.*, 369, 278-286.

Behboudnia, M. and Azizianekalandaragh, Y. (2007). Synthesis and characterization of CdSe semiconductor nanoparticles by ultrasonic irradiation. *Mater. Sci. and Eng.* B, 138, 65-68.

Bozanic, D. K.; Djokovic, V.; Bibic, N.; Sreekumari Nair, P.; Georges, M. K. and Radhakrishnan, T. (2009). Biopolymer-protected CdSe nanoparticles. *Carbohydrate Research*, 344, 2383-2387.

Bruchez, M.; Moronne, J. M.; Gin, P.; Weiss, S. and Alivisatos, A. P. (1998). Semiconductor nanocrystals as flourescent Biological labels. *Science,* 281, 2013-2016.

Brus, L. E. (1983). A simple model for ionization potential, electron affinity and aqueous redox potentials of small semiconductor crystallites. *J. Chem. Phys.*, 79, 5566-5571.

Brus, L. E. (1984). Electron-electron and Electron-Hole Interactions in Small Semiconductor Crystallites: "The Size Dependence of the Lowest Excited Electronic State" *J. Chem. Phys.*, 80, 4403.

Carrot, G.; Scholz, S. M.; Plummer, C. J. G.; Hilbron, J. G. and Hedrick, L. J. (1999). Synthesis and Characterization of Nanoscopic Entitles Based on Poly(Caprolactone)-Grafted Cadmium Sulfide Nanoparticles. *Chem. Mater.*, 11, 3571-3577.

Chan, W. C. W. and Nie, S. M. (1998). Quantum Dots Bioconjugates for Ultrasensitive Nonisotopic Detection. *Science*, 281, 2016-2018.

Coe, S.; Woo, W. K.; Bawendi, M. G. and Bulovi, V. (2002). Electroluminescence from single monolayers of nanocrystals in molecular organic devices. *Nature*, 420, 800-803.

Colvin, V.; Schlamp, M. and Alivisatos, A. P. (1994). Light-emitting diodes made from cadmium selenide nanocrystals and a semiconductor polymer. *Nature*, 370, 354-357.

Costa-Fernandez, J. M.; Pereiro, R. and Sanz-Medel, A. (2006). The use of luminescent quantum dots for optical sensing. *Trends in Anal. Chem.*, 25, 207-218.

Dabbousi, B. O.; Bawendi, M. G.; Onitsuka, O. and Rubner, M. F. (1995). Electroluminescence from CdSe quantum dot /polymer composite. *Appl. Phys. Lett.*, 66, 1316- 1318.

Dahl, J. A.; Maddux, B. L. and Hutchison, J. E. (2007). Towards Greener Nanosynthesis. *Chem. Rev.*, 107, 2228- 2269

Deng, D. W.; Qin, Y. B.; Yang, X.; Yu, J. S. and Pan, Y. (2006). The selective synthesis of water-soluble highly luminescent CdTe nanoparticles and nanorods: The influence of precursor Cd/Te molar ratio. *J. Cryst. Growth,* 296, 141-149.

Derfus, A. M.; Chan W. C. W. and Bhatia, S. N. (2004). Probing the cytotoxicity of Semiconductor Quantum dots. *Nano. Lett.*, 4. 11-18.

Diaz, D.; Rivera, M.; Ni, T.; Rodriguez, J. C.; Castillo-Blum, S. E.; Nagesha, D; Robles, J.; Alvarez-Fregoso, O. J. and Kotov, N. A. (1999). "Conformation of ethylhexanoate stabilizer on the surface of CdS nanoparticles" *J. Phys. Chem.* B, 45, 9854-9858.

Dragunski, D. C. and Pawlicka A. (2001). Preparation and characterization of starch Grafted with Toluene Poly (propylene oxide) diisocyanate. *Mater. Res.*, 4, 77-81.

Dzulkefly, K.; Koon, S. Y.; Kassim, A.; Sharif, A. and Abdullah, A. H. (2007). Chemical modification of SAGO starch by solventless esterification with fatty acid chlorides. *The Malaysian J. Analy. Sci.*, 11, 395-399.

Fang, J. M.; Fowler, P. A., Escrig, C.; Gonzalez. R.; Costa, J. A. and Chamudis, L. (2005). Development of biodegradable laminate films derived from naturally occurring carbohydrate polymers. *Carbohydrate polymers*, 60, 39-42.

Gaunt, J. A.; Knight, A. E.; Windsor, S. A. and Chechik, V. (2005). Stability and quantum yield effects of small molecules additives on solutions of semiconductor nanoparticles. *J. Colloid and Interface Science*, 290 (2), 437-443.

Green, M.; Harwood, H.; Barrowman, C.; Rahman, P.; Eggeman, A.; Festry, F.; Dobson, P. and Ng T. (2007). A facile route to CdTe nanoparticles and their use in bio-labelling. *J. Mater. Chem.*, 17, 1989-1994.

Guo, G.; Liu, W.; Liang, J.; He, Z.; Xu, H. and Yang, X. (2007). Probing the cytotoxicity of CdSe quantum dots with surface modification. *Mater. Lett.*, 61, 1641-1644.

Hines, M. A. and Scholes, G. D. (2003). Colloidal PbS Nanocrystals with Size-Tunable Near-Infrared Emission: Observation of Post-Synthesis Self-Narrowing of the Particle Size Distribution. *Adv. Mater.* 15; 1844-1849

Korgel, B. A. and Monbouquette, H. G. (1996). Synthesis of Size-Monodisperse CdS Nanocrystals Using Phosphatidylcholine Vesicles as True Reaction Compartments. *J. Phys. Chem.*, 100, 346-351.

Lee, S. M.; Jun, Y. W.; Cho, S. N. and Cheon, J. W. (2002). Single-Crystalline Star-Shaped Nanocrystals and Their Evolution: Programming the Geometry of Nano-Building Blocks. *J. Am. Chem. Soc.*, 124 (38), 11244-11245.

Lee, S. M.; Cho, S. N. and Cheon, J. W. (2003). Anisotropic shape control of colloidal inorganic nanocrystals. *Adv. Mater.*, 15, 441-444.

Li, J. H.; Ren, C. L.; Liu, X. Y.; Hu, Z. D. and Xue, D. S. (2007). "Green" synthesis of starch capped CdSe nanoparticles at room temperature. *Mater. Sci. Eng.* A, 458, 319-322.

Ma, X. D.; Qian, X. F.; Yin, J. and Zhu, Z. K. (2002a). Preparation and characterization of polyvinyl alcohol–selenide nanocomposites at room temperature. *J. Mater. Chem.*, 12, 663-666.

Ma, X. D.; Qian, X. F.; Yin, J.; Xi, H. A. and Zhu, Z. K. (2002b). Preparation and Characterization of Polyvinyl Alcohol-Capped CdSe Nanoparticles at Room Temperature. *J. Colloid and Interface Science,* 252, 77-81.

Malik, M. A.; Revaprasadu, N. and O' Brien, P. (2001). Air-stable single-source precursors for the synthesis of chalcogenide semiconductor nanoparticles. *Chem. Mater.*, 13, 913-920.

Mishra, P.; Yadav, R. S. and Pandey, A. C. (2009). Starch assisted sonochemical synthesis of flower-like ZnO nanostructure. *Digest. J. Nano. Bios.*, 4, 193-198.

Mondal, K.; Sharma, A. and Gupta, M. N. (2004). Three phase partitioning of starch and its structural consequences. *Carbohydrate polymers*, 56, 355-359.

Murphy, C. J. (2002). Optical Sensing with Quantum Dots. *Anal. Chem.*, 74, 520A-526A.

Murphy C.J. 2008. Sustainability as a Design Criterion in Nanoparticle Synthesis and Applications. *J. Mater. Chem.*, 18, 2173-2176.

Murray, C. B.; Norris, D. J. and Bawendi M. G. (1993). Synthesis and characterization of nearly monodisperse CdE(E=sulphur,selenium,tellurium) semiconductor nanocrystals. *J. Am. Chem. Soc.*, 115, 8706-8715.

Oluwatobi, S. O. and Revaprasadu, N. (2007). A Novel Method to Prepare Cysteine Capped Cadmium Selenide Nanoparticles. *Mater. Res. Soc. Symp. Proc. 0951*, E03-12.

Oluwafemi, O. S. and Revaprasadu, N. (2008). *'A new synthetic route to organically capped cadmium selenide Nanoparticles'*. *New J. Chem.*, 32, 1432-1437.

Oluwafemi, O. S., Revaprasadu, N. and Ramirez, A. J. (2008). A Novel One-Pot route for the Synthesis of Water-Soluble Cadmium Selenide Nanoparticles. *J. Cryst. Growth*, 310, 3230-3234.

Oluwafemi, O. S. and Revaprasadu, N. (2009). Study on Growth Kinetics of Hexadecylamine capped CdSe Nanoparticles using its electronic properties. *Physical B: Condense matter*, 404, 1204-1208.

Oluwafemi, O. S. (2009). A Novel 'Green' Synthesis of Starch-Capped CdSe Nanostructures. *Journal of Colloids and Surfaces B: Biointerfaces*, 73, 382-386.

Oluwafemi, O. S. and Revaprasadu, N. (2009). A Facile, 'Green" One - Step, Room Temperature Synthesis of a Series of monodispersed MSe(M = Cd or Zn)Water Dispersible Nanoparticles. *Mater. Res. Soc. Symp. Proc.*, 1138, FF 12-19.

Oluwafemi, O. S and Adeyemi, O. O. (2010). One -pot room temperature synthesis of biopolymer -capped ZnSe nanoparticles. *Journal of Materials Letters*, 64, 2310-2313.

Oluwafemi, O. S.; Revaprasadu N and Adeyemi O. O. (2010). A new synthesis of hexadecylamine-capped Mn-doped wurtzite CdSe nanoparticles. *Material Letter*, 64, 1513-1516.

Oluwafemi O. S., Revaprasadu N and Adeyemi O. O. (2010). A facile "green" synthesis of ascorbic acid-capped ZnSe nanoparticles. *Colloids Surf. B: Biointerfaces*, 79, 126-130.

Ozkan, M. (2004). Quantum Dots and other Nanoparticles: What can they offer to drug discovery?. *DDT*, 9, 1065-1071.

Patel, S.V.; Venditti, R. A. and Pawlak, J. J. (2010). "Dimensional changes of starch microcellular foam during the exchange of water with ethanol and subsequent drying," *BioResources*, 5, 121.

Peng, X. G. (2003). Mechanisms for the Shape-control and Shape-Evolution of Colloidal Semiconductor Nanocrystals. *Adv. Mater.*, 15, 459-463.

Pilen, M. P. (1997). Nanosized Particles Made In Colloidal Assembly. *Langmuir*, 13, 3266-3276.

Prasad, B. L. V.; Stoeva, S. I.; Sorensen, C. M. and Klabunde, K. J. (2002). Digestive Ripening of Thiolated Gold Nanoparticles: The Effect of Alkyl Chain Length. *Langmuir*, 28, 7515-7520.

Qi, L.; Colfen, H. and Antonietti, M. (2001). Synthesis and Characterization of CdS Nanoparticles Stabilised by Double Hydrophilic Block Copolymers. *Nano Lett.*, 1, 61-65.

Qu, L. and Peng, X. (2002). Control of Photoluminescence of CdSe Nanocrystals in Growth. *J. Am. Chem. Soc.*, 124, 2049-2055.

Raju, C. L.; Rao, J. L., Reddy, B. C. V. and Brahmam, K. V. (2007). Thermal and IR studies on copper doped polyvinyl alcohol. *Bull. Mater. Sci.*, 30, 215–218.

Raveendran, P.; Fu, J. and Wallen S. L. (2003). Completely"Green" Synthesis and Stabilization of Metal Nanoparticles. *J. Am. Chem. Soc.*, 125, 13940-13941.

Rockenberger, J.; Troger, L; Kornowski, A.; Vossmeyer, T.; Eychmuller, A.; Feldaus, J. and Weller, W. (1997). *J. Phys. Chem.* B, 101, 291

Rodriguez, P.; Munoz-Aguirre, N.; Martinez, E. S.; Gonzalez de la Cruz, G.; Tomas, S. A. and Angel O. Z. (2008). Synthesis and spectral properties of starch capped CdS nanoparticles in aqueous solution *J. Cryst. Growth*, 310, 160-164.

Rogach, A. L.; Nagesha, D.; Ostrander, J. W.; Giersig, M. and Kotov, N. A. (2000). "Raisin Bun"-Type Composite Spheres of Silica and Semiconductor Nanocrystals. *Chem. Mater.*, 12, 2676-2685.

Samia, A. C. S.; Hyzer, K.; Schlueter, J. A.; Qin, C. J.; Jiang, J. S.; Bader, S. D. and Lin, X-M. (2005). Ligand Effect on the Growth and the Digestion of Co Nanocrystals. *J. Am. Chem. Soc.*, 127, 4126-4127.

Sarangi, S. N. and Sahu, S. N. (2004). CdSe nanocrystalline thin films: composition, structure and optical properties. *Physica E*, 23, 159-167.

Selim, M. S.; Seoudi, R. and Shabaka, A. A. (2005).Polymer based films embedded with high content of ZnSe nanoparticles. *Mater. Lett.*, 59, 2650-2654.

Sondi, I.; Siiman, O.; Koester, S. and Matijevic E. (2000). Preparation of Aminodextran-CdS Nanoparticle Complexes and Biologically Active Antibody-Aminodextran-CdS Nanoparticle Conjugates. *Langmuir*, 16, 3107-3118.

Sooklal, K.; Hanus L. H.; Ploehn H. J. and Murphy, C. J. (1998). A Blue-Emitting CdS/Dendrimer Nanocomposite *Adv. Mater.*, 10, 1083-1087.

Spanhel, L.; Hasse, M.; Weller, H. and Henglein, A. (1987). *J. Am. Chem. Soc.*, 109, 5649.

Sweeting, O. J. (1968). *The science and technology of polymer films*, New York: Interscience Publishers

Tang, Z.; Kotov, N. A. and Giersig, M. (2002). Spontaneous Organization of Single CdTe Nanoparticles into Luminescent Nanowires. *Science*, 297, 237-240.

Taubert, A. and Wegner, G. (2002). Formation of Uniform and Monodisperse Zincite Crystals in the Presence of Soluble Starch. *J. Mater. Chem.*, 12, 805-807.

Trindade, T.; O'Brien P. and Zhang, X. (1997). Synthesis of CdS and CdSe Nanocrystallites using a novel single-molecule precursor approach. *Chem. Mater.*, 9, 523-530.

Trindade, T. and O' Brien, P. (1996). A single source approach to the synthesis of CdSe nanocrystallites. *Adv. Mater.*, 8, 161-163.

Trindade, T.; O'Brien, P. and Pickett, L. N. (2001). Nanocrystalline Semiconductors: Synthesis, Properties and Perspectives. Chem. Mater., 13, 3843-3858.

Wang, W.; Geng, Y.; Yan, P.; Liu, F.; Xie, Y. and Qian, Y. (1999). A novel mild route to nanocrystalline selenides at room temperature. *J. Am .Chem. Soc.*, 121, 4062-4063.

Wang, Q.; Pan, D.; Jiang, S.; Ji, X.; An, L. and Jiang, B. (2006). A solvothermal route to size- and shape-controlled CdSe and CdTe nanocrystals. *J. Cryst. Growth*, 286, 83-90.

Wei, Q. L.; Kang, S-Z. and Mu, J. (2004). "Green" synthesis of starch capped CdS nanoparticles. *Colloid. Surface. A*, 247, 125-127.

Xiong, S.; Huang, S.; Tang, A. and Teng, F. (2007). Synthesis and luminescence properties of water-dispersible ZnSe nanocrystalls. *Mater. Lett.*, 61, 5091-5094.

Yang, Q.; Tang, K.; Wang, C.; Qian, Y. and Zhang, S. (2002). PVA-Assisted Synthesis and Characterization of CdSe and CdTe Nanowires. *J. Phys. Chem.* B, 106, 9227-9230.

Yang, Y. J. and Xiang, B. J. (2005). Wet synthesis of nearly monodisperse CdSe nanoparticles at room temperature. *J. Cryst. Growth*, 284, 453-458.

Yochelis, S. and Hodes, G. (2004). Nanocrystalline CdSe Formation by Direct Reaction between Cd Ions and Selenosulfate Solution. *Chem. mater.*, 16, 2740-2744.

9

Alkylresorcinols Protect the DNA from UV-Damage *In Vitro* and *In Vivo* Models

Dmitry Deryabin, Olga Davydova and Irina Gryazeva
Orenburg State University, Department of Microbilogy
Russia

1. Introduction

Alkylresorcinols (AR) are a large group of secondary metabolites synthesized by bacteria [El-Registan et al., 2006], fungi [Zarnowski et al., 1999] and plants [Kozubek & Tyman, 1999]. In microorganisms AR control many aspects of a functional and morphological differentiation [Nikolaev et al., 2006], controlling transition to the dormant state [Mulyukin et al., 2003]. Such biological activity of AR is determined by their ability to interact with a broad spectrum of biopolymers, such as membrane lipids, proteins and nucleic acids. AR interaction with membrane lipids is determined by the amphiphilic properties of these molecules and leads to an increase of cytoplasmic membrane microviscosity, changing its permeability for monovalent ions [El-Registan et al., 1979; Reusch et al., 1983] and following cell dehydration [Kaprelyants et al., 1986]. The effect of AR interaction with the enzyme proteins appears in change of the rate of catalysis [Martirosova et al., 2004] and the spectrum of substrate specificity [Petrovskii et al., 2009] with a synchronous significant increase of protein globules stability to various extreme factors [Solyanikova et al., 2011]. Results of AR interaction with biopolymers and supramolecular structures in the whole cell are the control of metabolic processes, including their inhibition and the formation of cyst-like forms of microorganisms [Nikolaev et al., 2006], characterized by increased resistance to different stress factors [Stepanenko et al., 2004].

At the same time formation of anabiotic state and increase of stress resistance of bacterial cells are suggest the stabilization of the main genetic information storage biopolymer (DNA) with the obtaining of resistance to a wide range of abiotic and biotic factors [Azam & Ishihama, 1999; Setlow, 1995]. Due to chemical structure, alkylresorcinols are possible candidates for this role. Thus, the antimutagenic effect of AR is well described [Gasiorowski & Brocos, 2001], as well as effect of compactisation of the nucleoid in the dormant bacterial cells [Mulyukin et al., 2005], accompanied by a change of elastic and viscous characteristics of DNA.

In our previous works the fact of AR-DNA interactions, resulting in modifications of physicochemical properties of this biopolymer with formation of supramolecular complexes has been described [Davydova et al., 2005]. The AR-DNA interactions also leads to B $\rightarrow$ A transition of DNA, increase the thermostability of these complexes and improving the resistance of DNA to some external influences [Davydova et al., 2006, 2007].

Much less is known about the influence of AR on DNA sensitivity to the UV irradiation, though this problem is significant because UV is an important ecological factor and also finds wide application in various medical and biological technologies. The damaging effect of UV on DNA consists in its absorption by nitrogenuos bases with the formation of intra- and intermolecular cross-links [Lyamichev et al., 1990], while increasing the UV-dose causes deep degradation of the biopolymer caused by single- and double-stranded breaks [Cariello et al., 1988]. Potential effects of AR in this system are defined by their ability to absorb ultraviolet radiation in the range of 245-295 nm for different AR homologues, and to act as antioxidant similar to other photoprotectors [Fraikin et al., 2000]. According to accumulated data, AR are also capable to change the activity of stress regulons, including the SOS-response genes [Golod et al., 2009], that are realize an active DNA reparation at various damages.

In this reason the goal of this work is to investigate the complex effects of AR on UV sensitivity of DNA in molecular (*in vitro*) and cellular (*in vivo*) models.

2. Materials and methods

2.1 Materials

Chemically synthesized high-grade AR (99,9%), differing in the length and an arrangement of alkyl radical (table 1), including C_1-AR, C_5-AR and C_6-AR (Sigma, USA), and also C_3-AR and C_{12}-AR (Enamine, Ukraine) were used in this work. The DNA length markers λ/HindIII restricts (from 10000 down to 250 bp) and the plasmid pUC19 (2686 bp) were produced by SibEnzyme Ltd. (Russia). For PCR analysis DNA fragments (273 bp) of *Chlamydia trachomatis* were used (NPF "Liteh", Russia).

2.2 Bacterial strain

Escherichia coli K12 TG1 strain was used as a recipient for transformation. At studying of SOS-system activity the recombinant bioluminescent strain of *Escherichia coli recA'::lux* containing plasmid-borne fusions of the recA promoter-operator region to the Photorhabdus luminescens ZM 1 lux genes (GosNIIgenetika, Russia) was used. Increase of their luminescence in the presence of DNA damage factors [Rosen et al., 2000], were shown previously. Investigation of the luminescent response of this strain to UV radiation allows quantitatively estimate in a real time a SOS-system induction.

2.3 UV light irradiation

Preparations of DNA and bacterial suspensions irradiated using a germicidal bulb (Osram, Germany) through an optical filter of 254 nm with a power light exposure of 6.7 W/m^2 as determined by UV radiometer (TKA-PKM, Russia).

2.4 Agarose gel electrophoresis

DNA electrophoresis was performed in 0.8% agarose gel containing 0.5 μg/ml ethidium bromide in TBE buffer (pH 7.2) at field strength of 5 V/cm. The agarose gel was visualized on UV transilluminator (Vilber Lourmat, France) and photographed. The received digital images were processed with use of "ImageJ" software (NIH, USA).

AR-abbreviation	Structural formula	Molecular weight	Producer
C_1-AR	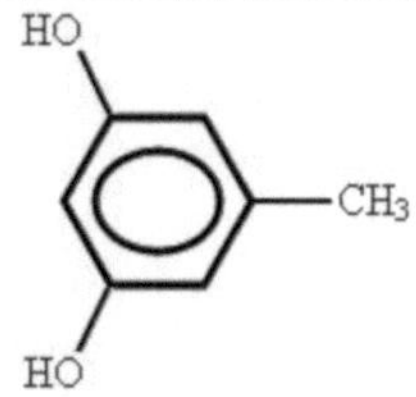	124	Sigma, USA
C_3-AR	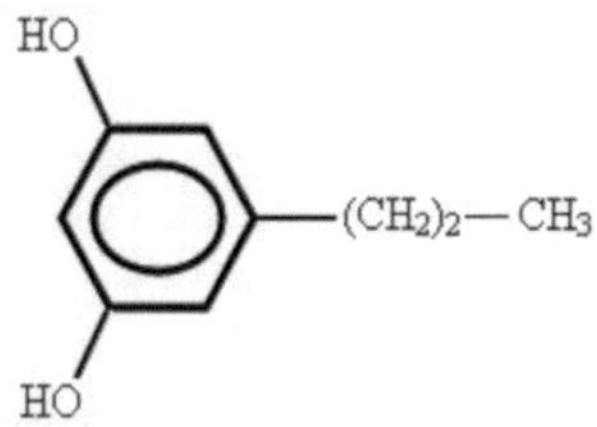	152	Enamine, Ukraine
C_5-AR	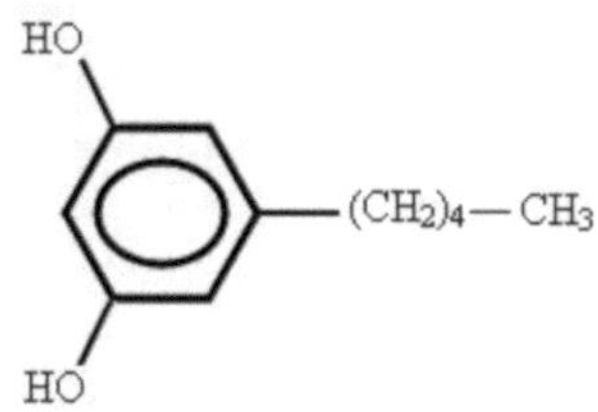	180	Sigma, USA
C_6-AR	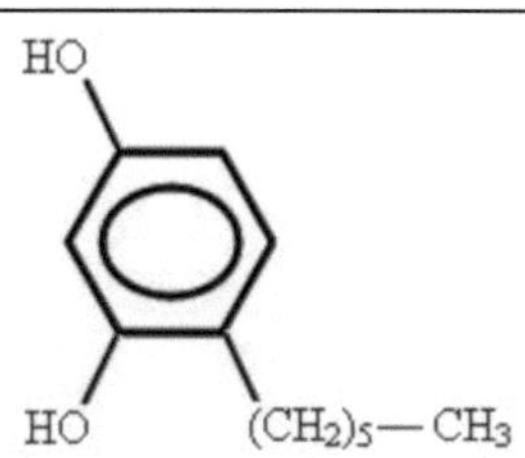	194	Sigma, USA
C_{12}-AR	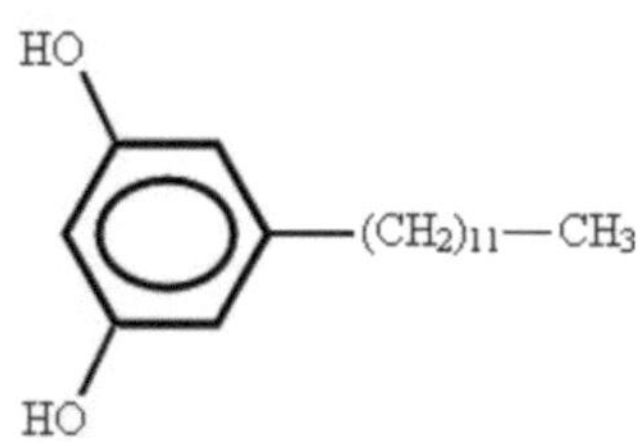	278	Enamine, Ukraine

Table 1. Chemical analogs of alkylresorcinols have been used in the study

2.5 Isolation of DNA and PCR-procedure

DNA was isolated from the gel block using the DNA extraction kit (Cytokine, Russia). DNA added to the tube containing the reagents for PCR (DNA-Technology, Russia) and amplified in 35 cycles (30s 93°C, 30s 59°C and 30s at 72°C). For subsequent analysis 5 µl of the product was analyzed by gel electrophoresis as described above.

2.6 Transformation

E.coli K12 TG1 were grown to log phase (up to OD_{600}=0.20-0.30) in Luria-Bertani (LB) broth, washed and ultimately concentrated 25 times in ice-cold 100 mM of $CaCl_2$. DNA was extracted from agarose gel after electrophoresis, added to 200 ml of competent cell and incubated at 0°C for 15 min. The cell-DNA complex was transferred to 42°C for exactly 90 s and was rapidly chilled in ice. Then 1000 ml LB-broth was added and the cells were incubated at 37°C for 60 min. 100 ml cells was spread on LB-agar with and without selective marker ampicillin (50 mg/ml), to obtain the number of transformants and viable cells respectively. Plates were incubated at 37°C for 18-24 h.

2.7 Bioluminescent and microbiological methods

E.coli recA'::luxCDABE strain were grown for 16-18 hours at 37°C in LB-broth in the presence of 20 μg/ml of ampicillin. Immediately before the experiment the culture was diluted 1:20 by fresh culture medium and incubated until early log-phase. The grown biomass was mixed with AR solutions in final concentrations of 10^{-5}, 10^{-4} и 10^{-3} M, with used for their dilution with distilled water (control) and incubated for 60 minutes. The luminescence intensity of UV-irradiated *E.coli recA'::lux* and intact specimens were registered by plate bioluminometer LM 01T (Immunotech, Czech Rep.) in a real time. The number of viable cells was determined from the colony-forming units (CFU) on a surface of a LB-agar after the subsequent incubation within 24 hours at 37 °C. A quantitative estimation of an induction of the SOS-system calculated on formula

$$F_i = \frac{luxA_i \cdot B_0}{luxA_0 \cdot B_i}, \tag{1}$$

where $luxA_0$ - light intensivity of not irradiated suspension of cells, $luxA_i$ - the light intensivity, the irradiated suspension of cells, B_0 - quantity of viable cells in not irradiated test, B_i - quantity of viable cells in the UV-irradiated test [Tsvetkova & Golyasnaya, 2007].

2.8 Statistical analysis

All studies were performed at least in triplicate and processed by variational statistic methods using "Statistica" software (StatSoft Inc., USA).

3. Results

3.1 AR directly protects DNA from UV damage

Electrophoresis of DNA phage λ linearized by HindIII restrictase, detected 7 fragments with a fixed length from 10 000 down to 250 bp (Fig. 1 lane 1). Three hours of UV exposure with a total dose of 3.64 J/m^2 significantly changed the pattern of electrophoresis, leading to the disappearance of the typical bands of DNA fragments and their transformation into solid track (Fig. 1 lane 2). The cause of these changes were the multiple breaks of double-stranded DNA with disintegration of each on irregular length of the fragments.

DNA preincubation with AR didn't lead to a change in electrophoretic mobility of fragments, but affected the sensitivity to UV radiation, which depends both on the chemical structure, and the used concentration of AR. For example, using 10^{-3} M of C_1-AR before the UV irradiation saves on 5.9 ± 0.6% of more DNA (Fig. 1, lane 4), although the use of 10^{-4} M of C_1-AR did not significantly change the electrophoretic pattern (Fig. 1 lane 3). On the other hand, the effect of long-chain C_6-AR was significantly higher, which led to the protective effect even at concentration of 10^{-4} M (Fig. 1 lane 5), and at concentration of 10^{-3} M it had been protected up to 22.4 ± 1.7% in the presence of C_6-AR under the same conditions (Fig. 1 lane 6).

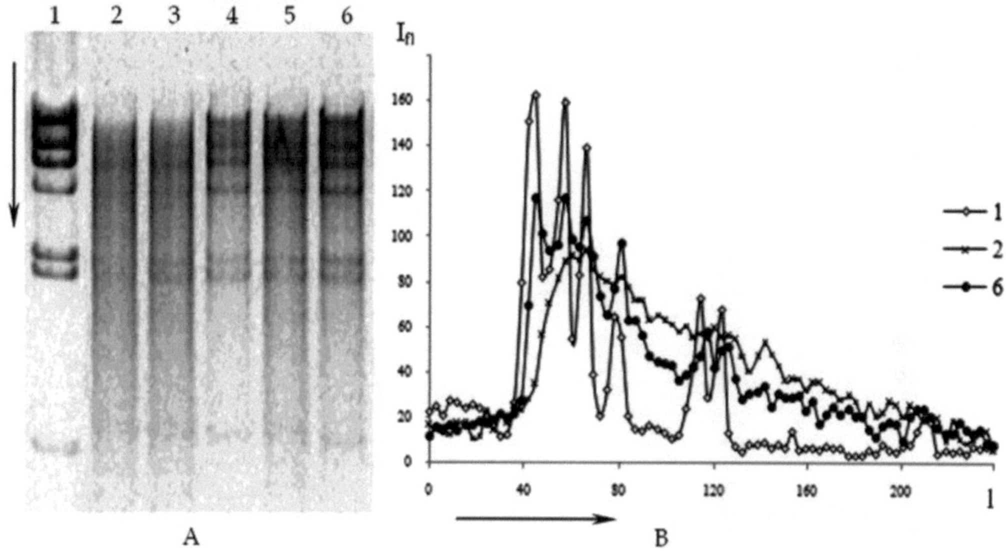

Fig. 1. Electrophoretic mobility of *Hind* III-*restrict* fragments of λ *DNA* after UV irradiation (A) and DNA electrophoregram trace (B): 1 - control, 2 - after UV irradiation for 3 hours (3.64 J/m^2), 3, 4 - the same in the presence of C_1-AR in concentrations of 10^{-4}M and 10^{-3}M, 5, 6 - the same in the presence of C_6-AR at concentrations of 10^{-4}M and 10^{-3}M. The arrow shows the migration vector.

The result of pUC19 plasmid DNA electrophoresis has allowed to detect the presence of two DNA conformations: supercoiled circular form (I) that accounts for 60.4% of the total DNA amount and relaxed circular form (II), accounts for 39.6% (Fig. 2 lane 1). UV irradiation of this plasmid in 30 minutes with a total dose of 0.61 J/m^2 changed the relative intensity of the typical bands, reflecting their conformational changes. It was exhibited by disappearance of a typical band I and parallel temporary increase of band II intensity that was the result of the transition of supercoiled plasmid DNA form into the relaxed supercoiled form caused by single-stranded breaks. The second effect was the appearance of the band III, which was interpreted as the result of a transition of supercoiled DNA molecules into the linearized form as a result of the formation one double-stranded breaks (Fig. 2 lane 2). The total number of electrophoretically detectable DNA after UV irradiation was 83.6 ± 4.1% of its initial content in the sample. Increasing of UV-dose up to 3.64 J/m^2 induced massive damage of plasmid DNA with the formation of a track from variable molecular weight linear fragments, as well as in the case of phage λ DNA (data is not shown).

Incubation of the pUC19 plasmid in the presence of AR prevented the formation of these effects, also depending on the chemical structure of used AR. So on the one hand, the amount of DNA after UV irradiation in the presence of 10^{-3}M C_1-AR was 91.9 ± 2.6%; in the presence of C_6-AR at the same concentration- 93.6 ± 3.1%, and in the presence of C_{12}-AR – 95.0 ± 3.4% if compared with samples of intact DNA molecules (Fig. 2B). On the other hand, on the conformational transitions of plasmid DNA in the presence of AR was also less expressed. In particular, long-chain AR: C_6-AR and C_{12} -AR almost completely prevents the appearance of band III (no more than 1.2% in comparision with irradiated DNA and kept the initial supercoiled conformation (lane I) as 87.3-98.7% of unirradiated DNA. Short-chain C_1-AR and C_3-AR slightly protected DNA from the formation of single- and double-stranded breaks so the DNA amount in the II and III bands corresponded to the irradiated DNA probes. Though they prevented the deeper DNA degradation with the preservation of quantity of initial supercoiled conformation (band I) in 1.1 times more in comparison with the UV-irradiated control.

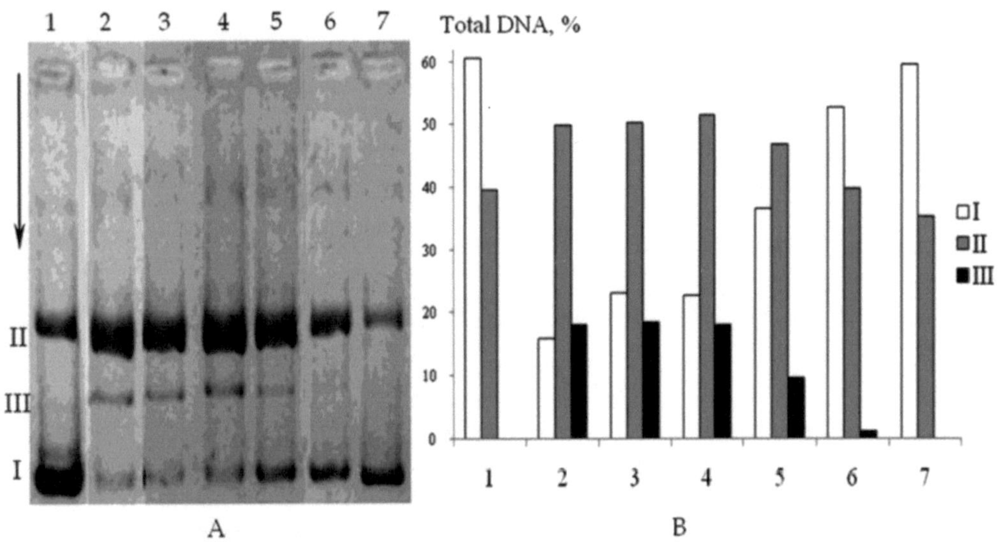

Fig. 2. Electrophoretic mobility of pUC19 DNA after UV irradiation (A) and relative DNA content (B) in the bands I-III for the lanes: 1 - control, 2 - after UV irradiation for 30 min (0.61 J/m²), 3 - then the same in the presence of 10^{-3}M C_1-AR, 4 - C_3-AR 5 - C_5-AR, 6 - C_6-AR, 7 - C_{12}-AR. The arrow shows the migration vector.

Thus, the AR directly protect DNA from the UV-damage. This is manifested in the saving of the total amount of this biopolymer, preventing its deep degradation due to formation of double-stranded breaks, as well as preventing the single -stranded breaks without transition of supercoiled into the relaxed circular form of DNA. Evidence of effects increased with length of the alkyl radical of the AR molecule and with AR concentration increase.

3.2 DNA protection by AR during agarose gel electrophoresis detection

The founded results formed the basis for the development of technology of DNA protection against the UV-damaging effect during the detection of gel electrophoresis results. The Actuality of this problem is the need to preserve the structural integrity of DNA for diagnostic studies using DNA technology, as well as saving of functional characteristics of the biopolymer for subsequent genetic engineering manipulations [Cariello et al., 1988; Hartman, 1991]. Known methods of DNA protection from UV radiation during the separation and detection are using such UV protectors as nucleosides [Grendemann & Schumig, 1996] or zinc-imidazole salts [Sosa et al., 1996] that are injected into the system for electrophoresis by different ways. Against this background, the identification of AR activity as photoprotectors determines the prospects for their use for this purpose. Availability of the approach is confirmed by well-known experience of AR using in the chemical industry as UV-protectors for rubber and plastics [Vagel & Roo, 2004].

As has been shown above photorotective effects are most expressed in long-chain C_6-AR and C_{12}-AR. However, some technological aspects are determined by low C_{12}-AR solubilization in water, demanded its preliminary dissolution in ethanol that gave additional difficulties for AR addition into different media for electrophoresis. For this reason, in the current paragraph of the study C_6-AR was used, photoprotective effects of wich were compared with C_1-AR, C_3-AR and C_5-AR.

The first DNA preparations in this part of the study was PCR product - DNA of *Chlamydia trachomatis* (273 bp), in the presence of a smaller by molecular mass internal control of human DNA. After migration the gel was exposured for 5, 30, 300 and 600 seconds by transilluminator Vilber Lourmat, equipped with 6 UV lamps with irradiance W = 0,24 W/m² and 254 nm filter. The degree of structural integrity loss of amplificated DNA was evaluated by the decrease of brightness intensity of the of the bands processed by using the tools of "ImageJ" computer program.

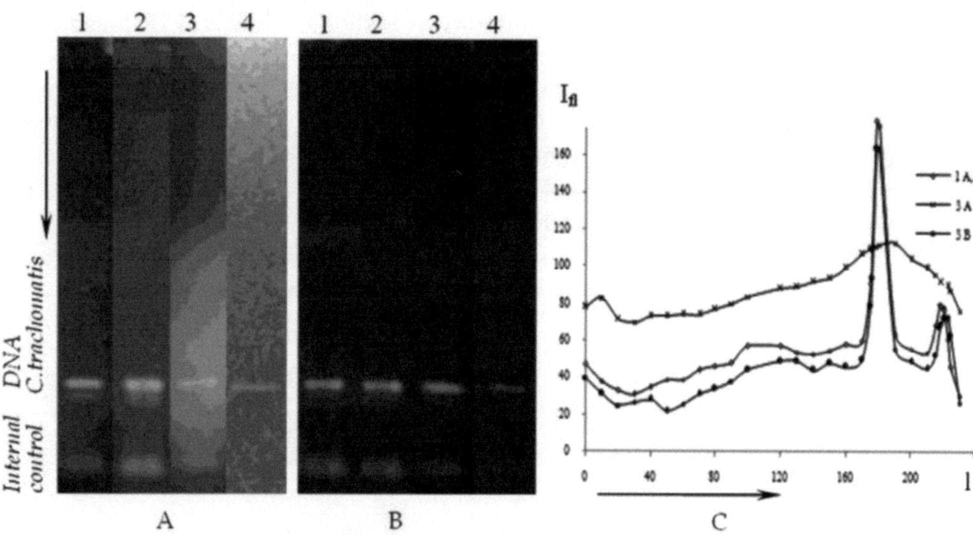

Fig. 3. Agarose gel after electrophoretic separation of amplified DNA of *Chlamydia trachomatis* without (A), and with the addition of C_6-AR (B) in the electrophoretic system at exposure by transilluminator at 254 nm for 5 (1), 30 (2), 300 (3) and 600 (4) seconds, and profiles of the electrophoretic mobility (C).

DNA exposure by transilluminator for 5-30 seconds has allowed to visualize the band that consist of amplified DNA fragments of *C. trachomatis* (Fig. 3 lane 1,2). The subsequent increasing of exposure up to 300 seconds resulted in a reduction of the relative emission intensity of this band (Fig. 3C) associated with photodamage both DNA molecules, and the development of luminescence of the gel, creating a backlight (Fig. 3B, lane 3). Finally, there were only trace amounts of this band after 600 seconds of irradiation (Fig. 3B, lane 4) which were not detected by the using software. In such situation, taking a right diagnostic conclusion was impossible.

On this background, the use of C_6-AR stabilized the gel pattern, preserving both DNA degradation and optical properties (transparency and coloration) of agarose. Thus, if at minimal exposure time of 5-30 seconds, there were no significant expressed differences of bands if compare with the control (Fig. 3B lane 1,2), then after 300 seconds, the luminescence of *C. trachomatis* DNA was much higher (up to 28.4%) than ones in control, and did not differ from baseline (88.9% - Fig.3C). On the other hand, another C_6-AR effect was optical properties saving while preserving the characteristics of the background (Fig. 3B lane 3). Moreover, with increasing of exposure time up to 600 seconds, registration of the results with software was still possible (Fig.3B lane 4).

The used variants of C_6-AR application were: adding to the wells of the gel to DNA, directly bringing into the agarose gel and the addition to electrophoretic buffer. The use of the latest way demonstrated its greatest efficiency by saving up to 1.63 times more DNA preparations if compared with the standard method of electrophoresis, while other ways showed 15.45% increase when C_6-AR was introduced into an agarose gel and 1.63%- when added to the DNA preparations.

A quantitative comparison of DNA band intensity at adding of different AR homologues into the buffer after 300 seconds of irradiation on the transilluminator has allowed to obtain a more detailed information about the structural integrity of DNA, depending on the AR concentration (Table 2). It was found that C_6-AR protected DNA greater than 1.5-fold in the range of 10^{-3}M and 5×10^{-3}M, with a maximum effect (163.5 + 15.2%) at concentrations of 10^{-3}M. On this background, C_1-AR and C_3-AR demonstrated poor photoprotective activity and C_5-AR showed a similar protective effect only at concentration of 10^{-3}M.

Concentration of AR, M	AR-abbreviation			
	C_1-AR	C_3-AR	C_5-AR	C_6-AR
10^{-4}	68.0±15.4	87.2±7.9	101.3±9.3	118.7±15.6
5×10^{-4}	97.1±9.8	109.1±10.6	128.1±13.3	135.0±13.7
10^{-3}	133.2±12.7	132.7±12.3	150.3±14.8*	163.5±15.2*
2×10^{-3}	124.2±12.3	129.1±13.2	109.3±10.1	153.5±13.2*
5×10^{-3}	126.9±13.5	119.2±10.8	90.2±8.5	142.6±14.8*

Table 2. DNA protection efficiency in comparision with the control (%) in the presence of various alkylresorcinols at UV-irradiation for 300 seconds. * – $P<0.05$.

C_6-AR also preserved the DNA functional properties for genetic engineering manipulations or amplification. For example, the efficiency of PCR-amplification of C_6-AR-treated DNA in comparison with UV-damaged control was significantly higher. Another result was the preservation of functional properties of DNA for transformation procedure. So, after the adding of the C_6-AR in the electrophoretic buffer in a concentration of 10^{-3}M and electrophoretic separation of pUC19 plasmid, its supercoiled form from agarose was extracted under visual control on the transilluminator for 30 seconds for transformation of bacterial *E. coli* cells. The transformation efficiency, expressed in a quantity of colonies of transformants was up 24.6% in comparision with the control.

On this base was developed a method for photoprotection of the DNA from the UV-damage (λ=254 nm) at detection of gel electrophoresis results, which consists in the fact that prior to the separation of DNA molecules in electrophoretic buffer is added C_6-AR at concentrations of 10^{-3}M. The main positive result of a this laboratory technology is the long saving of DNA from the damaging effects of UV radiation with a prolonged preservation of its structural integrity essential for the making of diagnostic deceision and functional activity of this biopolymer that is necessary for genetic engineering manipulations.

3.3 Bacterial cells UV protection by AR

The result about DNA protection *in vitro* models has given a study problem of photoprotective AR actions of the whole bacterial cells (*in vivo*). The model object for this study was *E.coli* strain carrying *precA'::luxCDABE-Amp*^{R} plasmids with a cassette of lux-genes of luminescent soil bacteria *Photorhabdus luminescence* ZM 1, cloned under the promoter of *recA E. coli* gene. Such a genetic design allows to characterize quantitatively the activity of SOS-reparation system after UV irradiation which are directly proportional to the intensity of DNA damage.

UV irradiation of this bacterial cells using a broadband mercury-quartz lamp through an interference filter (bandwith 254 nm), provided preferential DNA damage with minimal effects on other subcellular structures. The exposure time ranged from 0 to 180 minutes in increments of 60 minutes, which gave total dose of UV exposure 1.21, 2.43 and 3.64 J/m^2.

Thus, the dose of UV radiation in the range from 0 to 1.21 J/m^2 led to expressed growth of the absolute bioluminescence intensity of the probe with a maximum of 2.65 fold larger than background level. In turn, further increasing of dose resulted in a progressive decrease of the absolute bioluminescence intensity down to the initial values at a dose of 3.64 J/m^2 (Fig. 4). This effect depending on dose of UV-irradiation and developing in time effect has been caused by the loss of bacterial cells viability that has been shown in growth-test of analyzed probes on agar nutrient plates (Fig. 4).

On this background the dividing of absolute bioluminescence intensity to viable bacterial cells quantity gave the relative index F_i, showing proceeding activation of a DNA reparation with maximum value F_i = 282.85 at highest dose of the UV-irradiation (Fig. 4).

Another starting point was a data of the alkylresorcinols action on luminescense and viability of *E.coli recA'::lux* cells registered a partial dependence of this effects on AR chemical structure and concentrations.

Thus, none of the used AR homologues did not cause significant increase of the absolute level of bioluminescence, but causing its significant depression (Fig. 5). This effect increased in the range C_1-AR → C_3-AR → C5-AR → C_6-AR → C_{12}-AR, at the same time had direct concentration dependency (Fig. 5). In turn, the analysis of *E.coli recA'::lux*, incubated in contact with the AR shown the codirectionality between intensity of bioluminescence reduction and the CFU quantity, determined by the growth-repressed activity of these molecules (Fig.5B) [Grendemann & Schumig, 1996]. Finally, based on these data calculation of the relative index F_i allowed to reveal poorly expressed induction/inhibition of SOS-system as a result of AR exposure (Fig.5C). So, these factors at a concentration of 10^{-5}M, leading to the induction of stress response of bacterial cells [Golod et al., 2009], accompanied

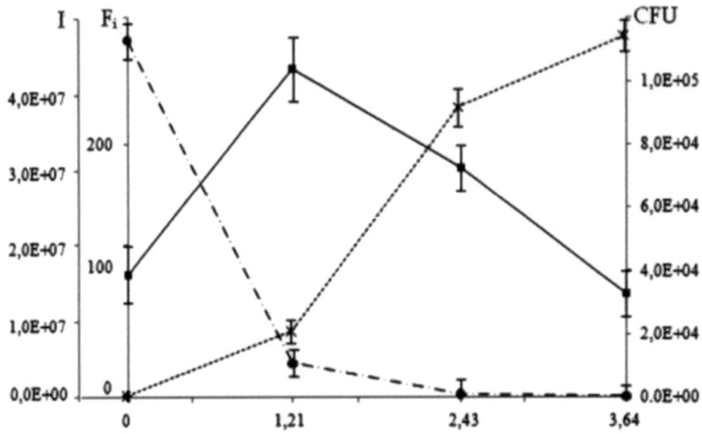

Fig. 4. Response of *E.coli recA'::lux* strain on UV exposure (on the horizontal axis - the exposure dose, J/m²): —■— the absolute intensity of bioluminescence (I); – ● – the number of viable cells (CFU); ---✳--- the relative index of induction SOS-response (F_i).

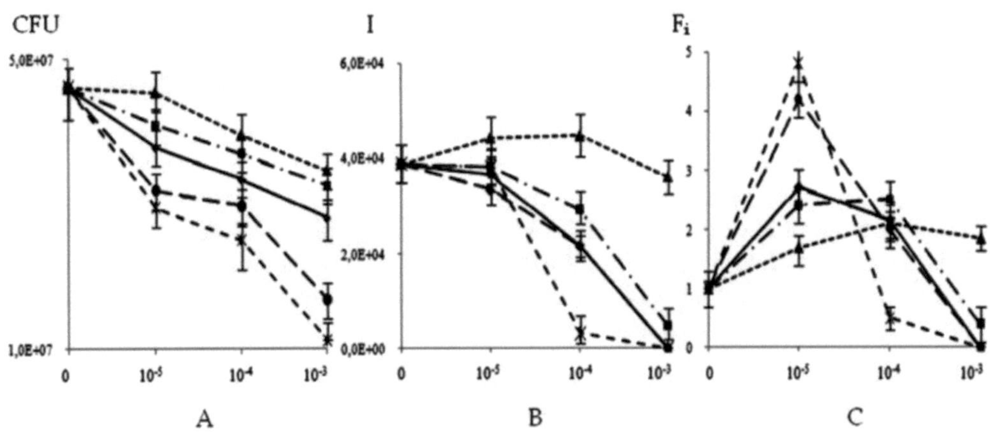

Fig. 5. *E.coli recA'::lux* strain response on AR action (horizontal axis - concentration, M): the number of survived cells (A), the absolute intensity of bioluminescence (B), the relative value of induction index of SOS-response (C). ---▲--- - C_1-AR, – ■ – - C_3-AR, —♦— - C_5-AR, –●– - C_6-AR, --✳-- - C_{12}-AR.

by the little activation of *E.coli recA'::lux* relative bioluminescence intensity. On the other hand, a high 10^{-3}M concentration, leading to the formation of the bacterial cell anabiotic state [Suzina et al., 2006], caused intensive SOS-response depression. The intermediate AR concentration 10^{-4}M caused a minimal change of this activity, making a best opportunity for the subsequent evaluation of the combined action of the studied factors.

The study of viability, bioluminescence and SOS-system activation index at UV exposure of *E.coli recA'::lux* pre-incubated has allowed to characterize some of interconnected effects characterizing the protective action of AR at UV irradiated cells. Thus, the number of *E.coli recA'::lux* cells, incubated in contact with the AR in a concentration of 10^{-4}M and saved their viability after subsequent UV exposure increased in the range C_1-AR → C_{12}-AR, at maximal dose of UV irradiation 2.0-14.4 times exceeded the value of control exposed - only to UV action (Table 3). In turn, reduction of AR concentrations down to 10^{-5}M, or an increase up to 10^{-3}M accompanied by a corresponding weakening or strengthening of such a UV-protective effect. These data are well corresponding with earlier findings, characterizing the alkylresorcinols as factors that protect the bacteria and yeast cells from the ultraviolet and ionizing radiation [El-Registan et al., 2005; Il'inskaya et al., 2002; Stepanenko et al., 2004].

On the other hand, comparative analysis of F_i variables gave the relative reduction of SOS-response in preincubated with AR bacterial cells (Table 3). So, repression SOS-response was proportionally to the length of the AR alkyl radical and was F_i = 3.7 (for C_6-AR) and F_i = 7.0 (for C_{12}-AR) that 76.43-40.40 fold lower than the SOS-system activation level in cells exposured only by UV radiation. With reduction of the AR concentration to 10^{-5}M is still observed statistically significant differences in the values of induction index (F_i) of the control and experimental samples, although the repression of SOS-response was less expressed. An increase of the AR concentration up to 10^{-3} M in the case of C_1-AR and C_3-AR led to some suppression, and for the C_5-, C_6- and C_{12}-AR to increase the values of F_i.

AR-abbreviation	The residual quantity of CFU (% of control: only the UV radiation)	Factor of SOS-system induction (F_i)
C_1-AR	0.6±0.1	234.9±21.0
C_3-AR	1.0±0.1*	141.9±13.3*
C_5-AR	1.9±0.1**	38.3±3.5**
C_6-AR	2.7±0.2**	3.7±0.3**
C_{12}-AR	4.3±0.2**	7.0±0.6**
Control	0.3±0.1	282.8±26.5

Table 3. Effect of AR concentration 10^{-4}M on the saving of *E.coli recA'::lux* cell viability and relative index of SOS-system induction at UV exposion (3.64 J/m^2). * - $P<0.05$;** - $P<0.01$.

This demonstrated that the AR realize a special mechanism of bacterial cells saving from UV-irradiation, connected with their action on "passive" and "active" DNA protection mechanisms. Probably it is based on direct interaction of AR with DNA, leading to the resistance of such comlexes to a wide range of stressors, with the simultaneous modification of the transcriptional activity of SOS-regulon. In tgis system a "passive" mechanism reduced the damage of DNA under UV irradiation and led to slower activation of SOS-system involving in "active" reparation of this biopolymer. This gave the opportunity to save viability of bacterial cells under UV irradiation at the minimal values of the repair processes activity. This results characterized AR induced resistance and SOS-repairation as two alternative "passive" and "active" mechanisms, evolutionary destined for the protection at different intensities of UV irradiation.

4. Conclusions

Deoxyribonucleic acid (DNA) is a very important biopolymer with the function of storage and transmission of genetic information. In this reason the protection of structural integrity and functional activity of DNA is essential for the viability of living systems, as well as the effectiveness of laboratory DNA-technics.

In turn, an important factor that can damage DNA in nature or at performing molecular genetic studies, is ultraviolet (UV) radiation that is absorbed by this biopolymer at bandwidth maxmum 254 nm. This led to the formation of different DNA photodamages, with increasing of the dose of UV radiation progressed from pyrimidine dimers to single- and double-stranded breaks [Cariello et al., 1988; Lyamichev et al., 1990].

In this paper we describe a new nature-determined and chemically-mediated DNA protection mechanism provided by a specific group of phenolic lipids - alkylresorcinols. These molecules are synthesized by bacteria, fungi and plants [El-Registan et al., 2006; Kozubek & Tyman, 1999; Zarnowski et al., 1999] and had different functions [Nikolaev et al., 2006], including the control of microorganisms transition in a dormant state [Mulyukin et al., 2003]. An important requirement for this process is the preservation and stabilization of a wide range of itracell biological macromolecules, including DNA [Solyanikova et al., 2011; Stepanenko et al., 2004]. This suggesting the role of AR in the protection of the biopolymer from UV exposure.

On the basis of previously determined AR-DNA interaction [Davydova et al., 2005] in this study the increasing of UV-resistance of this complex is shown. In particular, the interaction of AR with linear DNA molecules leads to biopolymer protection against deep degradation caused by UV exposure with double-stranded breaks prevention. The effect of long-chain C_6-AR if compared with C_1-AR was expressed and increased with AR concentration. In case of contact of AR with circular molecules, it is additionally shown prevention of the transition of supercoiled molecules into circular form as a result of single-stranded breaks. Thus simultaneous use of a wide range of AR homologs confirmed the maximum effect in long C_6- and C_{12}-AR.

This data led to the development of the original method of protecting the DNA from the damaging effects of UV radiation with a wavelength λ = 254 nm for detection of the results of gel electrophoresis. It is demonstrated the benefits of the adding of the C6-AR in the buffer solution compared to other ways of introducing an electrophoretic system (adding to the wells with DNA or agarose gel) and justify the optimal concentration of C_6-AR as 10^{-3} M, which provides maximum protection of the biopolymer. The achieved result was the stabilization of DNA electrophoretic separation pattern during the detection of PCR-result. This was accompanied by stabilization of the optical properties (transparency and coloration) of the agarose gel that was used for DNA separation. On the other hand, the introduction of the C_6-AR retained and functional characteristics of the shared DNA, which increased the effectiveness of subsequent *E.coli cells* transformation by these molecules.

Study of the action of AR on *E. coli precA'::luxCDABE*-Amp^R has confirmed it photoprotective effects and has shown features of such activity in live systems. Surprising was the interrelation between preservation of viability of AR-processed bacterial cells in the conditions of a long and intensive UV-irradiation and depression of activity their reparing SOS-systems. It has assumed AR action and the SOS-answer as alternative "passive" and "active" mechanisms for protection of bacterial cells DNA at various intensivity of UV-irradiation.

So this research reveals new mechanisms of bacterial autoregulation under extreme conditions, controlled by low weight molecules - alkylresorcinols. It applied aspects are defined by developing of methods for DNA protection in vitro and elongators of bacterial cells viability at UV exposure.

5. Acknowledgment

This work was supported by Federal Target Program ''Research and scientific-pedagogical personnel of innovative Russia" for 2009-2013 years (№ P327). We thank Hike Nikiyan for skillful technical assistance.

6. References

Azam, T., Ishihama, A. Twelve species of the nucleoid-associated protein from *Escherichia coli*. Sequence recognition specificity and DNA binding affinity. *The Journal of Biological Chemistry,* Vol.274, No.46, (November 1999), pp. 33105-33113, ISSN 0021-9258

Cariello, N., Keohavong, P., Sanderson, B., Thilly, W. DNA damage produced by ethidium bromide staining and exposure to ultraviolet light. *Nucleic Acids Research,* Vol. 16, No.9, (May 1988), pp. 4157-4161, ISSN 0305-1048

Davydova, O., Deryabin, D., Nikiyan, A., El-Registan, G. Mechanisms of interaction between DNA and chemical analogues of microbial anabiosis autoinducers.

Davydova, O., Deryabin, D., El-Registan, G. Influence of chemical analogues of microbial autorgulators on the sensitivity of DNA to UV radiation. *Microbiology,* Vol.75, No.5, (September 2006), pp. 654-661, ISSN 1350-0872

Davydova, O., Deryabin, D., El-Registan, G. IR spectroscopic research on the impact of chemical analogues of autoregulatory d1 factors of microorganisms on structural changes in DNA. *Microbiology,* Vol.76, No.3, (June 2006), pp. 266-272, ISSN 1350-0872

El-Registan, G., Duda, V., Kozlova, A., Duzha, M., Mityushina, L., Poplaurhina, O. Changes in constructive metabolism and ultrastructural organization of *Bacillus cereus* cells under the action of a specific autoregulatory factor. *Microbiology,* Vol.48, No.2, (February 1979), pp.240-244, ISSN 1350-0872

El-Registan, G., Mulyukin, A., Nikolaev, Yu., Stepanenko, I., Kozlova, A., Martirosova, E., Shanenko, E., Strakhovskaya, M., Revina, A. The role of microbial low-molecular-weight autoregulatory factors (alkylhydroxybenzenes) in resistance of microorganisms to radiation and heat shock. Advances in Space Research, Vol.36, No.9, (2005), pp. 1718-1728, ISSN 0273-1177

El-Registan, G., Mulyukin, A., Nikolaev, Yu., Suzina, N., Galchenko, V., Duda, V. Adaptogennye functions extracellular autoinducers microorganisms. *Microbiology,* Vol.75, No.4, (August 2006), pp. 446-456, ISSN 1350-0872

Fraikin, G., Strakchovskaya, M., Rubin, A. Light-induced processes of cell protection against photodamage. *Biochemistry,* Vol.65, No.6, (June 2000), pp. 865-875, ISSN 0006-2979

Gasiorowski, K., Brocos, B. DNA repair of hydrogen peroxide-induced damage in human lymphocytes in the presence of four antimutagens. A study with alkaline single cell gel electrophoresis (comet assay). Cellular & Molecular Biology Letters, Vol.6, (2001), pp. 897-911, ISSN 1425-8153

Golod, N., Loiko, N., Gal'chenko, V., Nikolaev, Y., El'-Registan, G., Lobanov, K., Mironov, A., Voieikova, T. Involvement of alkylhydroxybenzenes, microbial autoregulators, in controlling the expression of stress regulons. *Microbiology,* Vol.78, No.6, (September 2009), pp. 678-688, ISSN 1350-0872

Grendemann, D., Schumig, E. Protection of DNA during preparative agarose gel electrophoresis against damage induced by ultraviolet light. *BioTechniques,* Vol.21, No.5, (November 1996), pp. 898-903, ISSN 0736-6205

Hartman, P. Transillumination can profoundly reduce transformation frequencies. BioTechniques, Vol.11, No.6, (December 1991), pp. 747-748, ISSN 0736-6205

Il'inskaya, O., Kolpakov, A., Schmidt, M., Doroshenko, E., Mulyukin, A., El-Registan, G. The role of bacterial growth autoregulators (alkyl hydroxybenzenes) in the response of *Staphylococci* to stress. *Microbiology,* Vol.71, No.1, (January 2002), pp.23-29, ISSN 1350-0872

Kaprelyants, A., Suleimenov, M., Sorokina, A., Deborin, G., El-Registan, G., Stoyanovich, F., Lille, Yu., Ostrovsky, D. Structural-functional changes in bacterial and model membranes induced by phenolic lipids. *Biological membranes,* Vol.4, No.3, (March 1987), pp. 254-261, ISSN 0748-8653

Kozubek, A., Tyman, J. Resorcinolic lipids, the natural non-isoprenoid phenolic amphiphiles and their biological activity. *Chemical Reviews,* Vol.99, No.1. (January 1999), pp. 1-31, ISSN 0009-2665

Lyamichev, V., Frank-Kamenetskii, M., Soyfer, V. Protection against UV-induced pyrimidine dimerization in DNA by triplex formation. *Nature,* Vol.344, No. 6266, (1990), pp. 568-570, ISSN 1476-4687

Martirosova, E., Karpekina, T., El'-Registan, G. Enzyme modification by natural chemical chaperons of microorganisms. *Microbiology,* Vol.73, No.5, (August 2004), pp. 609-615, ISSN 1350-0872

Mulyukin, A., Soina, V., Demkina, E., Kozlova, A., Suzina, N., Dmitriev, V., Duda, V., El-Registan, G. Formation of resting cell by non-spore-forming microorganisms as a strategy of long-term survival in the environment. Proceeding of the SPIE. Vol.4939, (January 2003), pp. 208-218, ISSN 0277-786X

Mulyukin, A., Vakhrushev, M., Strazhevskaya, N., Shmyrina, A., Zhdanov, R., Suzina, N., Duda, V., El-Registan, G. Effect of alkylhydroxybenzens, microbial anabiosis inducers, on the structural organization of Pseudomonas aurantiaca DNA and on the induction of phenotypic dissociation. Microbiology, Vol.74, No.2, (March 2005), pp. 128-135, ISSN 1350-0872

Nikolaev, Yu., Mulyukin, A., Stepanenko, I., El-Registan, G. Autoregulation of the stressful answer of microorganisms. Microbiology, Vol.75, No.4, (June 2006), pp. 489-496, ISSN 1350-0872

Petrovskii, A., Loiko, N., Nikolaev, Yu., Kozlova, A., El'-Registan, G., Deryabin, D., Mikhailenko, N., Kobzeva, T., Kanaev, P., Krupyanskii, Yu. Regulation of the function activity of lysozyme by alkylhydroxybenzenes. Microbiology, Vol.78, No.2, (March 2009), pp. 144-153, ISSN 1350-0872

Reusch, R., Sadoff, H. Novel lipid components of the *Azotobacter vinelandii* cyst membrane. *Nature,* Vol. 302, (1983), pp.268-270, ISSN 1476-4687

Rosen, R., Davidov, Y., LaRossa, R., Belkin, S. Microbial sensors of ultraviolet radiation based on *recA'::lux* fusions. *Applied Biochemistry and Biotechnology,* Vol.89, No.2-3, (2000), pp. 151-160, ISSN 1559-0291

Setlow, P. Mechanisms for the prevention of damage to the DNA in spores of *Bacillus species*. *Annual Review of Microbiology,* Vol.49, (1995), pp. 29-54, ISSN 0066-4227

Solyanikova, I., Konovalova, E., El-Registan, G., Golovleva, L. Effect of alkylhydroxybenzenes on the properties of dioxygenases. *Journal of Environmental Science and Health,* Part B. Vol.45. (2011), pp. 810-818, ISSN 1001-0742

Sosa, A., Pupo, E., Casalvilla, R., Fernandez-Patron, C. Zinc-imidazole positive: a new method for DNA detection after electrophoresis on agarose gels not interfering with DNA biological integrity. *Electrophoresis,* Vol.17, No.1, (January 1996), pp. 26-29, ISSN 1349-9394

Stepanenko, I., Strakhovskaya, M., Belenikina, N., Nikolaev, Yu., Mulyukin, A., Kozlova, A., Revina, A., El'-Registan, G. Protection of *Saccharomyces cerevisiae* against oxidative and radiation-caused damage by alkylhydroxybenzenes. *Microbiology,* Vol.73, No.2, (March 2004), pp. 204-210, ISSN 1350-0872

Suzina, N., Dmitriev, V., Shorokhova, A., Duda, V., Mulyukin, A., Nikolaev, Y., Bobkova, Y. Barinova, E., Plakunov, V., El-Registan, G. The structural bases of long-term anabiosis in non-spore-forming bacteria. Advances in Space Research, Vol.38, No.6, (2006), pp. 1209-1219, ISSN 0273-1177

Tsvetkova, N., Golyasnaya, N. Induction of SOS-response in *Escherichia coli* under conditions of osmotic stress and in the presence of N-methyl-N'-nitro-Nnitrosoguanidine. *Microbiology,* Vol.76, No.4, (June 2007), pp. 448-455, ISSN 1350-0872

Vagel, A., Roo, E. Alkylresorcinols - rare chemicals available in bulk. *Innov. Pharm. Tech.* (2004), pp. 94-95

Zarnowski, R., Kozubek, A., Pietr, S. Effect of rye 5-n-alkylresorcinols on *in vitro* growth of phytopathogenic Fusarium and Rhizoctonia fungi. *Bulletin of the Polish Academy of Sciences,* Vol.47, No.2-4, (1999), pp. 231-235, ISSN 0239-7269

10

Polysaccharide-Specific Isoperoxidases as an Important Component of the Plant Defence System

Igor V. Maksimov, Ekaterina A. Cherepanova and Antonina V. Sorokan'
Institute of Biochemistry and Genetics, Ufa Science Centre, Russian Academy of Sciences
Russia

1. Introduction

The plant cell wall is a very complex and dynamic system, similar in importance to both the extracellular and intracellular processes which are recognised nowadays. The cell wall is a "vanguard" - an effective barrier in the way of different negative chemical and biotic factors, including pathogens and wounding. The defence functions of plant cell walls are associated with the construction of physical barriers consisting of lignin- and suberin-containing polymers on the path of pathogens inside a plant. This reaction develops more or less automatically, and barriers are only formed in the zone of pathogen penetration during active pathogen expansion into the host plant's tissues. Yet the mechanisms of these events are still unclear. It is well-known that peroxidases (PO) are key enzymes involved in lignification (Cosio & Dunand, 2009) and one of the few proteins secreted into the plant cell wall. However, POs have numerous applications in industry and one of the most important of these is the use of POs for lignin degradation. Therefore, both analytics and industry require a great volume of stable PO preparations of a high quality and at a low price, and the search for new methods and substrates for their extraction and purification has great commercial importance. Thus, the ability of plant oxidoreductases to interact with the biopolymers of the cell walls of plants and fungi has been studied for several decades (Siegel, 1957; McDougall, 2001). Moreover, it has been shown that plant POs can bind electrostatically with calcium pectate (Dunand et al., 2002) and chitin (Khairullin et al., 2000). Plants are likely to contain a whole subclass of these "polysaccharide-specific" isoPOs and their encoding genes. This subclass should be characterised by the ability to bind with polysaccharides and the defence function focused on strengthening the cell wall of the host and isolating the non-infected host tissues from the pathogen with the help of lignin. We suppose that the ability of plant POs to interact with some biopolymers without losing their activity can be applied for the isolation and purification of these enzymes. The possibility of the application of chitin in agriculture, biomedicine, biotechnology and the food industry has received much attention due to its biocompactibility, biodegradability and bioactivity. The low price and the ecological safety of this biopolymer define it as an available matrix for technological processes. As such, it may be possible to produce the high-quality preparations of POs that are needed for various fields of industry and analytical methods with the use of chitin (or other polysaccharide biopolymers) as a matrix. This article is focused on the biochemical and molecular features of POs binding with polysaccharides and their functions in plant defence reactions, and it covers some aspects of the application of polysaccharides for the purification of POs.

2. Molecular and biochemical features of plant peroxidases

PO (donor: H_2O_2 oxidoreductase, EC 1.11.1.7) belongs to a class of widespread and vital enzymes. These enzymes are used in enzyme immunoassays, diagnostic assays and industrial enzymatic reactions. The application of POs in the area of organic chemistry - especially when regio- and enantioselective oxidations are sought - are both numerous and appealing (Yoshida et al., 2003). Therefore, both analytics and industry require a great volume of stable PO preparations of a high quality and at a low price. Presently, the basic source of commercial POs is the horseradish roots (*Armoracia rusticana*). However, this PO has a poor stability under the different conditions required for industrial processes. Besides this, commercial POs often contain admixtures of other enzymes. One of the problems is purification of different isoPOs with different catalytic activity, since one of specific features of POs is the multiplicity of their molecular isoforms exhibiting various functions. Numerous of the genes encoding them are controlled by various *cis*-elements and *trans*-factors which are responsible for their different expression activity, depending on the environmental conditions and the stages of ontogenesis.

The rather high variety of PO genes together with the almost 90% homology of the functionally important sites responsible for their enzymatic activity was demonstrated by gene comparison, even for one plant species (Welinder et al., 2002; Duroux & Welinder, 2003; Cosio & Dunand, 2009). The inability of antibodies raised against the cationic isozymes of peanut PO to bind to the anionic PO of the same species indicates the differences in the structures of even those peroxidases belonging to a single species. At the same time, a high immune cross reactivity between the cationic isoPOs of horseradish, radish and carrot has been demonstrated (Conroy et al., 1982; Maksimov et al., 2010). POs are localised in cytoplasm, the plasma membrane or else secreted outside of the cell. In the cell wall, POs are present in the ionic-bound or covalent-bound fraction of proteins, and even freely circulate in apoplast. However, it is unclear which polysaccharides participate in PO anchoring within a plant cell wall.

PO performs vitally important functions in the plant cell and is mainly associated with the oxidation of phenolic compounds and with the formation and strengthening of the cell wall (Passardi et al., 2004). PO is involved in the oxidative transformation of molecules in growth-regulating or signalling activities and - as a result - can also perform regulatory functions in the cell. Plant POs are represented by genetically different proteins with the same enzymatic activity (Welinder et al., 2002).

These physiological functions of the enzyme are especially important in the case of cell damage due to exposure to various stress factors, including infection with pathogens (Almagro et al., 2009; Choi et al., 2007). The molecular and functional heterogeneity of isoPOs makes it possible to change the activity of different isoforms to the advantage of those that are most appropriate to the specific stress conditions of the environment, and an increase in PO activity can be suggested as a protective function of the organism. The functions of PO can be associated not only with the synthetic processes during cell differentiation and organogenesis, but also with the regulation of plant cell metabolism and the control of plant growth and development (Tian et al., 2003). However, it is still difficult to understand why the same isoPOs can be responsible both for normal physiological processes (Cosio et al., 2009) and for the oxidative burst during pathogenesis.

The available information points to the need for the determination of the role of oxidoreductases in the resistance of plants to adverse environmental factors and consideration of their role in the concentration, generation, and utilisation of ROS in the infection zone (Bindschedler et al., 2006). It is suggested that both quantitative and qualitative changes in the level and activity of oxidoreductases can lead to changes in the reactions of free radical oxidation. Therefore, ROS and the oxidoreductases involved in the system of their generation and degradation can be combined into the pro-/antioxidant system (Maksimov & Cherepanova, 2006). Nonetheless, one of the most common and widely discussed functions of POs, currently, is their participation in lignin synthesis in the cell wall (Marjamaa et al., 2009).

3. Polysaccharide-specific plant peroxidases

Cell walls have a complicated polysaccharide structure which serves as a "stumbling block" for plant biologists. Thus, about 15% of the genes of *Arabidopsis thaliana* take part in the formation and functioning of its cell wall (Carpita et al., 2001). Cellulose is the major polysaccharide of the plant cell wall, but there are also many other polysaccharides once called "hemicelluloses". Hemicelluloses include connecting glycans (xyloglycans, xylans, mannans and glucans (callose)) and pectins (polygalacturonic acid, rhamnogalacturonan, xylogalacturonan, arabinan and arabinogalactan) (Scheller et al., 2009). The rigidity and water-impermeability of cell walls is determined by the presence of phenolic polymer lignin. Lignin is considered to be a specific barrier in the way of fungi penetrating into a plant. It is believed that the main mechanism of plant defence against fungal plant pathogens is the synthesis and accumulation of lignin at the sites of penetration. Obviously, lignin molecules interact with the polysaccharide skeleton of cell walls. As such, a number of researchers agree that oxycinnamic acids bound with polysaccharides serve as primers for lignin polymerisation. Consequently, the lignin formation begins at the specific sites of polysaccharide molecules containing their oxycinnamic acids (Gorskova, 2007). Next, it was shown that the side chains of sugar beet (*Beta vulgaris*) pectins - which are mainly composed of arabinose and galactose residues - are esterified by ferulic acid units. Feruloyl esters may also be involved in the cross-linking of polysaccharides to lignins (Levigne et al., 2004). Recently, the genes involved in the feruloylation of arabinoxylan have been identified in rice (Piston et al., 2010). So, pectins can be an "anchor" for the initial step of lignin formation. The next step - polymerisation - requires the presence of PO for the oxidation of monolignols (Boerjan et al., 2003). Thus, researchers from Geneva University have shown the sorption of PO isolated from zucchini and *Arabidopsis* by pectate in the presence of Ca^{2+} (Dunand et al., 2002). The hypothesis about the ionic interaction of these POs with pectates was proved using PO with a deletion of the nucleotide sequence responsible for the translation of the polysaccharide-binding amino acid sequence and the subsequent transgenesis of the mutated genes of the protein into tobacco plants (Dunand et al., 2003). Zones with high electrostatic activity were found on the surface of some POs, and these zones can interact with cell wall pectins in the presence of calcium. Such isoPOs genes are activated in the zones of the formation of meristematic tissues and they display a strict tissue transcriptional activity (Carpin et al., 2001). It has been shown - with the help of electron microscopy - that in apical meristem of *Spinaceae* the majority of extracellular PO activity is found in middle lamella and cell corners (Crevecoeur et al., 1987), which coincides with the localisation of pectates in the cell wall.

In our investigations, we also detected the sorption of isoPO from potato, *Arabidopsis* and wheat, by calcium pectate. Moreover, we observed the binding with calcium pectate of potato PO from the fraction of proteins ionically bound with cell walls. It is likely that the ability of some PO isoforms to bind with pectin ensures the spatial proximity of these enzymes to the sites of the initiation of lignin synthesis and that these "pectin-specific" isoforms take part in this process.

The possible involvement of polygalacturonic acid-containing molecules in the defence reactions of tomato root cells against *Fusarium oxysporum* was suggested about 20 years ago by their accumulation at penetration sites. Since papillae are held to serve as a resistance mechanism to fungal penetration, it was assumed that the interrelation between pectin and other polymers - such as lignin - may contribute to enhancing the hardness of these newly formed structures (Benhamou et al., 1990).

Next, we observed that the activity of this isoPO was higher in calluses than in intact plants. Moreover, the significant activation of this isoform (pI~ 8.5) was shown in potato calluses infected by *Phytophthora infestans* (Fig. 1).

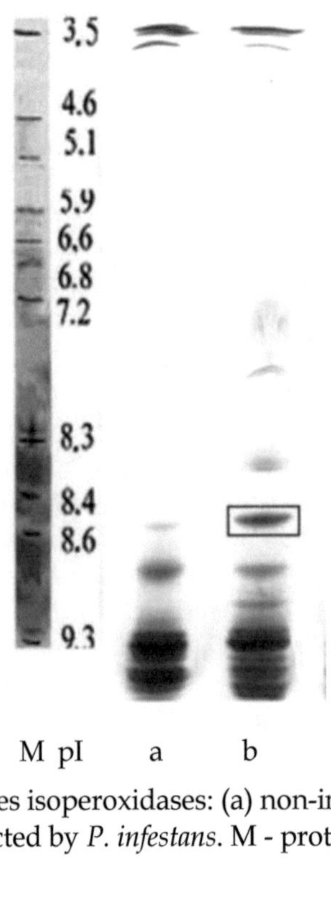

Fig. 1. Spectrum of potato calluses isoperoxidases: (a) non-infected plants protein extract, (b) protein extract from plants, infected by *P. infestans*. M - protein markers.

It may be that the presence of this isoform promotes the strengthening of calli cell walls through a special mechanism, since cultured cells have mainly undifferentiated cell walls missing lignification and suberin deposition, similar to meristematic cells in plants.

4. Peroxidases specific to the pathogen polysaccharides

It is interesting that Hammerschmidt and Kuc' (1982) observed the formation of lignin on the *Colletotrichum lagenarium and C. cucumerinum* fungal mycelium in the presence of PO extracted from cucumber plants, eugenol or coniferyl alcohol and hydrogen peroxide. Consequently, on the surface of these pathogens, a hypothetical "anchor" molecule for lignin synthesis initiation should be present, similar to pectin in the plant cell wall. Both plant tissues and pathogen hyphae may be lignified. The mechanisms of targeted lignification in the vicinity of infecting fungal structures, as well as the structures of plant cell walls, are poorly understood.

It has been shown that the cell walls of the majority of fungi contain chitin (Bowman, Free 2006). Chitin can be extracted in great quantities by industrial processes from crab and shrimp shells or the larvae cuticles of insects such as *Musca domestica* or *Galleria mellonella* (Ostanina et al., 2007). Some authors suggest a scheme of chitin extraction from the fungal mycelium of moulds (*Aspergillus sp., Mucor sp.*) (Wu et al., 2005). The low price and ecological safety of this biopolymer define it as an available matrix for technological processes. In this section, we would like to demonstrate the possibility of using chitin for the isolation of certain isoPOs from a great number of plants.

In 1958 Siegel showed that the addition of chitin to the reaction mixture for oxidation and polymerisation, with the help of the plant PO of the phenol compound eugenol present in lignin, increased the output of the reaction products and changed their structural quality as compared with the reaction without chitin. Based on these observations, it was proposed that chitin can be considered as a matrix for plant lignin formation. Using chitin as a biological matrix, we have developed an approach for isolating PO from the protein extracts of plants (Khairullin et al., 2000). Initially, we found that a fraction from wheat extract absorbed on chitin contained a dark-brown pigment which could be eluted with 1 M NaCl and which manifested the enzymatic activity of PO in the presence of H_2O_2. For the first time proteins with p*I* ~ 3.5 and a molecular weight of 37 kDa were found in plants, and these proteins could be absorbed by chitin due to ion exchange and manifest PO activity (Maksimov et al., 2005). Next, we detected POs with similar properties in a great number of plant species (Tab. 1) (Maksimov et al., 2003).

After this, we demonstrated the ability of wheat anionic POs to bind to the chitin of the cell walls of fungal pathogens. We called these POs "chitin-binding POs" (Maksimov et al., 2003). We were the first to demonstrate the binding of the anionic PO of wheat root to chitin (Maksimov et al., 1994). Besides this, we observed that in some species the activity of POs was increased in the unbound (*Armoracia rusticana, Lagenaria siceraria*) or eluted (*Pisum sativum, Galega orientalis, Brassica oleraceae*) fractions of proteins after interaction with chitin.

So, the sorption of PO on polysaccharides was not a classic ion-exchange interaction because the proteins were different in both isoelectric points and the molecular weights exhibited affinity for them. This conclusion was confirmed by the fact that the desorption of PO was facilitated by increasing NaCl concentrations as well as that isoPOs with a different p*I* can bind chitin, as has been shown earlier (Maksimov et al., 2003). Additionally, it was shown that the ion-exchange affinity of POs for polysaccharides is determined by the presence of zones with high electrostatic potential in the enzyme molecule (Dunand et al., 2002). Therefore, it can be assumed that the wheat POs adsorbed on chitin contain sites that can specifically interact with their acetyl residues.

N	Plant species	PO activity (U/mg protein)			isoforms in plant tissue	pI of chitin-specific POs	Chitin-specific isoPOs
		Crude extract	Unbound with chitin fraction	Binding with chitin fraction			
1	*Triticum aestivum*	10.5±1.2	4.3±0.5	11.0±3.0	9	3.5	1
2	*Avena sativa*	3.6±0.5	3.1±0.2	11.1±2.5	11	3.5; 3.6	2
3	*Oryza sativum*	1.7±0.2	4.9±0.6	0.5±0.2	15	3.5; 4.8; 6.2	3
4	*Zea mays*	1.5±0.3	2.9±0.7	0.3±0.1	7	3.6	1
5	*Allium cepa*	1.7±0.2	0.5±0.2	0.4±0.1	10	4.8; 5.8; 6.8	3
6	*A. porrum*	1.4±0.2	1.2±0.2	-	6	-	-
7	*A. sativum*	2.3±0.5	0.4±0.1	1.8±0.4	7	3.5; 5.8; 6.5; 7.0; 9.8	6
8	*Aloe vera*	1.0±0.2	1.6±0.2	0.7±0.1	4	5.5; 6.2; 6.5	3
9	*Lilium regale*	2.5±0.5	2.2±0.3	-	8	3.6; 7.0	3
10	*Hosta glauca*	2.6±0.5	2.8±0.3	0.3±0.1	11	7.2	1
11	*Phoenix dactylifera*	2.0±0.3	1.25±0.2	0.8±0.4	4	3.5; 5.8	2
12	*Armoracia rusticana*	7.0±1.0	19.2±2.3	2.0±0.4	8	3.5; 3.8	2
13	*Raphanus sativus*	9.0±0.7	11.3±0.9	2.9±0.3	6	3.6	1
14	*Brassica oleraceae*	5.6±0.2	6.2±0.3	15.5±0.8	6	8.0; 9.0	2
15	*Solanum tuberosum*	1.9±0.2	6.8±1.1	1.6±0.3	19	3.5; 3.7; 8,8; 9.1	4
16	*Nicotiana tabacum*	24.0±0.9	9.0±0.9	13.7±2.0	6	3.7	1
17	*Petunia hybrida*	5.2±0.5	3.7±0.6	23.9±2.2	6	3.5; 4.3; 8.0; 8.6	4
18	*Cucumis sativus*	9.8±0.5	1.9±0.1	13.5±0.9	11	3.4; 3.8; 7.2; 8.3; 8.5	7
19	*Cucurbita pepo*	6.2±0.6	1.7±0.2	5.1±0.8	4	3.5; 3.8; 9.8; 10	2 4
20	*Lagenaria siceraria*	2.4±0.3	4.6±0.4	-	8	-	-
21	*Arachis hypogaea*	8.2±1.1	1.3±0.1	6.1±0.9	13	from 3.4 to 5.5	9
22	*Pisum sativum*	3.8±0.3	2.9±0.5	26.1±3.9	4	3.5; 8.3; 10.2	3
23	*Galega orientalis*	3±0.2	3.9±0.3	44.2±0.2	8	3.5; 7.8; 9.0; 10.2; 10.8	6

Table 1. Analysis of chitin-specific isoperoxidases in plant species

We found that the PO activity in all the plants tested increased in the presence of chitin as compared with the activity in crude extracts. This activation was not only found in chitin-eluted fraction but also in that fraction not adsorbed on chitin. This data confirms the possibility of PO activation in the presence of polysaccharides (Gulsen et al., 2007). This increase in the activity is probably due to the conformational change of the enzyme molecules (Liu et al., 2005) or the clearing out from the extracts of various inhibitory factors possessing a higher affinity with chitin and remaining bound to it even after elution with 1 M NaCl (Maksimov et al., 2003).

In addition, such an increase in enzymatic activity could result from changes in the conformation of the enzymatic molecules due to the high electrostatic activity of chitin (Dunand et al., 2002; Ozeretskovskaya et al., 2002). It can be proposed that the PO sorption on chitin could not be considered to be a classic ion exchange process because both the anionic and cationic isoforms of the plant POs interact with chitin. Additionally, it contains 3 high anionic POs (3.5, 3.7, 4.0) but only 2 of them (3.5 and 3.7) adsorbed on chitin alongside with some cationic isoforms (Fig. 2).

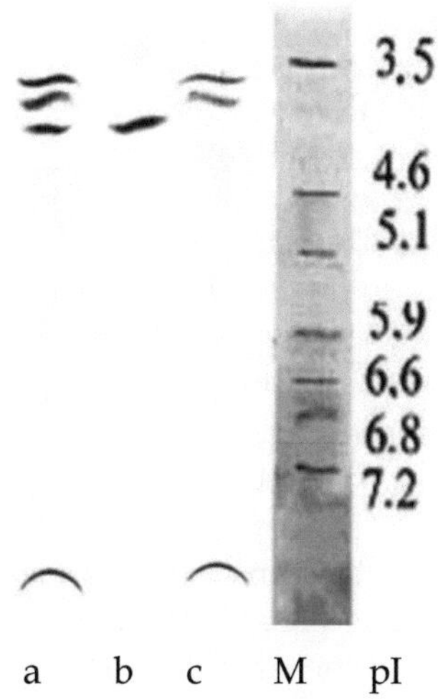

Fig. 2. Spectrum of anionic isoperoxidases isolated from potato: (a) crude protein extract, (b) protein fraction not adsorbed on chitin, and (c) chitin-specific peroxidises. M - protein markers.

In some cases, the anionic POs adsorbed on chitin have similar antigenic determinants, but the plants belonging to different families - and even members of the same family - could have polysaccharide-specific POs with different structures. Thus, the majority of investigated species had anionic chitin-specific peroxidises, and these isoforms from potato (*Solanaceae*) and horseradish (*Brassicaceae*) formed lines of precipitation with antibodies to wheat chitin-bound PO but not to anionic isoPO (Maksimov et al., 2000). However, protein extracts from several plants of *Brassicaceae, Cucurbitaceae* and *Fabaceae* formed precipitate with both the chitin-specific and anionic PO of wheat (Fig. 3). It was found that the greatest homology showed in plants and formed precipitation lines with the anionic PO of wheat (Tab. 2).

The analysis of the data on the presence of gene sequences encoding the polysaccharide-specific sites of PO in different plants was carried out. The fragment of the gene encoding amino-acid sequence 243-269 of wheat anionic isoPO TC151917 (tab. 2) homologous to the fragment of *Arabidopsis* gene encoding pectin-bound site of PO - was detected (Dunand et al., 2002).

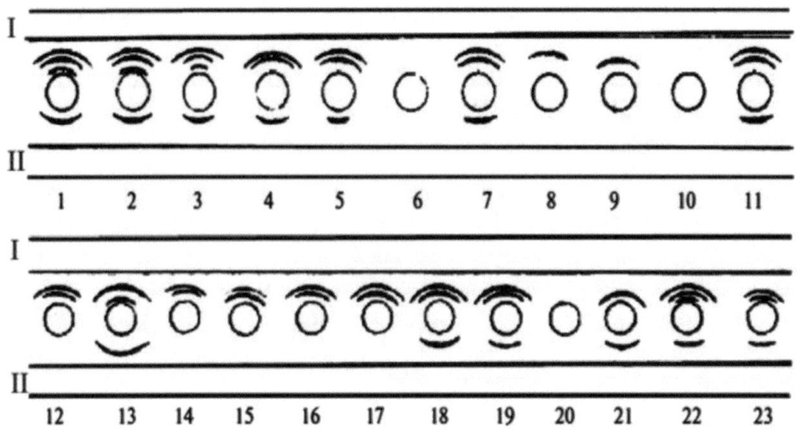

Fig. 3. The scheme of the precipitates formed by the crude protein extracts of plants of the groups monocotyledons (1-11, table 1) and dicotyledons (12-23 table 1) with antibodies against wheat chitin-binding proteins (I) and with antibodies against wheat anionic PO (II).

The characteristics of the POs to be activated by chitin and adsorbed suggest their involvement in the processes underlying two reaction-types which provide the plant with protection against pathogens. The first type includes the fast activation of the enzyme upon its contact with pathogen cell structures, as observed – most notably - with the rice, potato and horseradish POs interacting with chitin. The second reaction-type is comparable with the gradual accumulation of the enzyme molecules within a region of fungus location associated with the appearance of a specific "attracting" centre in the form of chitin-containing pathogen structures. Thus, it may be that due to these functions, lignification and the synthesis of other phenolic polymers resistant to the enzymatic attack of pathogenic fungi usually occur at the site where parasites are located (Simonetti et al., 2009; Kupriyanova, 2006).

The attachment of a pathogen to the surface of plant cells is the first step in formation of a multistep infection process. Unfortunately, it is difficult to observe these steps when fungi's growth and development are studied, and it is easier to study these steps using unicellular symbiotic bacteria. The A 75-kDa protein found in carrot is immunochemically similar to human vitronectin and the elongation factor eEF-1a and it is able to bind bacteria (Wagner & Matthysse, 1992). The authors indicated that these plant cells lost the ability to bind with bacterial cells after treatment with ionic detergents and that this proved the ionic character of these bonds. Proteins with a similar molecular weight and a high binding affinity with chitin were isolated from the microsomal fraction of rice cells (Ito et al., 2000; Yamaguchi et al, 2000). A vitronectin-like protein is involved in the adhesion of the plasma membrane to the cell wall and in extension of the fertilisation tube. In this connection, it is interesting that the PO activity was manifested by the animal protein peroxinectin, which is similar to vitronectin. The 76-kDa protein peroxinectin can bind phage particles and thus it is suggested as having lectin-like features. Both fibronectin and apolipoprotein E are found to possess a C-terminal heparin-binding domain (Kim et al., 2001). Unfortunately, the gene encoding the vitronectin-like protein has not been detected in plants, and this leaves open the degree of its molecular similarity to human vitronectin. The antibodies to the vitronectin-like protein and to the protein binding with bacterial rhicadhesin did not cross-react; however, they could competitively suppress the binding of bacterial cells with plant cells (e.g. with pea cells). Consequently, the above-mentioned proteins are different in molecular structure but their biochemical features are somewhat similar.

Plant species	Locus	Analyzed fragment	Reference
Triticum aestivum	TC151917 X56011	243fd**K**qyyhnlln**KK**glltsdq269 238fdnayytnlmsq**K**gllhsdq257	(http:// PO. isb-sib.ch/) (Rebmann et al., 1991)
Avena sativa	AF078872	249fdnsyynnllsq**K**gllhsdq259	(Cheng et al., 1997)
Oryza sativum	X66125 D84400 OS378734	243fdnayysnllsn**K**gllhsdq262 249fdnryyqnllnq**K**gllssdq268 255fdlgyf**K**nva**KRR**glfhsdg280	(Riemann et al., 1992) (Ito et al., 2000) (Chittoor et al., 1997)
Zea mays	AF037033 AY500792 ZM004710	175fdndyy**K**nllte**R**gllssdq194 265fdn**K**yyvgltnnlglf**K**sdv285 256fdn**K**yyfdlia**K**qglf**K**sdq279	(Padegimas et al., 2004) (Gullet-Claude et al., 2004) (Teichman et al., 1997)
Allium cepa	TC6261	240fdn**K**yyvdlln**R**qtlftsdq259	(http:// PO. isb-sib.ch/)
Armoracia rusticana	D90115	253fdn**K**yyvnl**K**en**K**gliqsdq273	(Fujiyama et al., 1990)
Raphanus sativus	X91172	258fdnsyf**K**nlmaq**R**gllhsdq277	(http:// PO. isb-sib.ch/)
Brassica oleraceae	75974310	260fdn**K**yyvnl**K**eh**K**gliqtdq280	(http:// PO. isb-sib.ch/)
Arabidopsis thaliana	ATg08770 NM101321	217fdn**K**yyvnl**K**en**K**gliqsd250 241fdnnyy**R**nlmq**KK**gllesdq261	(Dunand et al., 2002) (http: // ncbi.nlm.nih.gov)
Solanum tuberosum	M21334	271fd**K**vyydnlnnnqgimfsdq290	(Roberts et al., 1988)
Nicotiana tabacum	L02124	221fdndyftnlqsnqgllqtdq240	(Diaz-De-Leon et al., 1993)
Petunia hybrida	CV299755	22fdnmyf**K**nlq**R**g**R**glftsdq41	(http:// PO. isb-sib.ch/)
Cucumis sativus	DQ124871	183fdnnyf**K**nlvq**RK**glletdq203	(http: // ncbi.nlm.nih.gov)
Cucurbita pepo	DQ518906	239fd**K**nyytnlqan**R**glltsdq59	(Carpin et al., 1999)
Arachis hypogaea	71040666	252ldnnyy**R**nildn**K**glllvdh272	(Yan et al., 2003)
Pisum sativum	AF396465	16fdvgyf**K**qvv**KRR**glfesda36	(http:// PO. isb-sib.ch/)

Note: The amino acid residues of lysine (K) and arginine (R) which may be responsible for the binding of POs to polysaccharides are in bold. According to Dunand et al. (2002), the mutual substitution of these amino acids has no influence on the sorption properties of the ATg08770 PO of *Arabidopsis* with pectins, and the deletion of the fragment results in the loss of this function.

Table 2. Search and comparison of the regions of plant peroxidases homologous to the polysaccharide-binding site

So, among diversified plant proteins, we found POs which could be adsorbed on chitin, thereby preserving their enzymatic activity. An analysis of the isoenzymatic range and activity of chitin-binding POs revealed considerable differences between plant species. In particular, anionic isoPOs of practically all the examined species were adsorbed on chitin. This fact is of great importance because investigators accentuated some remarkable properties of anionic POs - high stability, resistance to temperature and pH changes and activity under high-oxidative conditions (Gazaryan & Lagrimini, 1996; Sacharov, 2004). Thus, the isolation of anionic POs with chitin matrices is likely to be very promising. Besides, practically every another polysaccharide-binding protein (such as chitinases) are cationic, and this fact makes chitin a more suitable matrix for POs isolation than - for example - pectates. However, in some species the constitutive activity of anionic POs is insignificant; therefore, one of the problems is how "to force" plants to synthesise more anionic isoPOs. The study of the physiological role of these proteins can help to solve this question.

5. Possible function of polysaccharide-binding plant peroxidases

We observed the activation of chitin-specific PO during infection with the causative agents of a number of diseases: in wheat under the influence of *Bipolaris sorokiniana* and the elicitors (Fig. 4), *Septoria nodorum* (Yusupova et al., 2006) and *Tilletia caries* (Khairullin et al., 2000); in potato infected by *Phytophthora infestans* (Maksimov et al., 2011), and in *Aegilops umbellulata* infected by *Septoria nodorum* (Maksimov et al., 2006).

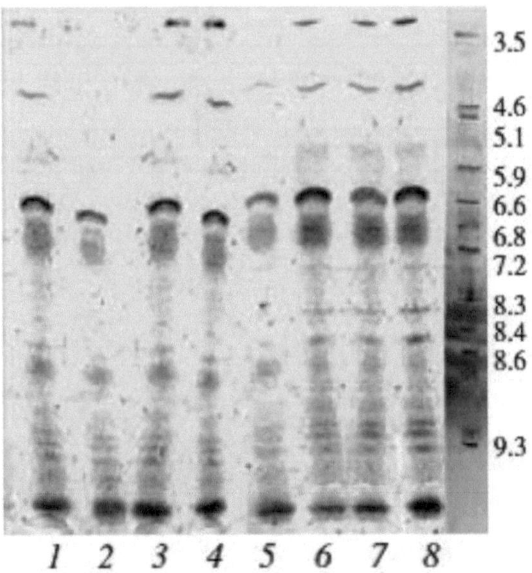

Fig. 4. PO pattern in a water-soluble protein fraction from the roots of wheat seedlings, contrasting in their resistance to the causal agent of the root rot *Bipolaris sorokiniana*, 48 h after infection (Burchanova et al., 2007). (1-4) - cultivar Znitsa (susceptible), (5-8) - cultivar Zarya (resistant). (1,5) control seedlings; (2,6) seedlings infected with *B. sorokiniana*; (3,7) seedlings treated with chitooligosaccharides; (4,8) - seedlings treated with chitooligosaccharides and infected with *B. sorokiniana*.

Akhunov et al. (2008) purified chitin-specific PO with fungicidal activity from cotton and observed the increase of its activity in plants, penetrated by *Verticillium dahliae*. Golubenco et al. (2007) showed the presence of the chitin-binding PO isozyme in *Hibiscus trionum*, which activated dramatically after inoculation by *V. dahliae*. The plants of *Nicotiana tabacum* overexpressing the anionic PO (chitin-specific according to our data) were more resistant to *Helicoverpa zea* and *Lasioderma serricorne* as compared with the wild-type (Dowd et al., 2006).

In this way, chitin-specific POs play an important role in the defence reactions of plants to microbial invasion.

These mechanisms are regulated substantially by the signalling molecule, inducing systemic resistance which is a form of long-lasting immunity to a broad spectrum of pathogens. For example, the accumulation of salicylic acid (SA) is often parallel to or else precedes the increase in the expression of PR genes and ROS accumulation needed for lignification (An & Mou, 2011). In our investigation, SA promoted the activity of wheat "chitin-specific" isoPOs, and SA-treated plants were more resistant to *Septoria nodorum* (Maksimov & Yarullina, 2007).

Thus, in wheat calli infected by *S. nodorum*, the activity of chitin-specific anionic PO with pI ~ 3.5 increased (Fig. 5). When the infected calli were treated with salicylic acid, the cytoplasmic anionic PO with pI ~ 3.5 (no. 1) was greatly activated as well as the one with pI~ 9.8 (no. 14), and was capable of interacting with the cell walls of fungi (presumably, "glucan-specific"). It is worth mentioning that all of the detected POs are involved in plant defence responses against pathogenic fungi.

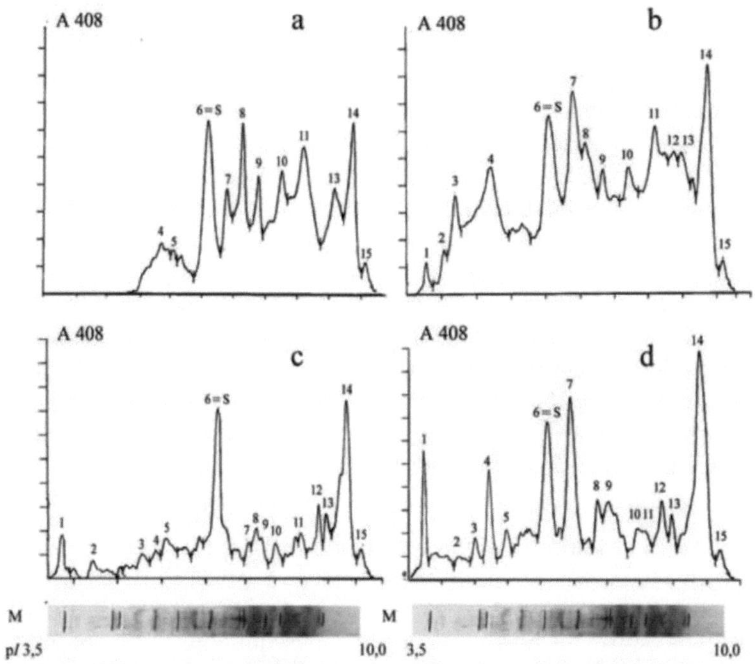

Fig. 5. Densitograms of the PAAG after IEF of water-soluble isoPOs isolated from wheat calli 10 days after infection. (a) – control, (b) calli, infected with *Tilletia caries;* (c) calli on the MS medium supplemented with 0.05 mM salicylic acid; (d) calli infected with *T. caries* on the MS medium supplemented with 0.05 mM salicylic acid; (M) marker (Maksimov & Yarullina 2007).

Therefore, the ability of certain POs to bind with chitin is a widespread phenomenon and - possibly - connected with the defence reactions of the organisms to pathogen attacks. Since it was shown that some biogenic molecules - such as chitooligosaccharides or salicylic acid - can activate an anionic POs, we might suggest that an application of these compounds optimises the process of anionic PO isolation with a chitin.

6. The degree of influence of the acetylation of polysaccharides on the sorption of peroxidases

Numerous polysaccharides in the cell walls of plants and fungi are subjected to cross-linking and are modified by methylation and acetylation in the cell wall (Scheller & Ulvskov, 2010). However, the functional role of the degree of the acetylation of polysaccharides is still unclear. According to our results, we suggest that the acetylation of polysaccharides promotes their binding with some PO isoforms, similarly to that of the chitin from fungal pathogens (Maksimov et al., 2005).

As such, the analysis of matrices capable of PO sorption revealed that PO interaction with chitin decreased with the deacetylation of the latter. It was suggested that the anionic PO of wheat should more actively interact with the acetylated derivatives of cellulose. In fact, acetyl cellulose adsorbed anionic PO, whereas cellulose did not adsorb it. This fact suggested the importance of the degree of polysaccharide acetylation for binding with PO. The binding coefficient of these oxidoreductases with acetyl cellulose was much higher relative to its sorption onto chitin. It is significant that acetylated chitooligosaccharides elicitors enhanced the defence reactions in *Arabidopsis* and wheat more effectively than the deacetylated ones (Cabrera et al., 2006). Besides, the effective concentration of deacetylated derivatives for triggering the defence reaction of soybean was higher than in those cases using highly acetylated oligomers (Shibuya, Minami, 2001). In our investigation, the ability of high-acetylated chitooligosaccharides to promote the transcription of wheat (Burchanova, 2007) and potato anionic PO was more significant than that of deacetylated ones. The investigations of El Guedarri et al. (2002) showed that the penetrating structures of the pathogens contained chitosan rather than chitin. This fact is argued by the absence of the interaction of the monoclonal antibodies specific to chitin following penetration and the appearance of specific reactions with chitosan antibodies. It should be noted that the aggressive race of phytopathogenic fungi has more active chitin deacetylase (Maksimov, Valeev, 2007). Our results show that microorganisms containing chitin (or its oligomers) are the targets of chitin-specific anionic plant POs performing a protective role. Thus, the defence mechanism in plants is specifically targeted and evolves where and when required. When chitin is deacetylated and transformed into chitosan, its ability to bind anionic PO declines (Fig. 6).

The ability of PO to interact with the acetyl residues of chitin allows us to compare them with monovalent lectins (i.e. extensins) which when binding with hemicellulose are only affected in a medium with a high ionic strength (Brownleader et al., 2006). As a rule, POs are bound with the plant cell wall and act as its modifiers. Some POs can form complexes with an extensin of cell walls (Brownleader et al., 2006). Consequently, chitin-specific sites that are capable of interacting with polysaccharides exist in the molecules of PO, and these sites can resemble the membrane receptor binding sites or else be similar to the domains of heparin-binding proteins (Kim et al., 2001).

As follows, during the plant-pathogen interaction a process of elicitor deacetylation takes place and it is possible that the effective penetration and colonisation of plant tissues in this case occurs due to avoiding "meeting" a specific plant PO in addition to the defence of fungus structures against plant chitinases. However, numerous of the polysaccharides of plants are also acetylated and it is not easy to suggest the cause for this fact. Thus, pectin acetylation with acetyl-CoA as the acetate donor has also been demonstrated *in vitro*. Recently, Scheller & Manabe (2010) have found that knockout mutants in one of the corresponding genes -

At3g06550 - are deficient in wall-bound acetate. It would be interesting to test their resistance, lignin depositions and other physiological parameters under the influence of pathogens. It may be that these experiments will make clear the role of polysaccharide acetylation.

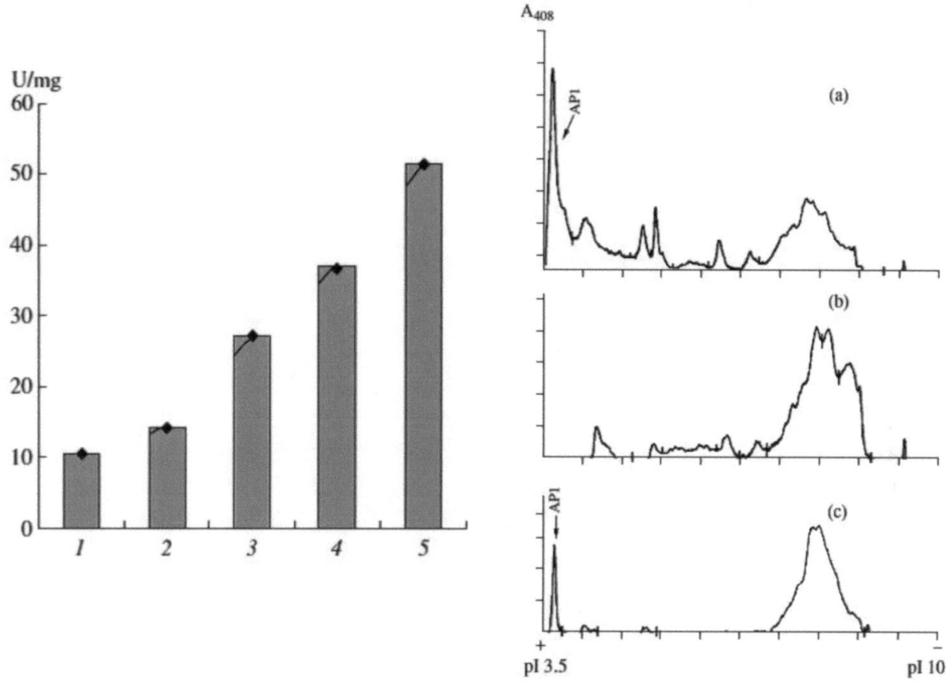

Fig. 6. Effect of the degree of chitin acetylation (%) on the interaction between chitin and chitin-specific wheat POs (A) (U/mg protein) (Maksimov et al., 2005); (B) PAAG after IEF of PO fractions from wheat roots (a) not bound to high-acetylated (b) and low-acetylated chitosan (c) (Khairullin et al., 2000). Designations: (1) 12%; (2) 23%; (3) 37%; (4) 45%; (5) 65%.

7. Can peroxidases bind with glucan-containing structures?

Earlier, we noted the ability of wheat PO to adsorb on purified chitin and suggested the possibility of the application of this polymer for POs' (in particular, for highly-stable anionic POs) purification. Chitin is a major component of fungi cell walls, but they also contain other polymers. Thus, using the spores of the bunt agent *Tilletia caries* as a sorbent for affinity chromatography we showed - as we predicted - the sorption on spores of the "chitin-specific" isoPO. However, in addition to this, significant content of PO with pI ~ 8.8 was observed (Fig.7) (Khairullin et al., 2000). Subsequently, through the chromatography of wheat POs on the cell walls of *Septoria nodorum* (Berk) we showed that only two cationic isoPOs bound with the cell walls of this pathogen and we didn't observe sorption of anionic PO.

It may be that this fact was associated with the rather complex composition of fungal mycelium, consisting of chitin enclosed in a glucan matrix (Bowman, Free, 2006). Therefore, mature saprophyte mycelium are completely covered by difficult-soluble glucans and the fraction of chitin in the apical cell wall is not sufficient. As such, we supposed that these cationic isoforms bound with another major component of the fungi cell wall - glucans. To check this hypothesis, we carried out an investigation of the sorption of the potato PO using the cell walls of the late blight pathogen *Phytophthora infestans* as a sorbent. Strictly speaking, *P. infestans* is not a fungal pathogen as it belongs to the class of *Oomycota* and its cell walls do not contain chitin, consisting of glucans, cellulose and some another components (Gaulin et al., 2006).

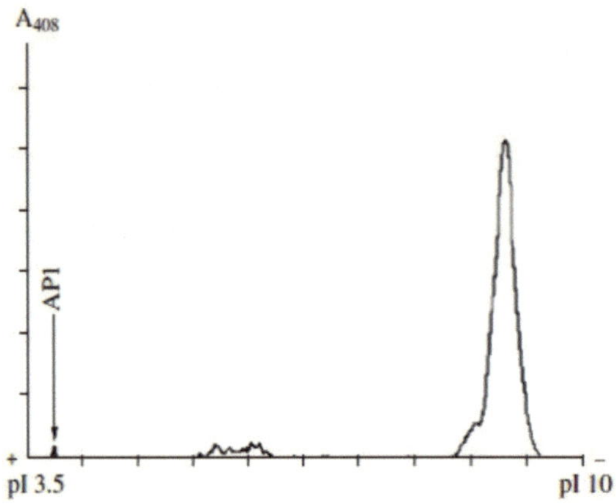

Fig. 7. PAAG densitogram after IEF of the PO fraction of wheat roots absorbed by the growing spores of *Tilletia caries*.

In fact, we observed the ability of cationic PO (pI 9.3) to bind with the purified cell walls of this pathogen (Fig. 8, A). This isoform was activated in the infected plants and under the influence of the stress hormone jasmonic acid, both individually and in combination with salicylic acid.

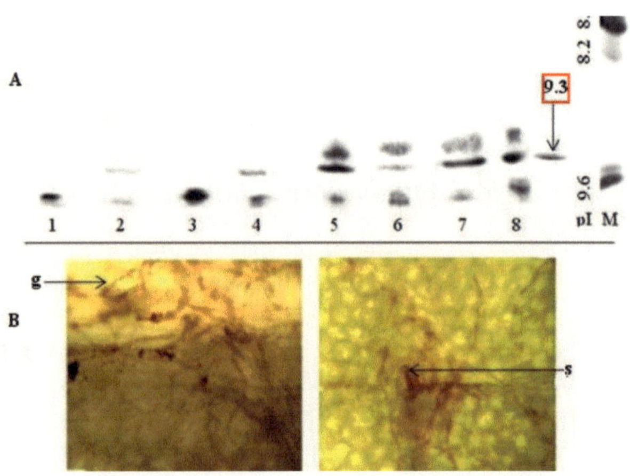

Fig. 8. Activation of the PO binding with *P. infestans* cell walls (glucan-specific?) under pathogen inoculation and treatment with salicylic (SA) and jasmonic (JA) acids (A); Peroxidase activity in stomata guard cells and intercellular spaces of adjoining epidermal leaf cells and on the surface of mycelium contacting with the stomata (B). (*1*) Non-treated control; (*2*) infection; (*3*) treatment with SA; (*4*) treatment with SA + infection; (*5*) treatment with JA; (*6*) treatment with JA + infection; (*7*) treatment with SA + JA; (*8*) treatment with SA + JA + infection; g – gifs of *P. infestans*; s – stomata guard cell. Specific to *P. infestans* cell walls, PO is highlighted.

Cytological experiments also demonstrated 2,2-diaminobensidine (DAB)-colouring of the infecting gifs on the surface of the leaves (Maksimov et al., 2011). Because DAB is intensively produced just on the surface of infectious products of a pathogen, this suggests focusing on them as ROS generators and their users, which are also likely to include POs (Fig. 8, B).

It is of interest that the activation of POs often takes place in stomata guard cells, since *P. infestans* mainly penetrates into plant tissues through stomata slits. The localisation of phenolic compounds - some of them seemingly being used by POs as a substrate - and PO activity was visible in guard cells. (Maksimov et al., 2011). As such, the immune reaction occurred in close proximity to pathogen structures.

8. Conclusion

It is well-known that POs can generate hydrogen peroxide and that it can act as a secondary messenger in defensive responses; besides this, they can oxidise numerous compounds - including phenols - and therefore catalyse a reaction that is directly associated with the lignification of the cell walls of plants and infectious fungal structures. These important physiological features of POs are found in the application of analytics and medical assays and industrial biocatalytic processes. Our results and our published data allow us to propose that a separate subclass of polysaccharide-binding isoPOs is present in plants and which take part in defence reactions against biotic stresses, including pathogen attacks and wounding (Carpin et al., 2001; Dunand et al., 2002). Unfortunately, currently, this subclass has yet to be characterised and its unique properties are not used in practice. It is possible that some of the POs of this class are functionally associated with the plant cell wall and contribute to its modification due to a high affinity for hemicellulose (most likely, to pectin). Some isoPOs can electrostatically bind with the components of the cell wall of pathogenic fungi and plants. These peroxidases probably facilitate the direct formation of lignin due to their ability to interact with polysaccharides. Therefore, it is possible that these isoenzymes play an important role in the defence reactions of plants against pathogens and wounding. The results obtained allow us to suggest the possibility of using polysaccharide biopolymers - chitin in particular - for some manipulations with POs.

Thus, the ability of PO to adsorb on chitin while preserving its enzymatic activity suggests the cooperative function of these enzymes in the defensive responses of wheat against chitin-containing pathogens and it opens up possibilities for using this biopolymer for the primary purification of chitin-specific proteins. It is worth noting that in the majority of cases the anionic POs have the ability to bind with chitin and according to some data (Gazaryan & Lagrimini 1996; Sacharov, 2004) these POs have higher thermal stability than cationic ones. Using these highly-stable, readily-produced POs can increase the quality of immunoblotting kits and stimulate the elaboration of new analytical methods.

It was shown that POs' binding with polysaccharides serves as a protective function in plants due to its immediate involvement in the action of the prooxidant and antioxidant systems. The possibility of regulating the PO-encoding genes' expression by the different regulators of plant resistance - including oligosaccharides - allows the determination of the role of the enzyme in plant immunity and it may also stimulate the production of POs (including anionic isoforms) by optimising their extraction. Besides this, we observed a unique feature of chitin in stimulating the POs activity. We suppose that this effect may be used for increasing the efficiency of obtaining PO preparations.

9. Acknowledgment

This study was supported by the grant of the Russian Federation Ministry of Education and Science P339.

10. References

Almagro, L.; Gomez Ros, L.V.; Belchi-Navarro S.; Bru R.; Ros Barcello A. & Pedreno M.A. (2009) Class III peroxidases in plant defence reactions // J. Exp. Botany. V. 60. P. 377-390.

An C. & Mou Z. (2011) Salicylic acid and its function in plant immunity // J. of Integrative Plant Biology. V. 53. P. 412–428.

Akhunov A.A.; Golubenko Z.; Khashimova N. R.; Mustakimova E.Ch. & Vshivkov S.O. (2008) Role of chitin-specific peroxidases in wilt-resistant cotton // Chem. Nat. Comp. V. 44. P. 493-496.

Benhamou N.; Chamberland H. & Pauze F.J. (1990) Implication of pectic components in cell surface interactions between tomato root cells and *Fusarium oxysporum* f. sp. *radicis-lycopersici* // Plant Physiol. V. 92. P. 995-1003.

Bindschedler L.V.; Dewdney J.; Blee K.A.; et al. Peroxidases -dependent apoplastic oxidative burst in *Arabidopsis* required for pathogen resistance // Plant J. 2006. V. 47. P. 851–863.

Boerjan W.; Ralph J. & Boucher M. (2003) Lignin Biosynthesis // Annu. Rev. Plant Biology. V. 54. P. 519-546.

Bowman S.M. & Free S.J. (2006) The structure and synthesis of the fungal cell wall // Bioessays. V. 28(8). P. 799-808.

Brownleader M.D.; Hopkins J.; Mobasheri A.; et al. (2002) Role of extension peroxidases in tomato (*Lycopersicon esculentum* Mill.) seedling growth // Planta. V. 210. P. 668-676.

Burhanova G.F.; Yarullina L. G. & Maksimov I. V. (2007) Effect of chitooligosaccharides on wheat defence responses to infection by *Bipolaris sorokiniana* // Russ. J. of plant physiology. V. 54. P. 104 -110.

Cabrera J. C.; Messiaen J.; Cambier P. & Van Cutsem P. (2006) Sise, acetylation and concentration of chitooligosaccharides elicitors determine the switch from defence involving PAL activation to cell death and water peroxide production in *Arabidopsis* cell suspensions // Physiologia plantarum. V. 127. P. 44-56.

Carpin S.; Crevecoeur M.; de Meyer M.; Simon P.; Greppin H. & Penel C. (2001) Identification of a Ca2+-pectate binding site on an apoplastic peroxidase // Plant Cell. V.13. P. 511-520.

Carpita N.; Tierney M. & Campbell M. (2001) Molecular biology of the plant cell wall: Searching for genes that define structure, architecture and dynamics // Plant Mol. Biol. V. 47. P. 1- 5.

Cheng E.H.; Kirsch D.G.; Clem R.J.; Ravi R.; Kastan M.B.; Bedi A.; Ueno K. & Hardwick J.M. (1997) Conversion of Bcl-2 to a Bax-like death effector by caspases // Science. V. 278. P. 1966–1968.

Chittoor J.M.; Leach J.E. & White F.F. (1997) Differential induction of a peroxidase gene family during infection of rice by *Xanthomonas oryzae pv. oryzae* // Mol Plant Microbe Inter. V. 10. P. 861–871.

Choi H.W.; Kim Y.J.; Lee S.C.; Hong J.K. & Hwang B.K. (2007) Hydrogen peroxide generation by the pepper extracellular peroxidase $CaPO_2$ activates local and systemic cell death and defense response to bacterial pathogens // Plant Physiology. V. 145. P. 890–904.

Conroy J.M.; Borzelleca D.C. & McDonell L.A. (1982) Homology of plant peroxidases. An immunochemical approach // Plant Physiol. V. 69. P. 28-31.

Cosio C. & Dunand C. (2009) Specific function of individual class III peroxidase genes // J. Exp. Botany. V. 60. P. 391-408.

Cosio C.; Vuillemin L.; De Meyer M.; Kevers C.; Penel C. & Dunand C. (2009) An anionic class III peroxidases from zuccini may regulate hypocotyl elongation through its auxin oxidase activity // Planta. V. 229. P. 823–836.

Crevecoeur M.; Pinedo M.; Greppin H. & Penel C. (1997) Peroxodase activity in shoot apical meristem from *Spinacia* // Acta Histochem. V. 99(2). P. 177-186.

Diaz-De-Leon F.; Klotz K.L. & Lagrimini L.M. (1993) Nucleotide sequence of the tobacco (*Nicotiana tabacum*) anionic peroxidase gene // Plant Physiol. V. 101. P. 1117–1118.

Dowd P.F. & Lagrimini L.M. (2006) Examination of the biological effects of high anionic peroxidases production in tobacco plants grown under field conditions. I. Insect pest damage. // Transgenic Research. V. 15. P. 197-204.

Dunand C.; de Meyer M.; Crévecoeur M. & Penel C. (2003) Expression of a peroxidases gene in zucchini in relation with hypocotyl growth. // Plant Physiology and Biochemistry. V. 41. P. 805–811.

Dunand C.; Tognolli M.; Overney S. et al. (2002) Identification and characterisation of Ca^{2+}pectate binding peroxidases in *Arabidopsis thailiana* // J. Plant Physiol. V. 159. P. 1165-1171.

Duroux L. & Welinder K. G. (2003). The peroxidases gene family in plants: a phylogenetic overview. // J. Molecular Evolution. V. 57. P. 397–407.

El Gueddari N.E.; Rauchhaus U.; Moerschbacher B.M. & Deising H.B. (2002) Developmentally regulated conversion of surface-exposed chitin to chitosan in cell walls of plant pathogenic fungi // New Phytologist. V. 156. P. 103–112.

Fujiyama K.; Takemura H.; Shinmyo A.; Okada H. & Takano M. (1990) Growth-stimulation of tobacco plant introduced the horseradish peroxidase gene prxC1a // Gene. V. 89. P. 163–169.

Gazaryan I.A. & Lagrimini L.M. (1996) Purification and unusual kinetic properties of a tobacco anionic peroxidases // Phytochemistry. V. 41. P. 1029-1034.

Gaulin E.; Dramé N.; Lafitte C.; Torto-Alalibo T.; Martinez Y. et al. (2006) Cellulose binding domains of a *Phytophthora* cell wall protein are novel pathogen-associated molecular patterns // Plant Cell. N. 18. P. 1766-1777.

Golubenco Z.; Achunov A.; Khashimova N.; Beresneva Y.; Mustakimova E.; Ibragimov F.; Abdurashidova N. & Stipanovic R. (2007) Induction of peroxidase as a disease resistance response in resistant (*Hibiscus trionum*) and susceptible (*Althea armeniaca*) species in the family *Malvaceae // Phytoparasitica.* T. 35(4). P. 401 – 413.

Gorshkova T.A. (2007) Plant Cell Wall as a Dynamic System (in Russian), Nauka, Moscow.

Guillet-Claude C.; Birolleau-Touchard C.; Manicacci D.; Rogowsky P.M.; Rigau J.; Murigneux A.; Martinant J.P. & Barriere Y. 2004. Nucleotide diversity of the ZmPox3 maize peroxidases gene: relationships between a MITE insertion in exon 2 and variation in forage maize digestibility // BMC Genet. V. 5. P. 19.

Gulsen O.; Shearman R. C.; Heng-Moss T. M.; Mutlu N.; Lee D. J. & Sarath G. (2007) Peroxidases gene polymorphism in buffalograss and other grasses // Crop Science. V. 47. P. 767-774.

Hammerschmidt R. & Kuc J. (1982) Lignification as a mechanism for induced systemic resistance in cucumber // Phisiol. Plant. Pathol. V. 20. P. 61 – 71.

Harholt J.; Suttangkakul A.; & Scheller H. V. (2010) Biosynthesis of pectin // Plant Physiol. V. 153. P. 384–395.

Ito H.; Higara S.; Tsugava H. et al. (2000) Xylem-specific expression of wound inducible rice peroxidases genes in transgenic plants // Plan Sci. V. 155. P. 85-100.

Khairullin R. M.; Yusupova Z. R. & Maksimov I. V. (2000) Protective responses of wheat treated with fungal pathogens: 1. Interaction of wheat anionic peroxidases s with chitin, chitosan, and thelyospores of *Tilletia caries* // Rus. J of plant physiology. V. 47. N. 1. P. 97–102.

Kim J.; Han I.; Kim Y. et al. C-terminal heparin-binding domain of fibronectin regulates integrin-mediated cell spreading but not the activation of mitogen-activated protein kinase // Biochem. J. 2001. V. 360. P. 239–245.

Kupriyanova E.V.; Ezhova T.A.; Lebedeva O.V. & Shestakov S.V. (2006) Intraspecific polymorphism in peroxidases genes located on *Arabidopsis thaliana* chromosome 5 // Biological Bull. N. 4. P. 437-447.

Levigne S. V.; Ralet M-C. J.; Quemener B. C.; Pollet B. N-L.; Lapierre C. & Thibault J.-F.J. (2004) Isolation from sugar beet cell walls of arabinan oligosaccharides esterified by two ferulic acid monomers // Plant Physiology. V. 134. P. 1173–1180.

Liu G. S.; Sheng X. Y.; Greenshields D. L.; Ogieglo A.; Kaminskyj S.; Selvaraj G. & Wei Y. D. (2005) Profiling of wheat class III peroxidases genes derived from powdery mildew-attacked epidermis reveals distinct sequence-associated expression patterns // Molecular Plant-Microbe Interactions. V. 18. P. 730–741.

Maksimov I.V. & Cherepanova E.A. (2006) The pro-/antioxidant system of plants and the resistance of plants to pathogens // Usp. Sovr. Biol. (on Russ.) V. 126. P. 250-261.

Maksimov I.V.; Cherepanova E.A.; Kuzmina O.I.; Yarullina L.G. & Achunov A.V. (2010) Molecular peculiarities of the chitin-binding peroxidases s of plants // Russian J of Bioorganic Chemistry. V. 13. P. 293–300.

Maksimov I. V.; Cherepanova E. A. & Khairullin R. M. (2003) "Chitin specific" peroxidases in Plants // Biochemistry (Moscow). V. 68(1). P. 111 - 115.

Maksimov I. V.; Cherepanova E. A.; Murtazina G. F. & Chikida N. N. (2006) The relationship between the resistance of *Aegilops umbellulata* Zhuk. seedlings to *Septoria nodorum* Berk. and peroxidases isozyme pattern // Biology Bulletin. V. 33. N. 5. P. 466–470.

Maksimov I. V.; Cherepanova E. A. & Surina O. B. (2010) Effect of chitooligosaccharides on peroxidase isoenzyme composition in wheat calli co cultured with bunt causal agent // Rus. J. of Plant Physiol. V. 57. P. 131-138.

Maksimov I. V.; Cherepanova E. A.; Yarullina L. G. & Akhmetova I. E. (2010) Isolation of chitin-specific oxidoreductases // Applied Biochem. & Mikrobiol. V. 41. P. 616-620.

Maksimov I. V.; Sorokan' A. V.; Cherepanova E. A.; Surina O. B.; Troshina N. B. & Yarullina L. G. (2011) Effects of salicylic and jasmonic acids on the components of pro/-antioxidant system in potato plants infected with late blight // Rus. J. of Plant Physiol. V. 58. N. 2. P. 299–306.

Maksimov I. V. & Yarullina L. G. (2007) Salycilic acid and local resistance to pathogens // Salycilic acid: a plant hormone, Springer-Verlag. Berlin. Heidelberg. P. 323 - 334.

Marjamaa K.; Kukkola E.M. & Fagerstedt K.V. (2009) The role of xylem class III peroxidases in lignification // J. of Exp. Botany. V. 60. P. 367-376.

Margalit H.; Fisher N. &Ben-Sasson S.A. (1993) Comparative analysis of structurally defined heparin binding sequences reveals a distinct spatial distribution of basic residues // J. Biol. Chem. V. 268. P.19228-19231.

McDougall G.J. (2001) Cell-wall proteins from stika spruce xylem are selectively insolubilised during formation of dehydrogenation polymers of coniferyl alcohol // Phytochem. V. 57. P. 157-163.

Ostanina E.S., Lopatin C.A., Varlamov B.P. (2007) The extraction of chitin and chitozan from *Galleria melonella // Biotechnology*. V. 3. P. 38 – 45.

Ozeretckovskaya O. L. (2002) Problems of specific immunity // Russian J. Plant Physiol. V. 49. P. 148-154.

Padegimas L. S.; Reichert N. A. Nematode-upregulated peroxidase gene promoter from nematode-resistant maize line Mp307 // USA Patent 6703541. 03.09.2004.

Passardi F.; Penel C.; Dunand C. (2004) Performing the paradoxial: how plant peroxidase modify the cell wall // Trends in Plant Sci. V.9. P. 534-540.

Piston F.; Uauy C.; Fu L. H.; Langston J.; Labavitch J. & Dubcovsky J. (2010) Down-regulation of four putative arabinoxylan feruloyl transferase genes from family PF02458 reduces ester-linked ferulate content in rice cell walls // Planta. N. 231. P. 677–691.

Rebmann G.; Hertig C.; Bull J.; Mauch F. & Dudler R. (1991) Complementary DNA cloning and sequence analysis of a pathogen-induced putative peroxidase from rice // Plant Mol. Biol. V. 16. P. 329–331.

Reimann C.; Ringli C. & Dudler R. (1992) Complementary DNA cloning and sequence analysis of a pathogen-induced putative peroxidase in rice // Plant Physiol. V. 100. P. 1611–1612.

Roberts E.; Kutchan T. & Kolattukudy P. E. (1988) Cloning and sequencing of cDNA for a highly anionic peroxidase from potato and the induction of its mRNA in suberizing potato tubers and tomato fruits // Plant Mol. Biol. V. 11. P. 15–26

Sacharov I. Yu. (2004) Palm tree peroxidases // Biochemistry (Moscow). V. 69. P. 823-829.

Shibuya N. & Minami E. (2001) Oligosaccharide signaling for defense responses in plant // Physiol. Mol. Plant Pathol. V. 59. P. 223-233.

Scheller H. V.; & Ulvskov P. (2010) Hemicelluloses // Annu. Rev. Plant Biol. V. 61. P. 263–289.

Siegel S.M. (1957) Non-enzymic macromolecules as matrices in biological synthesis. The role of polysaccharides in peroxidase catalyzed lignin polymer formation from eugenol // J. Amer. Chem. Soc. V. 79. P. 1628-1632

Simonetti E.; Veronico P.; Melillo M. T.; Delibes Á.; Andrés M.F. & López-Braña I. (2009) Analysis of class III peroxidase genes expressed in roots of resistant and susceptible wheat lines infected by *Heterodera avenae* // Mol. Plant-Microbe Interact. V. 22. P. 1081-1092.

Teichmann T.; Guan C.; Kristoffersen P.; Muster G.; Tietz O. & Palme K. (1997) Cloning and biochemical characterization of an anionic peroxidase from *Zea mays*// Eur. J. Biochem. V. 247. P. 826–832.

Tian M.; Gu Q. & Zhu M. (2003) The involvement of hydrogen peroxide and antioxidant enzymes in the process of shoot organogenesis of strawberry callus // Plant Science. V. 165. P. 701-707.

Wagner V. & Matthysse A. G. (1992) Involvement of a vitronectine-like protein in attachment of *Agrobacterium tumefaciens* to carrot suspension culture cells // J. Bacteriol. V. 174. P. 5999-6003.

Welinder K. G.; Justesen A. F.; Kjaersgard I. V.; Jensen R. B.; Rasmussen S. K.; Jespersen H. M. & Duroux L. (2002) Structural diversity and transcription of class III peroxidase POs from *Arabidopsis thaliana* // Europ. J. Biochem. V. 269. P. 6063–6081.

Wu T., Zivanovic S., Draugnon F. A., Conway W.S., Sams C.E. (2005) Physicochemical properties and bioactivity of fungal chitin and chitosan // J. Agric. Food Chem. V. 53. P. 3888-3894.

Yoshida K., Kaothien P., Matsui T., Kawaoka A. (2003) Molecular biology and application of plant peroxidase genes // Appl. Microbiol. Biotechnol. V. 60. P. 665 - 670.

Yusupova Z.R.; Akhmetova I.E.; Khairullin R.M. & Maksimov I.V. (2005) The effect of chitooligosaccharides on hydrogen peroxide production and anionic peroxidase activity in wheat coleoptiles // Rus. J. of Plant Physiol. V. 52. P. 209–212.

Yan J.; Wang J.; Tissue D.; Holaday A. S.; Allen R. & Zhang H. (2003) Protection of photosynthesis and seed production under water-deficit conditions in transgenic tobacco plants that over-express *Arabidopsis* ascorbate peroxidase // Crop Sci. V. 43. P. 1477–483.

Yamaguchi T.; Ito Y. & Shibuya N.(2000) Oligosaccharide elicitors and their receptors for plant defence responses // Tr. Glycisci. Glykotech. V. 12. P. 113-120.